The IMA Volumes in Mathematics and its Applications

Volume 155

T0205714

For further volumes:
http://www.springer.com/series/811

Institute for Mathematics and its Applications (IMA)

The Institute for Mathematics and its Applications was established by a grant from the National Science Foundation to the University of Minnesota in 1982. The primary mission of the IMA is to foster research of a truly interdisciplinary nature, establishing links between mathematics of the highest caliber and important scientific and technological problems from other disciplines and industries. To this end, the IMA organizes a wide variety of programs, ranging from short intense workshops in areas of exceptional interest and opportunity to extensive thematic programs lasting a year. IMA Volumes are used to communicate results of these programs that we believe are of particular value to the broader scientific community.

The full list of IMA books can be found at the Web site of the Institute for Mathematics and its Applications:

> http://www.ima.umn.edu/springer/volumes.html.

Presentation materials from the IMA talks are available at

> http://www.ima.umn.edu/talks/.

Video library is at

> http://www.ima.umn.edu/videos/.

Fadil Santosa, Director of the IMA

Stephen Childress • Anette Hosoi
William W. Schultz • Z. Jane Wang

Editors

Natural Locomotion in Fluids and on Surfaces

Swimming, Flying, and Sliding

 Springer

Editors

Stephen Childress
New York University
Courant Institute of Math Sciences
251 Mercer Street
New York, NY 10012
USA

William W. Schultz
Department of Mechanical Engineering
The University of Michigan
2027 Auto Lab
Ann Arbor, MI 48109
USA

Anette Hosoi
Massachusetts Institute of Technology
77 Massachusetts Avenue
Cambridge, MA 02139
USA

Z. Jane Wang
Cornell University
Mechanical and Aerospace Engineering
323 Thurston Hall
Ithaca, NY 14853
USA

ISSN 0940-6573
ISBN 978-1-4899-9916-0   ISBN 978-1-4614-3997-4 (eBook)
DOI 10.1007/978-1-4614-3997-4
Springer New York Heidelberg Dordrecht London

FOREWORD

This volume was based on a Workshop on Natural Locomotion in Fluids and on Surfaces: Swimming, Flying, and Sliding held at the Institute for Mathematics and its Applications (IMA) at the University of Minnesota, from June 1–5, 2010. We would like to thank all the participants for making this a stimulating and productive workshop.

In particular we would like to thank the organizers, Steve Childress, Anette Hosoi, William Schultz, Z Jane Wang for their role as workshop organizers and in organizing this volume.

We also take this opportunity to thank the National Science Foundation for its support of the IMA.

Series Editors

Fadil Santosa, Director of the IMA,

Jiaping Wang, Deputy Director of the IMA.

PREFACE

A Workshop on Natural Locomotion in Fluids and on Surfaces: Swimming, Flying, and Sliding was held at the Institute for Mathematics and its Applications (IMA) at the University of Minnesota, from June 1–5, 2010. This final workshop in the IMA 2009–2010 Thematic Year on Complex Fluids and Complex Flows brought together an exceptionally large and diverse group of scientists, which included engineers, mathematicians, biologists, and physicists. The subject matter ranged widely from observational data to theoretical mechanics, and reflected the broad scope of the workshop. In both the prepared presentations and in the informal discussions, the workshop engaged exchanges across disciplines and invited a lively interaction between modelers and observers.

The success of the Workshop depended on the efforts of many people. We begin with thanks to Charles Doering and the other members of the organizing staff of the Thematic Year, for recognizing the relevance of locomotion in the context of complex fluid flows, and for their help and encouragement in formulating workshop goals and in attracting a stellar turnout of experts in the field. We are indebted to IMA Director Fadil Santosa and his excellent staff, for easing the tasks of organization and implementation of a large meeting. We especially thank Chun Liu for his help with these preparations. Patricia Brick and Dzung Nguyen provided valuable assistance with the early organization of this volume during a time of transition of the publication process. Donna Chernyk of Springer US was a great help in finalizing the text. Finally, we thank the participants of the Workshop for their enthusiasm and engagement, and the authors of invited as well as contributed papers for helping to define the scope of the discussions and state of the field in 2010.

The papers in the volume provide a representative if necessarily incomplete account of the field of natural locomotion during a period of rapid growth and expansion. Driven by its myriad of applications in Nature, locomotion encompasses many fields. The fluid mechanics is especially complex, owing to the interaction between the fluid medium and the flexible, time-dependent locomoting body. The fluid motions are further coupled to the dynamics of the body, which in turn is responding to the fluid environment. Reasonable analytic models are not easily found, and numerical computations of time-dependent, three-dimensional fluid-body interactions are difficult and costly. The natural scientist is presented not only with large numbers of creatures capable of flying, swimming, and sliding, but also with a wide variety of *mechanisms of* locomotion, not to mention the *motivations for* locomotion. Given the need to go from A to

B, evolution has given us a vast panorama of examples, over a wide range of length scales, shapes, and environments.

The papers presented at the workshop, and the contributions to the present volume, can be roughly divided into those pertaining to swimming on the scale of marine organisms, swimming of microorganisms at low Reynolds numbers, animal flight, and sliding and other related examples of locomotion. The swimming of fish is characterized by the relatively high Reynolds number and the near neutral buoyancy of the organism. Lighhthill's pioneering analysis of the swimming of slender fish set the stage for a large literature on propulsion and maneuvering. Much of the recent literature has elaborated on the basic mechanisms and focused on specific instances of thrust generation and drag reduction as well as on the difficulty of separating out thrust and drag. This viewpoint is reflected in the contributions to the present volume, and was the subject of lively discussion in the workshop tutorials. The mechanisms used by sharks to reduce drag have received considerable attention, and the paper by Lang, Motta, Habegger, and Hueter addresses the role of the scales of the skin of the shark in controlling fluid separation. Zhu and Shoele discuss the fluid dynamics of caudal and pectoral fins using numerical modeling. It is shown how the structural properties of the fin can be essential to achieving high performance and efficiency.

The feeding of marine organisms often involves the generation of flow fields by movements akin to those of locomotion. The paper by Hamlet, Miller, Rodriquez, and Santhanakrishnan considers the feeding current generated by a jellyfish whose feeding arms extend upward. Experimental observations are combined with numerical modeling and analysis to study the implications of the flow patterns on particle capture and waste disposal. Several papers are concerned with the response of fish to their natural environment, with particular emphasis on the effects of turbulence and vortical eddies on their behavior. Aline Cotel and Paul Webb emphasize the importance of a collaborative effort between engineers and biologists to determine the effects of these natural flow fields on the distribution and density of fish populations. The interaction of a swimming fish with a free fluid surface is of interest as a means of tracking movement, as is shown in the paper by Rachel Levy and David Uminsky, which treats the patches of smooth water surface caused by the motion of the fluke of a whale. These are explained as the interaction of the ocean surface with eddies created by the fluke. A related interaction problem arises in the study of fish schooling and the swimming of a fish in the wake of another fish. A particularly interesting question is the role that such interactions might play in reducing the energy requirements of locomotion. Silas Alben considers this question and the importance of the arrangement and phase relationships of multiple swimming flexible bodies.

As a branch of theoretical mechanics, locomotion presents some intriguing questions concerning its appropriate mathematical setting. In fact locomotion can be viewed as a Lagrangian embodiment of kinematics, focusing on the displacement of the points of the locomoting body without regard to the dynamics utilized. The paper of Kelly, Pujari, and Xiong considers this geometric viewpoint in the case of an ideal irrotational fluid. This is then combined with a planar model allowing localized vortex shedding from a swimming fish, represented by a Joukowsky foil with time-dependent camber. In a similar vein, Babak Oskouei and Eva Kanso examine the motion of a passive body in the wake of a thrust producing locomoting body. They find unstable periodic orbits wherein the passive body extracts energy from the ambient vortices.

G.I. Taylor's seminal papers on the swimming of two-dimensional sheets in Stokes flow provided the foundation for the study of locomoting microorganisms, arguably the most active branch of our subject. Roughly a third of this volume concerns low Reynolds number locomotion, divided between studies of propulsion by a single organism and those of suspensions of many swimmers. There is also considerable interest in the interaction of locomoting cells with boundaries or an extracellular matrix. The paper of Desimone, Heltai, Alouges, and Lefebre-Lepot takes up the question of optimal low Reynolds number swimming, in the sense that a given displacement is realized with minimum energy expenditure. They study several model swimmers utilizing geometric control theory. Kiori Obuse and Jean-Luc Thiffeault investigate the movement of a treadmilling microswimmer near a semi-infinite wall, on which a no-slip condition is applied. Trajectories are calculated and the possibility of escape from the wall is determined. Janna, Kim, Yang, and Jung are similarly concerned with the influence of a boundary, in their case on the swimming behavior of *Paramecium* when placed in a microchannel. Susan Suarez discusses the motion of mammalian sperm in vivo and pinpoints key questions influencing successful fertilization. For cells which are essentially attached to a boundary or an extracellular matrix, swimming as such gives way to crawling. The paper by Qixuan Wang, Jifeng Hu, and Hans Othmer examines movement caused by "blebbing", a change of shape involving small ballooning protrusions, which may grow to take over the entire cell volume.

Suspensions of many swimming microorganisms are of great interest because of the collective behaviors that become possible through hydrodynamic interactions. Such "active" suspensions add a new dimension to the rheology of fluid-particle systems. David Saintillian provides a helpful summary of work in this area. He describes in detail the application of a kinetic model to problems of pattern formation and chemotaxis. Cisneros, Ganguly, Goldstein, and Kessler discuss the effect of cell concentration of the collective swimming of the bacteria *Bacillus subtilis*. Their observations reveal the emergence of anomalous speeds, both high and low, as the concentration affects the cell-cell interactions. A useful classification of these suspensions divides the swimmers into "pushers", which use organelles aft of the body center for propulsion, and "pullers" where the placement is forward of the center. Zhenlu Cui and Xiao-Ming Zeng study the rheology of active suspensions in the presence of weak steady shear. From a stability analysis of a continuum model they are able to treat systematically the pusher-puller duality.

Animal flight is distinguished by the need to lift the body weight as well as to locomote, and the manner by which this is achieved in flapping flight, allowing as well for the remarkable maneuverability of birds and insects, is one of the most interesting problems of locomotion in Nature. An explosion of activity in the field has been driven by modern techniques of high-speed photography, image analysis, and numerical modeling. The review paper by Ristroph, Bergou, Berman, Guckenheimer, Wang, and Cohen describes a major advance in our understanding of the dynamics and control of insect flight. Their study, which involves high-speed videography, dynamics modeling, and the reconstruction of three-dimensional wing and body motions, illuminates the mechanism of in-flight turns as well as the recovery of stable flight from an external disturbance. The role of wing flexibility on the efficiency of flapping flight is of great interest for the construction of robotic microflyers. The paper by Hassan Massoud and Alexander Alexeev explores numerically the flapping of elastic wings at resonance, and shows that flexibility can lead to enhanced lift. Sheng Xu considers the basic question of formulating a concise description of the dynamics of insect flight using a matrix form of Newton's laws of dynamics. And Sheng, Ysasi, Kolomentshiy, Kanso, Nitsche, and Schneider compare three numerical models for computing the two-dimensional vortex wake behind a flapping plate.

It should be clear that numerical simulation is a key component of locomotion studies, and two papers in the volume are devoted to specific numerical techniques. The immersed boundary method has provided a useful new approach to the computation of locomotion of a flexible body through a fluid. Anita Layton describes a two-dimensional formulation of the method for solving flow about an immersed interface with Dirichlet boundary conditions, and shows that the method exhibits second-order accuracy in time and space. Theoretical issues connected with the existence of solutions of the Navier-Stokes equations describing locomotion of a free body are taken up in the paper of Adam Boucher. He shows that a technique of elliptic regularization may be used to advantage in these problems, and outlines a proof for establishing classical solutions.

Finally, we turn to two papers which are closest to the "sliding" of our Workshop title. David Hu and Michael Shelley describe a new model for the locomotion of snakes. Their work, involving observation and numerical simulation, shows how anisotropic friction associated with scales of the snake's belly, combined with the flexibility of the body and the modulation of the weight of contact along its length, gives rise to many of the observed modes of movement across various types of surface. Locomotion on or through a granular material is the subject of the paper by Ding, Gravish, Li, Maladen, Mazouchova, Steinmetz, Umbanhowar, and Goldman. They review various modes of limb-ground interaction, and examine the sub-surface movement of the sandfish lizard using high-speed x-ray imaging. "Swimming through sand" is not inappropriate as a description of this interesting mechanism of locomotion.

Stephen Childress
Courant Institute of Mathematical Sciences
New York University
New York, NY, USA

Anette Hosoi
Department of Mechanical Engineering
Massachusetts Institute of Technology
Cambridge MA, USA
William Schultz
Department of Mechanical Engineering
University of Michigan
Ann Arbor, MI, USA

Jane Wang
Sibley School of Mechanical and Aerospace Engineering
Cornell University
Ithaca, NY, USA

CONTENTS

INVITED PAPERS

MODEL PROBLEMS FOR FISH SCHOOLING

SILAS ALBEN(✉)*

Abstract. We review recent work on model systems for body-vortex and body-body interactions in a fluid, related to fish schooling. The studies show a variety of structured and disordered dynamics which can occur when vortices interact with passive and actively-driven deformable bodies. The energy savings due to body-vortex interactions depend strongly on the geometric arrangement of multiple bodies and/or ambient vortices, the phases of the bodies' motions relative to those of the vortices and other bodies, and the rigidity of flexible appendages.

Key words. Flexible, vorticity, coupled, flow-body

AMS(MOS) subject classifications. 76Z10, 74F10, 74L15, 92C10

1. Introduction. This article reviews a collection of recent model problems which address different aspects of fish schooling. We focus on theoretical work, and the experiments which were their direct inspiration. In most cases the experiments can also be considered physical/biological models (in the sense of [1]) for the complicated phenomena of fish schooling. In particular, the experiments and theories described here mainly focus on the interactions of individual bodies with ambient vorticity, or two bodies coupled to each other through their shared fluid environment. Model "swimmers" such as rigid foils, passive elastic rods, or isolated fins driven at the leading edge, yield simpler problems which can be analyzed. Thus, important phenomena can be studied without the full complexity of real fish.

2. Bodies Interacting with Vorticity. Even though many aspects of thrust generation by individual flexible bodies remain mysterious, a number of recent studies have considered how multiple flexible bodies interact with each other through a fluid. The interaction occurs principally through vorticity that is shed from the edges of these bodies, and then collides with other bodies downstream. Thus a useful precursor to the problem of fish schooling is the study of how bodies interact with ambient vorticity.

The experiment of Liao et al. [2] increased interest in this field by showing how a single swimming trout alters its swimming motion in the presence of vorticity. When a D-shaped cylinder was placed in a steady flow, a "von Kármán" street of concentrated vortices of alternating sign was shed from the two sides of the cylinder. When placed in this vortex-street wake, the trout was able to hold its position in the flow with less energy, compared to the same flow situation without the upstream cylinder. The trout altered its usual swimming motion to increase its side-to-side

---

*School of Mathematics, Georgia Institute of Technology, Atlanta, GA 30332, USA, alben@math.gatech.edu. This work was supported by NSF-DMS 0810602 and 1022619

S. Childress et al. (eds.), *Natural Locomotion in Fluids and on Surfaces*, IMA 155, DOI 10.1007/978-1-4614-3997-4_1,
© Springer Science+Business Media New York 2012

motion, increasing its ability to interact with the vortices. The vortices emerged from this interaction significantly weakened, indicating that the trout had extracted significant kinetic energy from the vortices. The usual pattern of muscular contraction all along the trout body, in the absence of the cylinder, was replaced by a contraction concentrated in the upstream portion of the body. The remainder of the body was thereby left to interact with vortices more like a passive flexible element. Liao also reviewed the experimental literature on more general cases of swimming in turbulent flows [3–5].

An earlier theoretical study by Streitlien et al. [6] studied the motion of a rigid body in an array of point vortices, and found significant gains in propulsive efficiency due to the interaction of the body with the point vortices. However, no optimal motion could be identified, since the efficiency became arbitrarily large as the body became closer to the point-vortex singularities. In real flows, the vortices themselves deform significantly upon encountering a body, and their subsequent dynamics can depend sensitively on their spatial structure [7].

In [8], we considered a periodic flexible sheet swimming along the midline of a von Kármán street. The model is somewhat similar to Taylor's infinite swimming sheet at low Reynolds number [9]. In our work, the Reynolds number is large, and the flow can be solved analytically for small sheet deflections by posing a periodic vortex sheet along the swimming body. The body does not swim freely, but instead deforms transversely to itself, in a traveling wave which moves with the vortex street, in order to maximize the upstream thrust force acting on it. The optimal solutions are shown in Fig. 1a. When the vortex street width d is large relative to its streamwise periodic length l, shown schematically by the solid-line vortices, the optimal swimming shape is sinusoidal, also shown by a solid line. In the other limit, when the vortex street is narrow (d/l is small), shown by the dotted-line vortices, the optimal swimming shape consists of sharp peaks near the vortices, also shown by a dotted line. An intermediate case is shown for comparison. In all cases, the sheet has maximum slope adjacent to the vortices, which allows the pressure force from the vortices to act mostly in the upstream direction.

The analytical solutions yield closed formulae for the output power P_{out} on the sheet, per unit period length. P_{out} is the product of thrust force with the mean flow speed that the body encounters U, and is nondimensionalized by a product of the fluid area density ρ_f, the amplitude of the traveling wave A, the square of the strength of the vortices in the street Γ, and U. Figure 1b shows the output power for the full range of vortex street width parameter d/l. For narrow vortex streets, the output power diverges as a power law, $\sim (d/l)^{-5/2}$, while for wide vortex streets the output power decays exponentially, $\sim e^{-d/l}$. Wide vortex streets have an exponentially small central flow, so little pressure is induced on a body in the center of the vortex street.

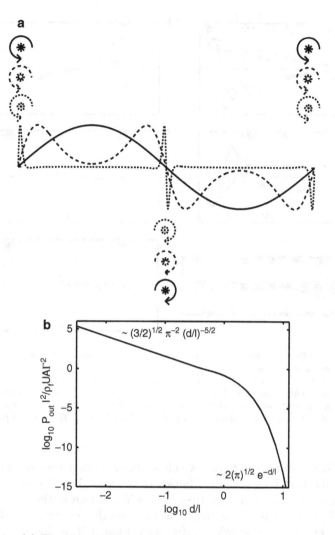

FIG. 1. *(a) Thrust-maximizing motions of a periodic swimming sheet immersed in a vortex street when the vortex street width is large (solid line), moderate (dashed line), and small (dotted line). (b) The output power P_{out}, nondimensionalized, versus the vortex street aspect ratio d/l (Adapted from [8])*

The periodic model in [8] is also reasonable for a body of finite length, when the body is long relative to the streamwise spacing between the vortices. In this case the trailing edge wake has relatively weak vorticity [10]. Excitation of passive bodies by vortices were studied in [11, 12]. More general cases of *finite* swimming bodies in a von Kármán street were studied in [10]. The study considered only relatively wide vortex streets, which are only slightly perturbed by the body's presence. This regime also simplifies

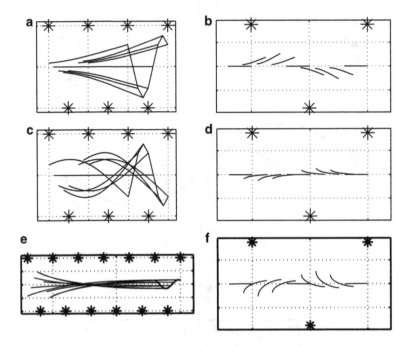

FIG. 2. *Efficiency-maximizing swimming motions for an elastic fin driven at the leading edge (a–d) and a whole body driven all along its length (e, f). Eight snapshots of the fin/body are shown over a period of motion, in a frame which moves with the vortex street. The stiffness and length of the elastic fin are varied as follows: a stiff, long fin (a); a stiff, short fin (b); a more flexible, long fin (c); and a more flexible, short fin (d). A long body is shown in (e), and a short body in (f) (Adapted from [10])*

the trailing vortex wake to one which is approximately straight, and with a sinusoidal-traveling-wave distribution of vorticity. In this case the entire problem has a single temporal frequency, which is that of the von Kármán street. The parameter d/l is thus removed from the problem, but the ratio of body length to streamwise vortex separation, L/l, is added. One can again identify body motions which maximize thrust force and efficiency. Some examples of body motions which maximize swimming efficiency are shown in Fig. 2.

Two different types of body forcing are considered in Fig. 2. In panels a–d, a leading edge heaving and pitching motion is applied to an otherwise passive elastic body, which is a model of a tail fin. Snapshots of the body motions are shown in the frame in which the vortex street is at rest, and the body swims upstream ("slaloms") through the vortex street. Panels a and b show a relatively stiff mode of bending, and panels c and d show a higher bending mode. Panels a and c show bodies which are long (relative to the streamwise vortex spacing) and panels b and d show bodies which are short. In general, the phase of maximum body deflection relative to the

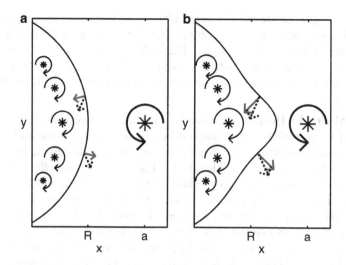

FIG. 3. *(a) Point vortex near a rigid circular wall, with induced flow at the wall shown by dashed and gray arrows. (b) Point vortex near a flexible circular wall, with a (now larger) induced flow shown by dashed and gray arrows (Adapted from [13])*

passage of the vortices can take on the full range from 0 to 2π depending on the parameters L/l and the body stiffness. Panels e and f show efficiency-maximizing motions for a body which is driven all along its length, similarly to an entire fish body. Here the thrust near the leading edge is important, and favors a large body curvature near the leading edge.

Many studies have considered interactions of vortices with rigid bodies relevant to engineering applications, principally aircraft [7, 14]. Interactions between a single vortex and *passive* flexible walls were studied in [13]. Traveling wave solutions were found for two simple wall geometries: an infinite straight line, and a circle, in the undeflected states. In these solutions, the vortex travels parallel to the wall, and the wall flexes outward towards the vortex and the outward bump moves with the vortex in a traveling wave of deflection.

For a circular wall, the force on the wall diverges as the inverse cube of the distance between the point vortex and the wall. The physical mechanism is illustrated in Fig. 3. In panel a, a portion of a rigid circular wall near the point vortex (at $x = a$) is shown. The smaller vortices along the inner boundary of the wall represent the bound vortex sheet induced on the wall, which prevents flow from penetrating the wall. The dashed arrows are the flow velocities induced by the point vortex at two points on the wall, and the gray arrows are their components normal to the wall. Panel b shows the same situation but for a flexible wall. The outward deflection of the wall brings it closer to the point vortex. Then the flow induced by the point vortex on the wall (dashed arrows) and their components normal to the wall (gray arrows) are larger. Hence the bound vortex sheet along the wall is stronger, and so is the pressure force on the wall, which leads to a larger deflection, and so on. This self-amplification of the wall deflection leads to the aforementioned inverse-cube force law, and the speed of the point vortex diverges as the inverse fourth power of distance to the flexible wall.

3. Multiple Bodies in Flows. Another level of complexity is introduced by considering *two* coupled bodies in a flow. Dong and Lu investigated numerically two bodies arranged side-by-side, which undergo a prescribed traveling wave motion [15]. They studied the effect of varying the relative phase between the bodies' traveling waves, and found that in-phase motions can enhance efficiency, while out-of-phase motions can increase the maximum thrust produced. Because the flow is incompressible, large pressures can develop when the bodies move out of phase and "squeeze" the fluid between them. In-phase motions, by contrast, decrease the relative speed between the body and the ambient fluid. The signature of these effects can be seen in the vortex streets produced by the foils. When the foils move out-of-phase, the vortices are strengthened, whereas for in-phase motions, the vortices are weakened. The resulting change in outward fluid momentum flux causes increased or decreased thrust on the bodies. For the in-phase motion, however, the decreased thrust is accompanied by increased efficiency.

These results are reminiscent of a recent study by Wang and Russell on the coupling of front and hindwing motions in dragonflies [16]. There the dragonfly switches from moving the (essentially rigid) wings in-phase for steady hovering, to out-of-phase, for take-off motions. The authors gave a simple interpretation for the increased lift force and efficiency in terms of the relative motion of the wing and the ambient fluid. In each of these cases, the vortex-laden viscous flows surrounding the wings are highly complicated. However, a quasi-steady drag law allows for a simple interpretation of the effect of one wing moving through the wake created by the other.

Other studies have considered coupled motions of two completely passive bodies. Zhu and Peskin used the immersed boundary method to study the effect of the lateral spacing between the flags in a two-dimensional flow [17], inspired by soap film experiments of Zhang et al. [18, 19]. They found a transition from out-of-phase flapping, when the flags are far apart, to in-phase flapping when the flags become sufficiently close. This fact connects well with the increase in efficiency noticed by Dong and Lu with in-phase swimming of side-by-side swimmers. The transition in flapping frequencies during these states were studied by Farnell et al. [20]. Jia et al. studied the same problem in a simpler model, which enabled a more thorough investigation of parameter space [21]. Important parameters are flag displacement-to-length ratio, dimensionless bending energy, and dimensionless flag density. In addition to in-phase and out-of-phase modes, they found more general states which fit into neither category, which they classified as "indefinite" modes. They also investigated different lengths of the flags, which alters the timing and strength of vortex shedding at the trailing edges of the flags through a complex interaction which is not understood on the basis of simple principles.

One of the most interesting problems has been the analysis of propulsion by *coupled* motions of fish fins arranged streamwise. Drucker and Lauder studied the relative motions of the Bluegill sunfish dorsal fin (located atop the body) and caudal (tail) fin [22]. Using a laser light sheet they measured the kinematics of the two fins and the fluid motion in a horizontal cross-section through both fins and the surrounding fluid. They found that vorticity shed from the dorsal fin collides with the caudal fin at different locations along the fin depending on the swimming gait. They found evidence for a constructive interaction between the vorticity shed by the two oscillating fins, which resulted in stronger net vorticity leaving the tail fin. A smaller sweep amplitude and phase advance of 30 degrees were reported for the dorsal fin relative to the caudal fin. Presumably these are among the important parameters characterizing how vorticity from the dorsal fin affects shedding from the tail fin. Due to the complex analysis required to visualize the flow fields, the results were confined to a small set of parameter values.

Ristroph and Zhang studied a related passive system, consisting of a streamwise array of flags in a soap film [23]. As the distance between the second flag's leading edge and the leading flag's trailing edge is varied, the drag on both flags can increase and decrease in a nonobvious manner. In particular, they reported an optimum spacing between the flags, leading to a minimum overall drag on the system. The phenomenon is different from "drafting" used to decrease drag in cycling. The amplitude of the flags' flapping is of the same order as the spacing between them, leading to an unsteady wake with a dominant frequency of oscillation which can only be determined by considering the interaction of the flags and the shed vortex sheets.

We have used a vortex sheet model to gain a more detailed picture of the two-flag system in parameter space [24]. Figure 4 shows snapshots which are representative of the typical dynamics for two flags side-by-side (panels a–c) and in the tandem configuration (panels d–h). For the side-by-side case, as the spacing between the flags is increased, we find a transition from nonperiodic, erratic flapping for closely spaced flags (panel a), to antiphase flapping (panel b), which lies at one end of a range of periodic flapping with monotonic variation of interflag flapping phase with separation distance. In-phase dynamics occur at a separation of about 2.5 flag-lengths (panel c). An experimental study of 2–4 side-by-side flags showed interesting higher-frequency modes of interaction [25]. A linear stability analysis was given by Michelin and Llewellyn Smith [26].

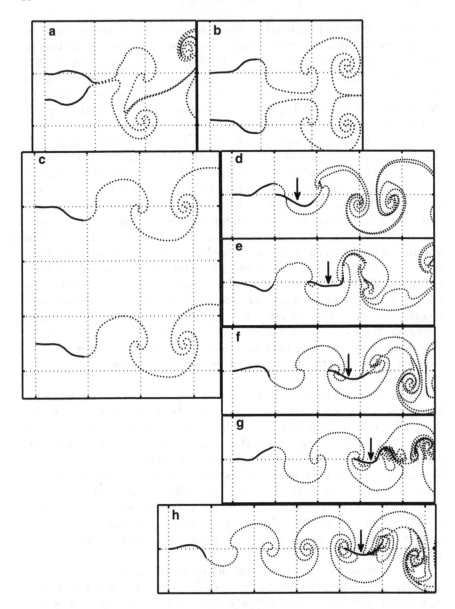

FIG. 4. *Snapshots of a model two flag system in side-by-side (**a–c**) and tandem (**d–h**) configurations. The trailing flags in the tandem configurations are indicated by arrows (Adapted from [24])*

The tandem case is somewhat richer, and shows an alternating sequence of synchronous (panels d, f, and h) and asynchronous (panels e and g) flapping states as separation distance increases. The drag force on the follower flag is increased in synchronous flapping, but not in asynchronous flapping. Synchronous flapping seems to occur at flag separations which allow the vortex streets of the leader and follower flag to nearly align, with vortices of the same sign rolling up together. Studies incorporating viscosity found similar results [27, 28].

4. Future Directions. We have discussed a set of recent model problems for understanding how bodies interact through flows. One of the future challenges is to move from model problems to an understanding of the fluid dynamical mechanisms at work in a school of fish. Weihs gave a simple model of the forces on individual fish in a school in terms of the relative spacings in a 2D array of fish, with fluid wakes modeled using far-field flows induced by point vortices [29]. More recent studies have modeled multiple swimmers using Hamiltonian systems, with omission or simplification of vortex shedding [30–32].

Biologists have challenged the hydrodynamic importance of schooling, with the observation that three species of schooling fish do not swim in appropriate positions to gain hydrodynamic advantage [33]. The matter remains unresolved, but the hydrodynamic interactions within a fish school are of sufficient intrinsic interest that a more detailed description of the flow in an array of swimmers is desirable [34, 35]. Rapid progress has been made recently on the understanding of collective swimming at low Reynolds numbers [36–39], and similar statistical and mean-field approaches may work well at high Reynolds numbers.

REFERENCES

[1] Koehl MAR (2003) Physical modelling in biomechanics. Philos Trans R Soc Lond Ser B Biol Sci 358(1437):1589

[2] Liao JC, Beal DN, Lauder GV, Triantafyllou MS (2003) Fish exploiting vortices decrease muscle activity. Science 302(5650):1566–1569

[3] Liao JC (2007) A review of fish swimming mechanics and behaviour in altered flows. Philos Trans R Soc B Biol Sci 362(1487):1973

[4] Newman JN (1975) Swimming of slender fish in a non-uniform velocity field. ANZIAM J 19(1):95–111

[5] Dabiri JO (2007) Renewable fluid dynamic energy derived from aquatic animal locomotion. Bioinspiration Biomim 2:L1

[6] Streitlien K, Triantafyllou GS, Triantafyllou MS (1996) Efficient foil propulsion through vortex control. AIAA J 34(11):2315–2319

[7] Doligalski TL, Smith CR, Walker JDA (1994) Vortex interactions with walls. Annu Rev Fluid Mech 26(1):573–616

[8] Alben S (2009) On the swimming of a flexible body in a vortex street. J Fluid Mech 635:27–45

[9] Taylor G (1951) Analysis of the swimming of microscopic organisms. Proc R Soc Lond Ser A Math Phys Sci 209(1099):447–461

[10] Alben S (2009) Passive and active bodies in vortex-street wakes. J Fluid Mech 642:95–125

[11] Eldredge JD, Pisani D (2008) Passive locomotion of a simple articulated fish-like system in the wake of an obstacle. J Fluid Mech 607:279–288

[12] Manela A, Howe MS (2009) The forced motion of a flag. J Fluid Mech 635:439–454

[13] Alben S (2011) Interactions between vortices and flexible walls. Int J Non-Linear Mech 46:586–591

[14] Rockwell D (1998) Vortex-body interactions. Ann Rev Fluid Mech 30(1):199–229

[15] Dong GJ, Lu XY (2007) Characteristics of flow over traveling wavy foils in a side-by-side arrangement. Phys Fluids 19:057107

[16] Wang ZJ, Russell D (2007) Effect of forewing and hindwing interactions on aerodynamic forces and power in hovering dragonfly flight. Phys Rev Lett 99(14):148101

[17] Zhu L, Peskin CS (2003) Interaction of two flapping filaments in a flowing soap film. Phys Fluids 15:1954–1960

[18] Zhang J, Childress S, Libchaber A, Shelley M (2000) Flexible filaments in a flowing soap film as a model for one-dimensional flags in a two-dimensional wind. Nature 408(6814):835–839

[19] Shelley MJ, Zhang J (2011) Flapping and bending bodies interacting with fluid flows. Annu Rev Fluid Mech 43(1):449

[20] Farnell DJJ, David T, Barton DC (2004) Coupled states of flapping flags. J Fluid Struct 19(1):29–36

[21] Jia L-B, Li F, Yin X-Z, Yin X-Y (2007) Coupling modes between two flapping filaments. J Fluid Mech 581:199–220

[22] Drucker EG, Lauder GV (2001) Locomotor function of the dorsal fin in teleost fishes: experimental analysis of wake forces in sunfish. J Exp Biol 204(17):2943–2958

[23] Ristroph L, Zhang J (2008) Anomalous hydrodynamic drafting of interacting flapping flags. Phys Rev Lett 101(19):194502

[24] Alben S (2009) Wake-mediated synchronization and drafting in coupled flags. J Fluid Mech 641:489–496

[25] Schouveiler L, Eloy C (2009) Coupled flutter of parallel plates. Phys Fluids 21:081703

[26] Michelin S, Llewellyn Smith SG (2009) Linear stability analysis of coupled parallel flexible plates in an axial flow. J Fluids Struct 25(7):1136–1157

[27] Zhu L (2009) Interaction of two tandem deformable bodies in a viscous incompressible flow. J Fluid Mech 635:455–475

[28] Kim S, Huang W, Sung HJ (2010) Constructive and destructive interaction modes between two tandem flexible flags in viscous flow. J Fluid Mech 661:511–521

[29] Weihs D (1973) Hydromechanics of fish schooling. Nature 241(5387):290–291

[30] Kanso E, Marsden JE, Rowley CW, Melli-Huber JB (2005) Locomotion of articulated bodies in a perfect fluid. J Nonlinear Sci 15(4):255–289

[31] Kelly SD, Xiong H (2006) Controlled hydrodynamic interactions in schooling aquatic locomotion. In: Decision and control, 2005 and 2005 European control conference. CDC-ECC'05. 44th IEEE conference on, Seville. IEEE, pp 3904–3910

[32] Nair S, Kanso E (2007) Hydrodynamically coupled rigid bodies. J Fluid Mech 592:393–411

[33] Partridge BL, Pitcher TJ (1979) Evidence against a hydrodynamic function for fish schools. Nature 279(5712):418

[34] Fauci LJ (1990) Interaction of oscillating filaments: a computational study. J Comput Phys 86(2):294–313

[35] Fish FE (1995) Kinematics of ducklings swimming in formation: consequences of position. J Exp Zool 273(1):1–11

[36] Saintillan D, Shelley MJ (2007) Orientational order and instabilities in suspensions of self-locomoting rods. Phys Rev Lett 99(5):58102

[37] Ishikawa T, Pedley TJ (2008) Coherent structures in monolayers of swimming particles. Phys Rev Lett 100(8):88103

[38] Saintillan D, Shelley MJ (2008) Instabilities and pattern formation in active particle suspensions: kinetic theory and continuum simulations. Phys Rev Lett 100(17):178103

[39] Pedley TJ (2010) Collective behaviour of swimming micro-organisms. Exp Mech 50(9):1293–1301

THE CHALLENGE OF UNDERSTANDING AND QUANTIFYING FISH RESPONSES TO TURBULENCE-DOMINATED PHYSICAL ENVIRONMENTS

ALINE J. COTEL(✉)* AND PAUL W. WEBB†

Abstract. The natural habitats of fishes are characterized by water movements driven by a multitude of physical processes of either natural or human origin. The resultant unsteadiness is exacerbated when flow interacts with surfaces, such as the bottom and banks, and protruding objects, such as corals, boulders, and woody debris. There is growing interest in the impacts on performance and behavior of fishes swimming in "turbulent flows". The ability of fishes to stabilize body posture and their swimming trajectories is thought to be important in determining species distributions and densities, and hence resultant assemblages in various habitats. Understanding impacts of turbulence and vorticity on fishes is important as human practices modify water movements, and as turbulence-generating structures ranging from hardening shorelines to control erosion, through designing fish deterrents, to the design of fish passageways become common. Collaboration between engineers and biologists is essential in order to generate adequate and sustainable solutions. Previous work on fish responses to turbulent perturbations is discussed and new theoretical concepts/framework are proposed to quantify fish-eddy interactions.

Key words. Fish/eddy interaction, vorticity, turbulence, fish responses

1. Introduction. Turbulence is so ubiquitous and almost so familiar that often little attention is paid to it. Generations of aquatic biologists have been introduced to the river continuum, in which running water systems are divided into various zones by current speed [1–3]. These zones are also characterized by differences in turbulence, but this is implicitly assumed as an unspecified correlate of flow velocity. Current speeds and turbulence are high in headwaters, with slopes >1/20, associated with recognizable fish assemblages of species capable of breasting strong and variable flows. In the most extreme high flows, such as in the Himalayas, this ichthyofauna has earned its own moniker of "torrential" fauna comprised of a small number of species with adaptations to attaching to substrate or refuging behaviors to avoid areas of extreme flow [4]. In slower, but still high-current high-turbulence habitats, salmonids and cottids dominate, well adapted to such conditions. At intermediate gradients, flowing waters are characterized by riffle-pool systems, populated by grayling, dace, darters and suckers, and at the lower-slopes of such systems, by shiners and other minnows. As the current in a river system becomes sluggish, fish diversity increases and suckers, bullheads and catfishes, various minnows and shiners, pike and centrarchids become abundant, many with forms that are

---

*University of Michigan, Department of Civil and Environmental Engineering, Ann Arbor, MI 48109-2125, USA, acotel@umich.edu.

†University of Michigan, School of Natural Resources and Environment, Ann Arbor, MI 48109-1041, USA, pwebb@umich.edu.

S. Childress et al. (eds.), *Natural Locomotion in Fluids and on Surfaces*, IMA 155, DOI 10.1007/978-1-4614-3997-4_2,
© Springer Science+Business Media New York 2012

not adapted for turbulent conditions. Differences in turbulence in these river sections are clear, but have not been well described or quantified.

Similarly, coral reefs experience variable flows and turbulence, and the structure of protruding reef structures and organisms would be expected to be major sources of elevated turbulence. It is well known that fishes behave in ways that would avoid flow-turbulence extremes, for example using refuges on peak ebb and flow tides [5, 6]. Turbulence also varies across reefs, depending on exposure to waves, which in turn affects fish assemblages. In this situation, a crude measure of flow variances has been determined from rates of dissolution of gypsum balls [7–10]. On the Great Barrier Reef, the abundance of small, largely benthic, cryptic gobies and blennies in sand/rubble habitats decreased with increasing exposure. Shallow, wave-exposed reef zones had fewest species and lowest abundance [11]. On the sheltered side of the reef, also in sand/rubble habitats, fishes were more diverse and abundant. Among pelagic fishes, larger fishes and more powerful swimmers were more common in exposed habitats while less strong swimmers became more abundant in sheltered reef zones [9, 10].

Thus, there has been an *implicit* recognition that turbulence is an important ecohydraulic factor affecting fish distributions and abundance, and hence impinging on fundamental questions in ecology. However, in considering factors affecting habitat suitability for various species, simple hydraulic parameters such as current speed and depth have been the common metrics. While turbulence is often correlated with discharge, variations in turbulent conditions have had little role in habitat assessment in spite of being an important ecological factor in its own right.

We do not offer a solution to the vexing question on how to define turbulence. Instead, we explore approaches to transfer ideas on turbulence to the biological realm, considering new developments in concepts and methods, as well as future directions for evaluating turbulence in both the field and laboratory. Especially important are conceptual approaches recognizing structure in turbulence flows, evaluating such flows in terms of eddy structures and vorticity rather than statistical measures derived from point measurement techniques. Modeling has also become an integral part of ubiquitous turbulence in fish habitat, reproducing a wider range of conditions than physically possible in the laboratory. However, such models need to be carefully calibrated with field and laboratory data.

2. The Nature of Turbulence. At its simplest, numerous factors create velocity gradients and therefore vorticity (as the curl of the velocity vector) in turn resulting in curvature of streamlines leading to the formation of eddies (regions of finite vorticity) that characterize turbulent flows. Eddies vary in size from large gyres that can span ocean basins, such as the North Pacific Gyre, to very small Kolmogorov length scale eddies, so small that they are damped by viscosity dissipating their energy as heat. Vorticity can facilitate swimming when a fish can sense and respond to an eddy that boosts it on its way or reduces transport costs [12]. Some eddies facilitate contacting food items [13]. Most commonly, however, flow fluctuations have negative impacts on fishes, notably reducing swimming speed, increasing energy expenditure and destabilizing fishes [14–16], impacts that are postulated to underlie broader ecological effects of turbulence.

Historically, turbulent flows have been investigated using point measurement techniques, such as Hot Wire Anemometry, Acoustic Doppler Velocimetry, etc. High temporal resolution was achieved and some understanding of turbulent flow fluctuations was gained using various statistical tools. However these techniques have provided little reliable information regarding spatial gradients in the flow. Hypotheses, such as Taylor's assumption of frozen, advected turbulent eddies, have provided some information on length scales, such as the integral length scale (see below), but do not lead to an accurate, physically quantifiable "picture" of the flow at different instants across a range of eddies of various sizes and strengths. Optical techniques such as Particle Image Velocimetry are providing new means of obtaining instantaneous representations of turbulent flows. These methods have lower temporal resolution than point measurements, which could be a limitation in very high speed flows, but most probably not an issue in applications related to ecohydraulics. At this point in time, it seems that a combination of both point-measurement and optical techniques provide the most productive path towards a reliable quantitative description of turbulent flows with adequate temporal and spatial resolution.

3. Linking the Flow to Fishes Using Point Measurements. The first step in any new or developing field is assessing metrics that are important in quantifying responses of fishes to turbulent flows, ideally starting in the field with data on both fish swimming and flow characteristics. These should subsequently lead to complimentary laboratory experiments. In practice, the historical but implicit recognition of the importance of turbulence in the field has lead to laboratory experiments without adequate characterization of the flow. As a result, results from laboratory experiments have been contradictory [17]. Therefore, we first consider concepts and results for tools useful in both field and laboratory situations.

The most common tool in field and laboratory experiments is the Acoustic Doppler Velocimeter (ADV). An ADV measures fluctuations in the magnitude and direction of water speed in a very small volume – essentially one point – as a function of time. The sampling frequency typically varies from a few Hz to 50 Hz depending on manufacturer and instrument. Higher frequencies are now available, up to 200 MHz. ADV data consist of temporal signals of velocity in three directions, u, v and w. Various statistical descriptions relating to flow are calculated from the instantaneous u, v and w values. Engineers interested in turbulence most commonly use the root mean square (rms) velocity, u_{rms}, as a measure of mean flow, recognizing potential effects of velocity variation, and hence of turbulence. The root mean square velocity, u', is calculated from the mean square deviation in resultant velocity from the mean, numerically the statistic variance. The instantaneous resultant velocity, u_{inst}, is be calculated from u, v and w. Average velocities, $\bar{u}$, $\bar{v}$, $\bar{w}$ and u_{inst} are calculated from a time series of measurements over a long period, commonly two or more minutes.

The mean speed and variance are often combined in a simple non-dimensional derived turbulence parameter, **Turbulence Intensity**, TI [14, 18, 19], where:

$$TI = u'/(\bar{u}^2 + \bar{v}^2 + \bar{w}^2)1/2 \tag{1}$$

TI takes into account a speed-dependent aspect of stability. The velocity variance challenges stability while the momentum of a fish contributes to damping disturbances. Thus, TI may be considered a ratio of disturbance to damping. Pavlov and his collaborators have amassed an impressive range of results on performance of fishes in turbulent flows [14, 30]. As current speed increases through very low values, fish first orient to the flow (rheotaxis), which reduces drag and promotes station holding [21, 22]. Threshold current speeds at which roach, *Rutilus rutilus*, first oriented to the flow were generally lower when turbulence intensities increased [14]. As current speed increased above speeds stimulating rheotaxis, fish holding station on the bottom eventually slipped, this being the limit for the gait transition from station holding to free swimming [21]. Also gudgeon, *Gobio gobio*, slipped at lower speeds as TI increased [14]. Maximum cruising speed also decreased as TI increased for perch, *Perca fluviatilis*, and roach. Similarly, prolonged swimming speeds were reduced with increasing TI for perch, roach and gudgeon [14]. Of particular interest in these experiments was the finding that performance of fishes from still-water habitats was reduced more with TI than fishes from populations from streams and rivers. Maximum burst speeds also were reduced with increasing TI for roach and again, the effects of TI were larger for fishes from quieter-water habitats [14]. Finally, Reynolds stresses >30 N.m^{-2} for 10 min in turbulent flows that visibly buffeted fishes, reduced startle responses of hybrid bass (striped bass, *Morone saxatilis* × white bass, *M. chrysops*), and Atlantic salmon parr, *Salmo salar* [19].

There is also an interaction between current speed and TI, such that fishes tend to avoid high TI conditions at both low and at high $\overline{u_{ires}}$, but choose higher TI at intermediate current speeds [14]. This dome-shaped response pattern was also found for the ability of fishes to flow-refuge in the turbulent wake behind vertical and horizontal cylinders [23, 24], and in feeding efficacy [13, 25]. Such dome-shaped relationships probably reflect trades-off in control capabilities and energy costs at various speeds. The low momentum and kinetic energy of fishes at low speeds reduce a fish's ability to control stability, such that energy costs of swimming in turbulent regions may be higher than for swimming in less turbulent areas. Under these circumstances, fish would be expected to avoid high TI [23, 26, 27].

At higher $\overline{u_{ires}}$, the energy costs for translocation will be high, perhaps leaving little surplus for other activities, such as controlling stability, especially if energy costs are as large as many studies suggest [15, 28–35].

Unfed fish also select higher levels of TI than fed fish [14]. This can be attributed to the potential for higher levels of turbulence to increase the probability of bringing food to fish through such mechanisms as increased shear stress displacing prey organisms [35]. Furthermore, contagion rates of food also increase at intermediate turbulence levels, lower turbulence levels cause lower feeding, probably through reduced delivery rates, and high turbulence levels interfere with feeding because of the cost of swimming in such environments [13, 14, 25].

Similar results have been found where fishes have a wider range of choices of positions in complex flows. Locations chosen by juvenile rainbow trout, *Oncorhynchus mykiss*, in a flume were recorded for several hydraulic options created by various combinations of bricks [36]. The juvenile trout chose locations with below-average current speeds, lower TI and also smaller integral lengths as determined from ADV time-series velocity data. When current speeds were high, trout chose regions with below-average current speeds, in locations where TI tended to be higher. In addition, the density of rainbow trout supported in flumes with various bricks also varied with turbulence levels [36].

While most studies to date have shown reduced swimming performance in all gaits with increasing TI, there are some notable exceptions of **neutral effects**. For example, elevated turbulence had no effect on cruising speeds of bluefish, *Pomatomus saltatrix*, [37] and similar results were observed for sprints by inanga, *Galaxias meculatus* [38]. Startle responses of juvenile rainbow trout, *Onchorhynchus mykiss*, were unaffected by turbulence levels that did affect this behavior in hybrid striped bass and Atlantic salmon parr [19]. The authors of these studies have suggested that differences relate to eddy size relative to fish size, such that eddies experienced by fishes were too small to create displacements large enough to require conscious control. Alternatively, differences might result from offsetting effects of different eddy sizes in the incident water movements, some facilitating swimming, while others interfering with swimming. The discrepancies between these studies and those showing negative effects of turbulence underscore the importance of improved methods flow visualization, and the need for more experiments on fish performance in known incident flow fields [24].

An important component of behavior of fishes in turbulent flows is the **deployment of control surfaces**. There is increased use of various median and paired fins by fishes faced with perturbing flows [39, 40]. Tail-beat frequencies of salmonids swimming in the field are often irregular compared to swimming in a flume, and the difference has been attributed to turbulence [31, 39]. McLaughlin and Noakes also described deployment of the pectoral fins by young-of-the-year brook trout, *Salvelinus fontinalis*, swimming in turbulent flows. Paired fin use declined with increasing swimming

speed, consistent with the expectation that stability control is easier at
higher swimming speeds when flow is higher over control surfaces and momentum is greater to damp perturbations. These additional fin motions
and stabilizing behaviors are expected to increase energy costs of swimming compared to those in steady flows. The only direct measure of increased costs for fish swimming in turbulence were obtained for Atlantic
salmon, *Salmo salar*, swimming in a water tunnel with imposed wave-like
pulses [15]. These experiments are unusual in that the flow apparently was
intended to create surge perturbations. The shear forces developed in the
system, however, would undoubtedly have calved eddies, probably creating a flow dominated by a limited size range of eddies. A power spectrum
based on ADV measurements of the incident flow reveals some dominant
flow structures, but PIV would be needed to determine their nature. The
approach deserves further study as a possible method to impose more controlled flow perturbations on fishes than most current methods. Costs
of swimming of fish in the surge-generated incident water movements increased by 30% as TI increased from 0.28 to 0.36 when swimming at an
average speed of $\sim$18 cm s^{-1}, and 1.6-fold as TI increased from 0.21 to
0.30 when swimming at $\sim$23 cm s^{-1} [15]. Metabolic rates for the salmon
swimming in turbulent flows was 1.9- to 4.2-fold larger than expected from
data obtained by others using water tunnels with typical micro turbulent
flow conditions.

In considering costs of swimming in flows that challenge control of stability, such control involves inertial corrections. Hence stability is mechanically similar to maneuvers, and energy costs associated with maneuvers
illustrate the potential for increased energy costs involved in controlling stability. Maneuvers can increase energy expenditure compared to traveling at
the same average translocation speed by over tenfold [15, 20, 28, 29, 31–35].

Another parameter commonly used to quantify turbulent flows is the
Turbulent Kinetic Energy (TKE), which measures the increase in kinetic energy due to turbulent fluctuations in the flow. Then:

$$\text{TKE} = 0.5\,(\sigma_u^2 + \sigma_v^2 + \sigma_w^2) \tag{2}$$

TKE was a good predictor of juvenile rainbow trout densities in flumes [36].

The energy associated with any velocity occurring at a given length
scale is proportional to the square of the magnitude of the velocity. In the
flow, the rate of energy dissipation in the inertial sub-range for an energy
cascade from large to smaller and smaller eddies is a constant [41]. The
relationship between the energy in the flow and fish swimming kinematics
is a topic of active research as the energy in the flow is obviously coupled
to the energy expended by the fish. No simple relationship is apparent
and a functional relationship is unknown. However, TKE proved the best
predictor of density, with highest numbers at intermediate TKE in the case
of trout swimming in a flume with various hydraulic regimes and geometric features [42]. In the field, low current speeds may provide less usable
habitat than higher flows, while high flows displace fish and also increase

energy costs of holding position [9, 10, 15, 35, 36, 42]. Observations in a natural trout stream showed that brown trout, *Salmo trutta,* occupy habitat with intermediate levels of turbulence compared to those available [43]. These situations also have dome-shaped relationships between turbulence levels chosen versus current speed, consistent with the behavior of individual fishes in flumes.

In contrast to negative impacts of swimming in more turbulent flows, positive effects can occur, **improving performance**. For example, there is a potential for fish to extract energy from turbulent flow elements and thereby decrease energy consumption. In flume experiments, swimming speeds of sockeye salmon, *Oncorhynchus nerka,* migrating through a turbulent river were 1.4–76 times higher than those expected based on tail-beat frequencies of fishes swimming in rectilinear, steady incident flow [44]. Muscle activity was reduced in rainbow trout using the Kármán gait, slaloming between the vortices in the Kármán vortex street downstream of a cylinder with a D-shaped cross-section [45, 46].

4. Using Spatial Correlations of Point Measurements. It has been suggested that rather than absolute measures of flow variation, gradients are more important. Spatial gradients would be associated with larger asymmetries than point effects when integrated along a linear dimension of a fish, and would be expected to amplify destabilizing forces. Thus du/ds is thought to affect location preferred by salmonids [47, 48], where u is the local velocity in a given direction and s the spatial coordinate in the same direction. $dTKE/ds$ affects salmonid density at a stream-reach level [49].

Arrays of ADVs have been used in field situations to provide a spatial correlation between individual instantaneous velocity signals [50]. However flow perturbations created by the presence of a large number of instruments is an issue.

Acoustic Doppler Profilers (ADP) provide another tool to evaluate spatial patterns of velocity fluctuations in the field, with receivers placed along a transect line providing instantaneous velocity information at many locations at once. These instruments are usually deployed in deep waters such as oceans and lakes, and the spacing between receivers does not always provide enough resolution to evaluate fine-scale features of the flow. Recent developments in this area are beginning to remedy such limitations. Nevertheless, ADP still provides only 1-D information, so that eddy structure remains elusive with such methods.

Statistical analysis of time-series data of velocity has also been used to estimate a length scale. This is obtained using autocorrelation analysis to determine the "integral length", which can be considered an estimate of average eddy size. Thus, [14, 30] estimated the average eddy size for their flume experiments as 0.66 L, where L was the total length of the fish being studied. We suggest that conceptual and measurement problems are sufficient to make estimations of integral length of little utility in relating turbulent flow structure to embedded organisms. For example: (1) an autocorrelation analysis provides some average length scale for any flow, although the size range averaged is not clear. In practice, specific eddy sizes within the range of length scales are expected to be biologically relevant. (2) Interpretation of time-series data by autocorrelation to obtain the length scale over a large spatial domain includes assumptions such as Taylor's frozen turbulence, i.e. the flow has not changed significantly between each datum sample. This is often likely to be an overly optimistic assumption in natural flows occupied by fishes. (3) In terms of methodology, autocorrelation analyses must be applied over a domain that is large relative to the integral scale to be estimated, which is usually not practical, or practiced [51].

Although consideration of spatial gradients in flow fluctuations is undoubtedly an improvement over point measures alone, a more promising approach builds from basic concepts recognizing that turbulent flows are typically comprised of eddies. The first step is visualizing those eddies for which a growing range of new techniques is emerging; for example Particle Image Velocimetry (PIV), Particle Tracking Velocimetry. With PIV, successive positions along the trajectories of neutrally buoyant microspheres or naturally occurring suspended particles are recorded as they pass through a laser sheet. The video feed or the laser is pulsed at known rates. From successive positions of particles, their velocity – magnitude and direction – can be determined, and hence the flow region can be mapped in terms of streamlines, fields of velocity vectors or vorticity. From these data, the size of eddies can be determined and then related to fish size [16, 45, 52]. PIV has been almost exclusively used in the laboratory, however recent developments of under-water PIV using handheld lasers and cameras in waterproof casings are beginning to allow for acquiring data in the field [53, 54]. It is now possible to obtain the relevant physical parameters in a natural environment. At this time, information on the spatial distribution of turbulent flow features relevant to ecohydraulics is extremely sparse. Neither the composition of the turbulent flows over the domain of a passageway nor eddy characteristics along the paths chosen by a fish are known. Consequently, although structures in the incident turbulent flow are implicated as causing disturbances that affect behavior and fish ecology, definitive evidence in lacking. A notable exception are laboratory studies on a novel use of eddies whereby fish "surf" the Kármán Vortex Street shed by D-cylinders [45, 46]. Liao and colleagues not only recorded the behavior of the fish, but also obtained unique data on muscle use and energetics, while using PIV to visualize eddies.

The observations by Liao and colleagues are also exemplary of an important aspect of behavioral interactions with turbulent flow elements whereby fish extract energy from the incident flow. There has long been discussion on the likely ability of migrants to extract energy to decrease migration times or energy costs [15, 44]. Fish in schools also may be able to extract energy from thrust-eddies shed by adjacent fishes [55]. Strict geometric position relationships are necessary among school members to take maximize advantage of flow components in eddies that have a velocity component in the direction of motion of the benefiting fishes. In addition, not all fish benefit from thrust-eddies at the same time, such that competition would seem likely for preferred conditions. Experimental evidence for appropriate spacing of school members is equivocal [56], but does suggest that some fishes of some species benefit from energy extraction from eddies at some times [57, 58]. In addition, energy may be extracted from eddies shed by the body and fins of a fish by downstream propulsor surfaces [44, 59–65].

5. Going One Step Further: The Physical Basis of Characterizing Turbulent Flow Using Eddies. Eddy-dominated flow is undoubtedly characteristic of fish habitat. It is this eddy structure that is revealed by flow visualization, and which is the basis for positive and negative impacts of turbulence on fishes. Eddies are areas of the flow where the trajectories of water particles, or streamlines, curve, leading to circular motions, such as a whirlpool. Vorticity is defined as the curl of the velocity vector, ω, i.e. a form of angular velocity. Therefore, eddies are directly linked to vorticity and in fact are best defined as regions of finite vorticity. Modern measurement techniques such as PIV provide direct measurement of vorticity, which can be used to define the physical limits and hence sizes of eddies [52, 66].

Most vorticity in fish habitats will derive from viscous forces and shear layers due to the gradient of velocity, the formation of boundary layers, and local stretching of vortex lines. Baroclinic torque increases the amount of vorticity in situations where stratification is present. The most important sources of velocity gradients occur where flows interact with each other or with physical structures in a flow. The overall bathymetry of stream beds, ponds, lakes, estuaries, oceans etc. variously constrict flow, create expansion areas, and cause waves to break, all affecting eddy formation and their subsequent growth as eddies travel downstream. Objects protruding into currents are another important source of eddies commonly encountered by fishes in their habitats. Such protuberances include substratum ripples, corals and macrophytes, rocky materials, woody debris, sunken ships and many structures used in stream and lake improvement projects.

Turbulent flow is comprised of numerous eddies varying in size and vorticity (Fig. 1) as intuitively pictured in Da Vinci's drawings in the late fifteenth century and further prescribed by Kolmogorov [41]. Eddies affecting fishes are in the "inertial sub-range". The largest eddy size in the inertial sub-range, is defined as δ. Over time, and further downstream from the source, the eddy composition of a flow develops finer- and finer-scale turbulence, until the smallest eddy size, λ_o, reaches the Kolmogorov eddy size [41]. Within the inertial range, energy is passed from one eddy to the next of smaller size in an inviscid fashion (i.e. no energy is lost due to

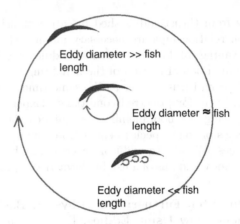

FIG. 1. *Schematic of eddy sizes with respect to a single fish size. Eddy sizes are a function of geometry and Reynolds number*

viscous effects) until the Kolmogorov size is reached. At this lower limit of the eddy size range, eddies are affected by viscosity, so that Kolmogorov eddies are eventually damped by viscosity and their energy is dissipated as heat. Eddy composition of fully developed flow can be described as a frequency distribution of eddy sizes, with eddy size decreasing logarithmically from many small-sized eddies to few large-size eddies (Fig. 2). The ratio of the largest to smallest eddy sizes is a function of Reynolds number based on δ, and $\lambda_o/\delta = \mathrm{Re}^{-0.75}$ [41]. Although data on the distribution of eddy sizes in natural fish habitats are lacking, it will vary locally depending on Reynolds number, upstream conditions, and details of local physical structures and bathymetry, fishes in natural habitats will encounter eddies of sizes within the limits of λ_o and δ.

Understanding turbulence is a major challenge for engineers, physicists and mathematicians, so it should be noted that the inertial sub-range idea is a theoretical construct. It has worked very well in engineering applications by providing a framework to compare eddy characteristics and to quantify turbulent flows. Therefore, this approach to turbulence is by no means a closed subject. Nevertheless, it is proving useful to classify turbulent situations to which fish react [17, 24]. By analogy, the influential classic work in the 1960s and 1970s also used the theoretical construct of vortex sheets to develop ideas on fish swimming [67–69].

Given the lack of information on the composition of natural flows and specific observations and how fish interact with such flow variations, a generalizable mechanistic explanation for the destabilizing properties of turbulent flow is a major need in ecohydraulics. The need for a common analytic framework is illustrated by the sometimes conflicting results on fish performance from various studies that have used widely different methods

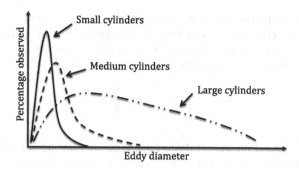

FIG. 2. *Distribution of eddy diameters for the different configurations tested in the laboratory [16]. A wider range of eddy sizes is obtained for the large cylinder diameter case*

to induce unsteadiness into the flow. Turbulence is commonly induced in experimental studies by in-current screens, bricks and similar objects [14, 42, 50]. A variant was used in which turbulence was induced by corrugated plastic walls and blocking some openings in an upstream grid [38]. Flow visualization using methods such as PIV would provide a basis for common descriptions of flow composition in these disparate situations.

Flow visualization also provides the information necessary to understand fish-eddy interactions. Here, an important basic concept has been proposed [14], one used by most subsequent studies seeking explanations for their results. This concept recognizes that relative size of eddies and fishes are important. Thus destabilizing disturbances in eddy-dominated flow are related to the relative size of eddies compared to the size of a fish (Fig. 1). Thus large gyres, such as the North Pacific Gyre, and other eddies very much larger than fish size are certainly important in dispersal and migration, but it is likely that fish respond to them in the same way as they would for uniform flow. At the opposite end of the spectrum, eddies that are very much smaller than fish length are undoubtedly ignored, although they may contribute to mixing and maintaining high dissolved oxygen levels and similar water characteristics important to the physiology and ecology of fishes. In contrast, eddies of intermediate size, apparently of the order of 0.5–1 fish lengths are associated with negative effects such as reduced swimming performance, and can cause displacements triggering stability failure [14, 17, 24].

At this time, the analysis of the characteristics of eddies relative to those of a fish are rare, and essentially limited to length scale as a measure of size. Length alone is insufficient to define the attributes of eddies that challenge posture and trajectory control. Instead fish-turbulence interactions should be sought in terms of momentum and forces that determine the strength of perturbations. These work against the damping and corrective properties of a fish, the interaction determining the magnitude and direction of displacements. A useful starting point for such evaluation of turbulent-flow impacts on a fish is circulation, which is measured by integrating the vorticity, ω, over a surface area, A, or by integrating the velocity, V, around a certain region of the flow, which here can be represented by an embedded fish. Thus circulation, Γ, is defined as:

$$\Gamma = - \oint_A \omega.dA = - \oint \vec{V}.dl \qquad (3)$$

The absolute value of circulation per unit area has been used to quantify flow complexity within a study reach or within meso-scale habitats such as pools, eddies, riffles and transverse flows because positive and negative circulation could potentially cancel out [49]. The minus sign appears because of the convention of positive line integrals to be counterclockwise whereas positive circulation is typically defined in the clockwise direction.

We propose using circulation and vorticity as starting points for quantifying the impact of turbulent eddies on fish swimming and stability through a series of non-dimensional parameters. Normalizing Γ by the product of fish velocity, V_{fish} and fish size, L_{fish}, gives a non-dimensional measure of the impact of eddy circulation on a fish. When this ratio is small, resulting displacements of the fish will be small, and may be small enough to neglect. Intermediate values will require active control behaviors, while large values will overwhelm these control systems and fish posture or trajectories will become highly irregular. In addition, other significant parameters are proposed whose relative importance will depend on the type and duration of interactions between fish and eddies. These are primarily: (1) linear or angular impulse, (2) linear or angular momentum (Table 1) both directly related to eddy circulation [70]. Impulse is related to the amount of force produced by eddies (i.e. the integral of thrust or force over a given time) and to momentum flux by definition. For example, momentum flux data derived from PIV measurements was used to estimate the amount of force produced by the tail of a robot lamprey [71]. Furthermore, velocity and pressure can be back-calculated from these quantities and the flow surrounding the embedded body estimated. Pressure can be integrated along the body, for example, and lift and drag forces calculated (i.e. horizontal and vertical forces), while taking into account added-mass effect [72]. This step in analyzing fish-turbulence interactions has yet to be made, but has been used in other situations [73].

We estimated the values of the different parameters for three turbulent flow conditions created by cylinder arrays of different sizes, small (0.4 cm in diameter), medium (1.6 cm) and large cylinders (8.9 cm) (Table 2). Swimming performance of creek chub, *Semotilus atromaculatus*, total length 12.2 ± 0.9 cm, was measured for fish swimming in the turbulent wakes created downstream of vertical and horizontal cylinder arrays [16]. PIV was used to measure eddy diameters, which showed that successively larger cylinder arrays essentially added successively larger eddies to

TABLE 1

Non-dimensional parameters for considering interactions between fishes and eddies to evaluate conditions where eddies should be expected to negatively affect swimming and thresholds for stability failure

Parameters	Incident flow Eddy diameter L_e	Fish Fish length L_f	Dimensionless variables
Length scale			L_e/L_f
Circulation	$\Gamma_e \propto V_e.L_e$	$\Gamma_f \propto V_f.L_f$	Γ_e/Γ_f
Linear impulse	$I_e \propto \rho_e \Gamma_e L_e{}^2$	$I_f \propto \rho_f V_f L_f{}^3$	I_e/I_f
Momentum flux	$M_e \propto \rho_e V_e{}^2 L_e{}^2$	$M_f \propto \rho_f V_f{}^2 L_f{}^2$	M_e/M_f

TABLE 2

Non-dimensional measures for fish-eddy interactions based on observations [16] on creek chub swimming in turbulent flow created by arrays of small (diameter 0.4 cm), medium (diameter 1.6 cm) and large (diameter 8.9 cm) cylinders. Gaps between cylinders were equal to cylinder diameters. Calculations were made for the 95th percentile eddy diameters

Flow	L_e/L_f	Γ_e/Γ_f	I_e/I_f	M_e/M_f
Small cylinders	0.09	0.01	0.0008	0.0001
Medium cylinders	0.25	0.09	0.006	0.0087
Large cylinders	0.69	1.32	0.64	1.76

the flow. The large cylinders added eddies with diameters of the order of 8.4 cm, resulting in ∼10–20% reductions in swimming performance and frequent failures of posture and trajectory, seen as failure to hold position in the flow. This simple analysis clearly identified flow conditions that create significant challenges for fish swimming and stability, i.e. the case of large cylinders is the only one creating conditions for which all the dimensionless parameters are on the order of unity. Simultaneous images of the turbulent incident flow and fish responses were not possible as fish maneuvered through a large three-dimensional space [16]. As a result, the actual eddies triggering failure were not known, which contributes to the variability in the magnitude of these metrics.

We are cognizant that analytical steps to go from vorticity or circulation to perturbing forces are not simple. Irrespective of these challenges, our analysis of fish-eddy interactions clearly indicates that simultaneous flow-fish observations are vital [45] if ecohydraulics is to be provided with a strong conceptual basis. It might seem attractive to use velocity measurements in the wake to obtain drag force, analogous to attempts to use such semi-empirical, quasi-static approaches to model undulatory propulsion [74]. However, there were few alternatives when these early studies were performed. The uncertainties concerning appropriate drag coefficients, and accommodation of unsteady effects make such an approach even less suitable today.

6. Numerical Tools. The practical problems of explicitly determining incident-flow-embedded-body forces suggest that modeling may be a desirable complement, or alternative in some cases, to direct observations. In particular, Computational Fluid Dynamics (CFD) has the capability to describe turbulent flow components in complex flows, and evaluate the consequences for objects embedded at various places in the flow. CFD models have been widely used in many engineering applications. Such models use a variety of techniques, from Reynolds Averaged Navier-Stokes Equations (RANS), to Large Eddy Simulations (LES), to Direct Numerical Simulations (DNS), etc. For example, the flow behind a large rectangular block in a large aspect-ratio channel was analyzed [75], and slow moving large-scale structures were found to interact with the shear layer behind the block. This type of flow dynamics is important in designing river restoration projects and understanding fish behavior behind such structures. Fish swimming behind an obstacle integrating flow features and fish kinematics was computed [76]. Others use a combination of inviscid and viscous approaches to compute the locomotion of fish-like bodies [77]. This is an active area of research with significant recent developments. As with other flow measurement and flow visualization empirical methods, all have advantages and constraints. Some of the limitations are due to the lack of data available for calibration of such models.

7. Future Directions. New developments in conceptual approaches, instruments and numerical tools are very promising and will help pave the future in this area. The need for simultaneous flow and fish visualization has been emphasized above. Current technology is essentially limited to 2-D observation, but 3-D technology is now appearing. Other technologies are needed to rapidly assay streams and other situations, to provide guidance for science-based management, and also to refine questions and variables for exacting laboratory studies. For example, ultrasound could be used in streams to directly evaluate the circulation of vortices shed behind obstacles. Ultrasound has been used on airport runways to estimate the circulation of large wing tip vortices and help optimize the time delay between take-offs. The principle is that the presence of turbulence in the flow affects the delay in ultrasound traveling through the water. Sound emitters and receivers in a square formation are used, and average vorticity and circulation within a sample volume, which can be as small as $2\,cm^2$ can be determined. This method has not to our knowledge been used in studying fish, but transducers are fairly robust and could be deployed in the field and laboratory [78]. CFD will also be vital in the lexicon of tools to model fish-flow-habitat interactions, and will be especially important for rapid assays to focus laboratory and field studies, and for designing management options.

One of the key factors for success is the continued and increased communication between engineers and biologists so that the data obtained in both laboratory and field settings can be used to calibrate and validate both analytical and numerical models.

REFERENCES

[1] Lagler KF, Bardach JE, Miller RR, Pasino DRM (1977) Ichthyology. Wiley, New York
[2] Bond CE (1996) Biology of fishes. Saunders, New York
[3] Allan JD, Castillo MM (2007) Stream ecology: structure and function of running waters, 2nd edn. Springer, Dordrecht

[4] Hora SL (1935) Ancient hindu concepts of correlation between form and locomotion of fishes. J Asiat Soc Bengal Sci 1:1–7

[5] Popper D, Fishelson L (1973) Ecology and behavior of anthias squamipinnis (Peters, 1855) (Anthiidae, Teleostei) in the coral habitat of eilat (Red Sea). J Exp Zool 184:409–424

[6] Hopson ES (1974) Feeding relationships of teleostean fishes on coral reefs in Kona, Hawaii. US Fish Bull 7:915–1031

[7] Jokiel PL, Morrissey JI (1993) Water motion in coral reefs: evaluation of the clod-card technique. Mar Ecol Prog Ser 93:175–181

[8] Fulton CJ, Bellwood DR (2002) Ontogenetic habitat use in labrid fishes: an eco-morphological perspective. Mar Biol Prog Ser 236:255–262

[9] Fulton CJ, Bellwood DR (2005) Wave-induced water motion and the functional implications for coral reef fish assemblages. Limnol Oceanogr 50:255–264

[10] Fulton CJ, Bellwood DR, Wainwright PC (2005) Wave energy and swimming performance shape coral reef fish assemblages. Proc R Soc London, Ser B 272:827–832

[11] Depczynski M, Bellwood DR (2005) Wave energy and spatial variability in community structure of small cryptic coral reef fishes. Mar Biol Prog Ser 303:283–293

[12] Liao JC (2007) A review of fish swimming mechanics and behavior in perturbed flows. Philos Trans R Soc Biol Sci 362:1973–1993

[13] McKenzie B, Kiorobe T (1995) Encounter rates and swimming behavior of pause-travel and cruise larval fish predators in calm and turbulent laboratory environments. Limnol Oceanogr 40:1278–1289

[14] Pavlov DS, Lupandin AI, Skorobogatov MA (2000) The effects of flow turbulence on the behavior and distribution of fish. J Ichthyol 40(2):S232–S261

[15] Enders EC, Boisclair D, Roy AG (2003) The effect of turbulence on the cost of swimming for juvenile Atlantic salmon (Salmo salar). Can J Fish Aquat Sci 60:1149–1160

[16] Tritico HM (2008) The effects of turbulence on habitat selection and swimming kinematics of fishes. Dissertation submitted in partial fulfillment of the requirements for the degree of doctor of philosophy, University of Michigan, Ann Arbor

[17] Webb PW, Cotel AJ, Meadows LA (2010) Waves and eddies: effects on fish behaviour and habitat distribution. In Domenici P and Kapoor BG (eds), Fish Locomotion: An Eco-Ethological Perspective Science Publishers, Enfield, NH, pp. 1–39

[18] Sanford LP (1997) Turbulent mixing in experimental ecosystem studies. Mar Biol Prog Ser 161:265–293

[19] Odeh M, Noreika JF, Haro A, Maynard A, Castro-Santos T, Cada GF (2002) Evaluation of the effects of turbulence on the behavior of migratory fish. Final report 2002, report to Bonneville Power Administartion, Contract no., 00000022, Project no. 200005700, pp 1–55

[20] Puckett KJ, Dill LM (1984) Cost of sustained and burst swimming of juvenile coho salmon (Oncorhynchus kisutch). Can J Fish Aquat Sci 41:1546–1551

[21] Arnold GP, Weihs D (1978) The hydrodynamics of rheotaxis in the plaice (Pleuronectes platessa). J Exp Biol 75:147–169

[22] Webb PW (1989) Station holding by three species of benthic fishes. J Exp Biol 145:303–320

[23] Webb PW (1998) Entrainment by river chub, nocomis micropogon, and smallmouth bass, micropterus dolomieu, on cylinders. J Exp Biol 201:2403–2412

[24] Tritico HM, Cotel AJ (2010) The effects of turbulent eddies on the stability and critical swimming speed of Creek chub, Semotilus atromaculatus. J Exp Biol 213:2284–2293

[25] Galbraith PS, Browman HI, Racca RG, Skiftesvik AB, Saint-Pierre J (2004) Effect of turbulence on the energetics of foraging in Atlantic cod Gadus morhua larvae. Mar Ecol Prog Ser 201:241–257

[26] Webb PW (2002) Control of posture, depth, and swimming trajectories of fishes. Integr Comp Biol 42:94–101

[27] Webb PW (2006) Stability and maneuverability. In: Shadwick RE, Lauder GV (eds) Fish physiology. Elsevier Press, San Diego, pp 281–332

[28] Blake RW (1979) The energetics of hovering in the mandarin fish (Synchropus picturatus). J Exp Biol 82:25–33

[29] Weatherley AH, Rogers SC, Pinock DG, Patch JR (1982) Oxygen consumption of active rainbow trout, Salmo gairdneri Richardson, derived from electromyograms obtained by radiotelemetry. J Fish Biol 20:479–489

[30] Lupandin AI (2005) Effect of flow turbulence on swimming speed of fish. Biol Bull 32:558–565

[31] Puckett KJ, Dill LM (1985) The energetics of feeding territoriality in juvenile coho salmon (Oncorhynchus kisutch). Behaviour 92:97–111

[32] Webb PW (1991) Composition and mechanics of routine swimming of rainbow trout, Oncorhynchus mykiss. Can J Fish Aquat Sci 48:583–590

[33] Boisclair D, Tang M (1993) Empirical analysis of the swimming pattern on the net energetic cost of swimming in fishes. J Fish Biol 42:169–183

[34] Krohn MM, Boisclair D (1994) Use of a stereo-video system to estimate the energy expenditure of free-swimming fish. Can J Fish Aquat Sci 51:1119–1127

[35] Boisclair D (2001) Fish habitat models: from conceptual framework to functional tools. Can J Fish Aquat Sci 58:1–9

[36] Smith DL, Brannon EL, Shafii B, Odeh M (2006) Use of the average and fluctuating velocity components for estimation of volitional rainbow trout density. Trans Am Fish Soc 135:431–441

[37] Ogilvy CS, DuBois AB (1981) The hydrodynamics of swimming bluefish (Pomatomus saltatrix) in different intensities of turbulence: variation with changes in buoyancy. J Exp Biol 92:67–85

[38] Nikora VI, Aberle J, Biggs BJF, Jowett IG, Sykes JRE (2003) Effects of size, time-to-fatigue and turbulence on swimming performance: a case study of Galaxias meculatus. J Fish Biol 63:1365–1382

[39] MacLaughlin RL, Noakes DL (1998) Going against the flow: an examination of the propulsive movements made by young brook trout in streams. Can J Fish Aquat Sci 55:853–860

[40] Webb PW (2004) Response latencies to postural disturbances in three species of teleostean fishes. J Exp Biol 207:955–961

[41] Kolmogorov AN (1941) Local structure of turbulence in an incompressible viscous fluid at very high Reynolds numbers. Dolk Akad Nauk SSSR 30:299. Reprinted in Usp Fix Nauk 93:476–481 (1967) (Trans: in Sov Phys Usp 10:734–736 (1968))

[42] Smith DL, Brannon EL (2005) Response of juvenile trout to turbulence produced by prismatoidal shapes. Trans Am Fish Soc 134:741–753

[43] Cotel AJ, Webb PW, Tritico H (2006) Do trout choose habitats with reduced turbulence? Trans Am Fish Soc 135:610–619

[44] Standen EM, Hinch SG, Rand PS (2004) Influence of river speed on path selection by migrating adult sockeye salmon (Oncorhynhus mykiss). Can J Fish Aquat Sci 61:905–912

[45] Liao JC, Beal DN, Lauder GV, Triantafyllou MS (2003) The Kármán gait: novel body kinematics of rainbow trout swimming in a vortex street. J Exp Biol 206:1059–1073

[46] Liao JC (2004) Neuromuscular control of trout swimming in a vortex street: implications for energy economy during the Kármán gait. J Exp Biol 207:3495–3506

[47] Fausch KD, White RJ (1986) Competition among juveniles of coho salmon, brook trout, and brown trout in a laboratory stream, and implications for great lakes tributaries. Trans Am Fish Soc 115:363–381

[48] Hayes JW, Jowett IG (1994) Microhabitat models of large drift-feeding brown trout in three New Zealand rivers. North Am J Fish Manag 14:710–725

[49] Crowder DW, Diplas P (2002) Vorticity and circulation: spatial metrics for evaluating flow complexity in stream habitats. Can J Fish Aquat Sci 59(4):633–645

[50] Roy AG, Buffin-Bélanger T, Lamarre H, Kirkbride A (2004) Size, shape and dynamics of large-scale trubulent flow structures in a gravel-bed river. J Fluid Mech 500:1027

[51] O'Neill PL, Nicolaides D, Honnery D, Soria J (2004) Autocorrelation functions and the determination of integral scale with reference to experimental and numerical data. In: 15th Australasian fluid mechanics conference, Sydney

[52] Drucker EG, Lauder GV (1999) Locomotor forces on a swimming fish: three-dimensional vortex wake dynamics quantified using digital particle image velocimetry. J Exp Biol 202:2393–2412

[53] Tritico HM, Cotel AJ, Clarke J (2007) Development of small scale submersible PIV system. Meas Sci Technol 18(8):2555–2562

[54] Katija K, Dabiri JO (2008) In situ field measurements of aquatic animal fluid interactions using a self-contained underwater velocimetry apparatus (SCUVA). Limnol Oceanogr Method 6:162–173

[55] Weihs D (1973) Hydromechanics of fish schooling. Nature (London) 241:290–291

[56] Partridge BL, Pitcher TJ (1980) Evidence against a hydrodynamic function for fish schools. Nature (London) 279:418–419

[57] Herskin J, Steffensen JF (1998) Energy savings in sea bass swimming in a school: measurements of tail beat frequency and oxygen consumption at different swimming speeds. J Fish Biol 53:366–376

[58] Svendsen JC, Skov J, Bildsoe MJ, Steffensen F (2003) Intra-school positional preference and reduced tail beat frequency in trailing positions in schooling roach under experimental conditions. J Fish Biol 62:834–846

[59] Lighthill J (1969) Hydromechanics of aquatic animal propulsion. Ann Rev Fluid Mech 1:413–45

[60] Triantafyllou GS, Triantafyllou MS, Gosenbaugh MA (1993) Optimal thrust development in oscillating foils with application to fish propulsion. J Fluids Struct 7:205–224

[61] Weihs D (1993) Stability of aquatic animal locomotion. Contemp Math 141:443–461

[62] Nauen JC, Lauder GV (2000) Locomotion in scombrid fishes: morphology and kinematics of the finlets of the chub mackerel Scomber japonicas. J Exp Biol 203:2247–2259

[63] Nauen JC, Lauder GV (2001) Locomotion in scombrid fishes: visualization of flow around the caudal peduncle and finlets of the chub mackerel Scomber Japonicas. J Exp Biol 204:2251–2263

[64] Beal DN, Hover FS, Triantafyllou MS, Liao J, Lauder GV (2006) Passive propulsion in vortex wakes. J Fluid Mech 549:385–402

[65] Alben S, Madden PG, Lauder GV (2006) The mechanics of active fin-shape control in ray-finned fishes. J R Soc London Interface. DOI 10.1098/rsif.2006.0181

[66] Adrian RJ, Christensen KT, Liu Z-C (2000) Analysis and interpretation of instantaneous turbulent velocity fields. Exp Fluids 29:275–290

[67] Lighthill J (1975) Mathematical bio fluid dynamics. Society for Industrial and Applied Mathematics, Philadelphia

[68] Newman JN, Wu TY (1975) Hydromechanical aspects of fish swimming. Symp Swim Fly Nat 2:615–634

[69] Wu TY (1977) Introduction to scaling of aquatic animal locomotion. In: Pedley TJ (ed) Scale effects of animal locomotion. Academic, New York, pp 203–232

[70] Saffman PG (1992) Vortex dynamics. Cambridge University Press, Cambridge/New York

[71] Hultmark M, Leftwich M, Smits AJ (2007) Flow measurements in the wake of a robotic lamprey. Exp Fluids 43:683–690

[72] Dabiri J (2005) On the estimation of swimming and flying forces from wake measurements. J Exp Biol 208:3519–3532

[73] Schultz WW, Webb PW (2002) Power requirements of swimming: do new methods resolve old questions? Integr Comp Biol 42:1018–1025

[74] Gray J (1968) Animal locomotion. Weidenfeld & Nicolson, London

[75] Paik J, Sotiroupolos F (2005) Coherent structure dynamics upstream of a long rectangular block at the side of a large aspect ratio channel. Phys Fluids 17:104–115

[76] Borazjani I, Sotiropoulos F (2008) Numerical investigation of the hydrodynamics of carangiform swimming in the transitional and inertial flow regimes. J Exp Biol 211:1541–1558

[77] Eldredge J (2010) A reconciliation of viscous and inviscid approaches to computing locomotion of deforming bodies. J Exp Mech 50:1349–1353

[78] Johari H, Durgin WW (1998) Direct measurement of circulation using ultrasound. Exp Fluids 25:445–454

THE FLUID DYNAMICS OF FEEDING IN THE UPSIDE-DOWN JELLYFISH

CHRISTINA HAMLET(✉)*, LAURA A. MILLER†,
TERRY RODRIGUEZ†, AND ARVIND SANTHANAKRISHNAN‡

Abstract. The jellyfish has been the subject of numerous mathematical and physical studies ranging from the discovery of reentry phenomenon in electrophysiology to the development of axisymmetric methods for solving fluid-structure interaction problems. In the area of biologically inspired design, the jellyfish serves as a simple case study for understanding the fluid dynamics of unsteady propulsion with the goal of improving the design of underwater vehicles. In addition to locomotion, the study of jellyfish fluid dynamics could also lead to innovations in the design of filtration and sensing systems since an additional purpose of bell pulsations is to bring fluid to the organism for the purposes of feeding and nutrient exchange. The upside-down jellyfish, *Cassiopea* spp., is particularly well suited for feeding studies since it spends most of its time resting on the seafloor with its oral arms extended upward, pulsing to generate currents used for feeding and waste removal. In this paper, experimental measurements of the bulk flow fields generated by these organisms as well as the results from supporting numerical simulations are reviewed. Contraction, expansion, and pause times over the course of many contraction cycles are reported, and the effects of these parameters on the resulting fluid dynamics are explored. Of particular interest is the length of the rest period between the completion of bell expansion and the contraction of the next cycle. This component of the pulse cycle can be modeled as a Markov process. The discrete time Markov chain model can then be used to simulate cycle times using the distributions found empirically. Numerical simulations are used to explore the effects of the pulse characteristics on the fluid flow generated by the jellyfish. Preliminary results suggest that pause times have significant implications for the efficiency of particle capture and exchange.

Key words. Animal locomotion, fluid dynamics of feeding, jellyfish, fluid-structure interaction, immersed boundary methods

1. Introduction. The current dogma in the biology literature is that jellyfish propulsion represents the simplest model system of animal locomotion [7], integrating one of the simplest nervous and muscular systems [16, 17, 33]. Due in part to their relatively simple design, a large body of work in comparative biomechanics is focused on understanding the flows generated by pulsing jellyfish with the goal of relating their morphology to their ecology. Another goal of this work is to use jellyfish as a case study for unsteady propulsion using the philosophy that improved understanding of such systems will lead to innovations in the design of underwater vehicles. The fluid dynamics and scaling effects of this unsteady, biomechanical system have been investigated using mathematical modeling [8, 9, 12],

---

*Department of Mathematics, North Carolina State University, 27695, Raleigh, NC, USA

†Department of Mathematics, University of North Carolina, 27599, Chapel Hill, NC, USA

‡Department of Biomedical Engineering, Georgia Institute of Technology, 30332, Atlanta, GA, USA

S. Childress et al. (eds.), *Natural Locomotion in Fluids and on Surfaces*, IMA 155, DOI 10.1007/978-1-4614-3997-4_3,
© Springer Science+Business Media New York 2012

experiments [6, 10, 24] and numerical simulations [15, 18, 23, 31] with a focus on the generation of propulsive forces. The focus of this paper is to describe the fluid dynamics of jellyfish feeding with some discussion on how feeding currents may be coupled to or uncoupled from propulsion.

1.1. Swimming in Pelagic Jellyfish. Jellyfish propulsion may generally be divided into the rowing or paddling mode of swimming used by oblate medusae (such as *Aurelia auriata* [10]) and jet propulsion used by prolate medusae (such as *Nemopsis bachei* [11]). In the rowing mode of swimming, a starting vortex ring is generated during bell contraction, and a stopping vortex ring is generated during expansion. Vortex rings separate from the bell margin when the motion reverses and then travel away from the bell in pairs. Each traveling pair of vortex rings form a lateral vortex superstructure. Adjacent lateral vortex superstructures pull fluid into the wake and downstream of the jellyfish, enhancing forward propulsion. In the case of prolate medusae, a single vortex ring is generated during the contraction that is quickly swept downstream. The complicated vortex superstructures consisting of oppositely spinning starting and stopping vortices observed in oblate medusae are not present for the prolate species.

1.2. Coupled Swimming and Feeding. In addition to propulsion, another benefit of having a vortex-laden wake is that it can be used to enhance the efficiency of prey capture in medusae that use cruising modes of predation. For example, computational studies suggest that the tentacles of *Aurelia victoria* are positioned within the starting and stopping vortices so as to maximize the possibility of capturing prey throughout the entire pulsing cycle. In addition, the prey are less likely to escape due to the presence of these rotational regions in the flow field. In contrast, vortex shedding by the prolate medusa *S. tubulosa* ejects fluid far from the bell presenting little opportunity for localized foraging. This has implications on their ecological role as ambush predators, where there is an increased demand for short-time impulses of large thrust generation for prey capture and not necessarily for efficient swimming [5].

1.3. Feeding in the Benthic Jellyfish. The genus *Cassiopea*, the upside-down jellyfish, presents a unique opportunity to uncouple the fluid dynamics of feeding and swimming. *Cassiopea* is an oblate scyphozoan medusa that belongs to the order Rhizostomeae and is found in shallow protected marine environments saturated with sunlight. These medusae seldom swim, and instead prefer to rest their bell upside-down on a substrate directing their oral arms to the free surface. *Cassiopea* medusae have been examined in previous studies for their ecological roles in influencing benthic oxygen and inorganic nutrient fluxes [38], ability to thrive in eutrophic environments [2], potential to accumulate trace metals from urban environments [35], and ability to function as a biomonitor for environmental phosphates in marine coastal systems [36]. Through periodic contractions and relaxations of the bell margin, *Cassiopea* medusae drive

water into and awayfrom the subumbrellar cavity and through their oral
arms. This volume of entrained fluid is used to sample for particulate prey
consisting of copepods and other zooplankton, and the fluid flow is also
used for the exchange of oxygen, nutrients and the excretion of organic
fecal material [1, 21, 37]. Unlike the commonly studied *Aurelia* spp. that
are of the Order Semaeostomeae, *Cassiopea* do not possess tentacles or a
central primary mouth. Instead, the upside-down jellyfish have prominent
oral arms that can extend to lengths larger than the maximum bell radius (see Fig. 1). Contained within each of the frilly oral arm branched
vesicles are multiple secondary mouths. More details on the anatomy and
development of *Cassiopea* can be found elsewhere [3].

In this paper, the nature of the fluid flow generated by *Cassiopea* is
described with the use of experimental measurements of the fluid flow fields
and numerical simulations of an idealized, pulsing jellyfish. The results are
compared to the observed fluid flows generated by free-swimming oblate
medusae.

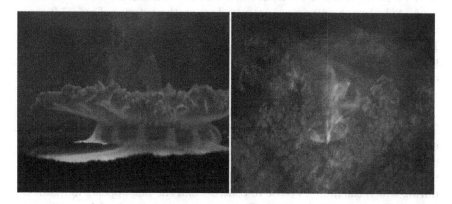

FIG. 1. *Dye visualization of flow driven by pulsations of the bell of the upside down
jellyfish, Cassiopea xamachana. The left picture shows a side view of the organism, and
the right picture shows a top view of the flow through the oral arms*

2. Methods.

2.1. Experimental Methods. An individual medusa was placed
into a 10 L glass aquarium that was 20.7 × 40.3 cm at the base and 26 cm
in height (see Fig. 2). Black sand was used in the observation tank to
reduce glare in the optical measurements. Videos of the medusa were
filmed using a Canon XH A1 camcorder at a rate of 30^{-1} s. To quantify
the time-averaged flow generated by *Cassiopea*, measurements were performed using two-dimensional digital particle image velocimetry (DPIV).
The laser sheet for the PIV measurements was generated by a 50 mJ
double-pulsed Nd:YAG laser manufactured by Continuum Inc., which
emits light at a wavelength of 532 nm. Phase-averaged data was ac-

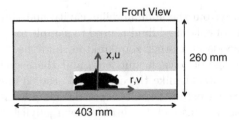

FIG. 2. *Particle Image Velocimetry (PIV) setup: 50 mJ double-pulsed Nd:YAG laser cut through the axis of symmetry of the jellyfish. The motion of illuminated particles is used to construct the flow field*

quired using a 14 bit CCD camera (Imager Intense, LaVision Inc.) with a $1{,}376 \times 1{,}040$ pixel array at a camera frame rate of 15 Hz. For the time-averaged data, 340 individual images were recorded for processing resulting in 170 velocity vector fields from which flow information and statistics were generated. Image pairs were processed using the software Davis 7.0 provided by LaVision Inc. Each DPIV realization consisted of an array of 100×75 vectors, each separated by a width of 12 pixels. Further details of the experimental setup and analysis may be found in Santhanakrishnan et al. [32].

2.2. Temporal Structure of Pulse Cycle and Markov Model.

The motion of the bell may be broken down into the following stages: (1) contraction, (2) first pause, (3) expansion, and (4) second pause. The contraction of the bell was defined as the time during which the sides of the bell move toward the central axis. The first pause is defined as the time during which the muscles cease to contract and the bell appears motionless before beginning its expansion. The expansion is defined as the motion of the bell out from the central axis of the organism. Finally the second pause is defined as the time that the bell appears motionless before the start of the next contraction. The total pulse period is defined as the time from the beginning of a bell contraction to the end of the rest period prefacing the next contraction.

The contraction, first pause, and expansion times were well described by a Gaussian distribution with relatively small variance (see Results). The length of the rest periods between expansion and contraction varied greatly from cycle to cycle. The bimodal distribution of the pause times suggests that a simplifying assumption that the lengths of the pauses can be described by the Markov Property might be appropriate. The Markov Property is such that given a system that exhibits a particular state i at time t, the probability that the system transitions to state j at time $t+1$ is independent of past behavior. Using this assumption, a two state discrete time Markov chain (DTMC) is created, and the transition probabilities are estimated from data collected on individual medusae.

2.3. Numerical Method. The goal of this part of the study is to numerically solve the Navier-Stokes equations coupled to a moving boundary representing the jellyfish bell and a porous layer representing the oral arms. One popular method for simulating such problems is the Immersed Boundary Method (IBM) first introduced by Peskin [28] and later applied to a variety of problems in animal locomotion [13, 18, 26]. The following outline describes the two-dimensional formulation of the immersed boundary method. For a complete description of the exact numerical scheme used in this paper, see Peskin and McQueen [29]. The equations of fluid motion are given by the Navier–Stokes equations:

$$\rho(\mathbf{u}_t(\mathbf{x},t) + \mathbf{u}(\mathbf{x},t) \cdot \nabla \mathbf{u}(\mathbf{x},t)) = \nabla p(\mathbf{x},t) + \mu \nabla^2 \mathbf{u}(\mathbf{x},t) + \mathbf{F}(\mathbf{x},t) \quad (1)$$
$$\nabla \cdot \mathbf{u}(\mathbf{x},t) = 0 \quad (2)$$

where $\mathbf{u}(\mathbf{x},t)$ is the fluid velocity, $p(\mathbf{x},t)$ is the pressure, $\mathbf{F}(\mathbf{x},t)$ is the force per unit area applied to the fluid by the immersed boundary, ρ is the density of the fluid, and μ is the dynamic viscosity of the fluid. The independent variables are the time t and the position $\mathbf{x}$.

The interaction equations between the fluid and the boundary are given by:

$$\mathbf{F}(\mathbf{x},t) = \int \mathbf{f}(q,t)\delta\left(\mathbf{x} - \mathbf{X}(q,t)\right) dq \quad (3)$$

$$\mathbf{X}_t(q,t) = \mathbf{U}(\mathbf{X}(q,t)) = \int \mathbf{u}(\mathbf{x},t)\delta\left(\mathbf{x} - \mathbf{X}(q,t)\right) d\mathbf{x} \quad (4)$$

where $\mathbf{f}(q,t)$ is the force per unit length applied by the boundary to the fluid as a function of Lagrangian position and time, $\delta(\mathbf{x})$ is a two-dimensional delta function, $\mathbf{X}(q,t)$ gives the Cartesian coordinates at time t of the material point labeled by the Lagrangian parameter q. Equation 3 applies force from the boundary to the fluid grid and adds an external force term, and Eq. 4 evaluates the local fluid velocity at the boundary. The boundary is then moved at the local fluid velocity, and this enforces the no-slip condition. Each of these equations involves a two-dimensional Dirac delta function δ, which acts in each case as the kernel of an integral transformation. These equations convert Lagrangian variables to Eulerian variables and vice versa.

The force $\mathbf{f}(q,t)$ is specific to the problem. In recent studies, such forces have included active elastic forces due to muscle contraction [19, 29], cohesive forces between boundaries or cells [14], the action of dynein molecular motors [22, 27], and thermal fluctuations [30]. In studies that focus on feeding efficiency, it is simpler to specify the motion of the boundary. In the case presented in this paper, the jellyfish bell is moved using target points [25].

The oral arms of the jellyfish are modeled as a simple porous layer using the porous immersed boundary method of Stockie and others [20, 34]. In this method, an infinitely thin porous layer is represented by modifying the movement of the boundary relative to the fluid. According to Darcy's law [4], the flow through the boundary is proportional to the pressure jump across the boundary. Since the pressure jump is equivalent to the force per unit area normal to the boundary, this force can be used directly to calculate the flow through the boundary. The flow is then modeled as a slip between the fluid and the boundary specified by the equation

$$\mathbf{X}_t(q,t) = \mathbf{U}(\mathbf{X}(q,t)) + \lambda(\mathbf{f}(q,t) \cdot \mathbf{n})\mathbf{n} \tag{5}$$

where λ is the porosity and $\mathbf{f}(q,t) \cdot \mathbf{n}$ is the force per unit length acting normal to the boundary.

2.4. Mathematical Model. The sides of the jellyfish bell are modeled as hyperellipsoids, and the aboral surface of the jellyfish bell lies against the substrate and is modeled as a straight line. The oral arms are modeled as a porous line drawn above the bell (see Fig. 3). The lengths of the major and minor axes are changed in time in order to create bell pulsations. Specifically, two reference configurations were defined as a completely contracted state and a completely expanded state. The curve of each of these configurations is defined by

$$(x,y)(\theta) = \frac{L}{A}(cos(C\pi + 2\pi\theta), Bsin(C\pi + 2\pi\theta)) \tag{6}$$

$A = 16$, $B = 2$, and $C = 1$ for the expanded sides of the bell, and $A = 4$, $B = -1$, and $C = 2$ for the contracted sides of the bell. θ denotes the angle between the line drawn from the center of the ellipse to the boundary point and the major axis. To determine intermediate conformations between fully expanded and fully contracted configurations, the position is linearly interpolated using the calculated motion times from the video footage. Figure 4 shows the position of the bell margin in the fully contracted and expanded states. Complete details of the mathematical model can be found in Hamlet et al. [15].

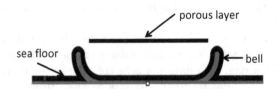

FIG. 3. *A simplified mathematical model of the upside down jellyfish. The aboral surface of the jellyfish bell is modeled as a straight line. The sides of the jellyfish bell are modeled as pieces of an ellipse. The lengths of the major and minor axes are changed in time in order to create bell pulsations. The oral arms are modeled as a porous line*

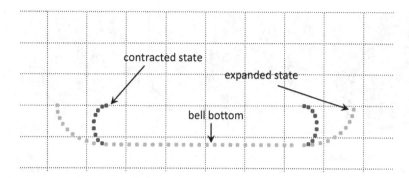

FIG. 4. *Contracted state of the model bell margin (blue) and expanded state (green). The bottom of the bell appears as a straight dotted green line and rests against the substrate*

3. Results.

3.1. Experiments. The PIV results show that fluid moves towards the *Cassiopea* medusa along the floor during the entire pulse cycle and is then ejected upwards and through the oral arms. The effect of this flow pattern is that new fluid is continually transported through the oral structures. Spatially resolved flow velocities averaged over the entire pulse cycle are shown for a 6 cm diameter medusa in Fig. 5. Starting vortex rings are generated during the power stroke or contraction phases, and counter-rotating stopping vortices are generated during bell expansion (see Fig. 6). As the pairs of vortex rings are advected upward and through the oral

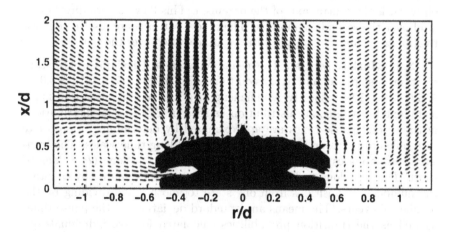

FIG. 5. *Phase-averaged velocity vector field of the flow induced by a Cassiopea medusa whose position is shown by the black silhouette. Note that fluid moves along the floor of the tank towards the jellyfish and is then pushed through the oral arms and up into the water column*

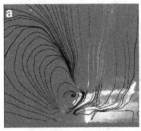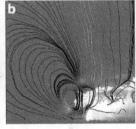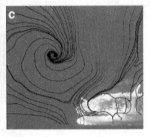

FIG. 6. *Instantaneous streamlines of the flow generated by pulsing of a Cassiopea medusa of 6 cm maximum bell diameter. (a) shows the phase of pulsing immediately prior to completion of the power stroke, (b) shows the pulsing phase corresponding to full contraction of the bell and (c) shows the phase corresponding to full relaxation of the bell*

arms, the rings are broken up and significant mixing occurs over the oral surface.

It is likely that the pulsations of the bell act as a suction pump drawing nutrient-rich waters towards the oral arms and projecting nutrient-poor waters up and away from the medusa. This flow pattern is suggested by the advection of dye seen in Fig. 1. Peak radial velocity is at the height of the bell margin and decreases as the horizontal distance increased from the medusa. This provides highly directed flow towards the oral arms and advection into a vertical jet. As a consequence, teach new volume of water is entrained horizontally and is least contaminated by flow entrained by the previous cycle. The distribution of time-averaged axial velocities shows that the majority of flow transported through and above the oral arms occurs in the region directly above the outer half of the medusa and tapers off towards the central axis of the organism. This flow pattern directs new fluid over the secondary mouths. These flow patterns are also in contrast to *Aurelia victoria* where the vortices are centered over the tentacles.

3.2. Markov Model. The means and standard deviations of the contraction, first pause, and expansion times are 0.62 ± 0.052 s, 0.13 ± 0.04 s, and 0.69 ± 0.12 s, respectively. The lengths of the second pause times typically exhibit a bimodal distribution as seen in Fig. 7. The distributions of the second pause times are approximated by determining appropriate thresholds to partition the data into short and long time scales. The transition probabilities from short to short (P_{ss}), short to long (P_{sl}), long to short (P_{ls}), and long to long (P_{ll}) pause times are estimated by counting the number of transitions from one state to another and dividing by the number of events. The means and standard deviations of the pause times as well as the transition probabilities calculated for five individuals are reported in Table 1.

Pulsing dynamics are simulated by starting the jellyfish in either a long or short pause state and randomly selecting a pause time from the appropriate distribution. The transition matrix is then used to determine

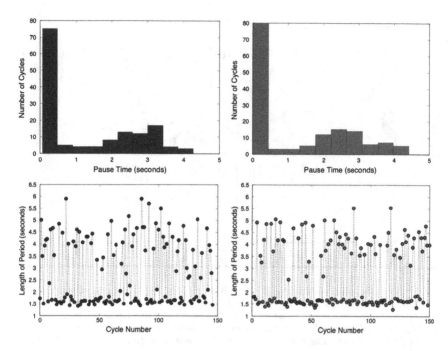

FIG. 7. *Experimentally observed and simulated times for the lengths of the pauses between bell expansion and the subsequent bell contraction. A histogram showing the frequency of the total cycle lengths (**a**) recorded over 150 cycles for a medusa and (**b**) simulated using the Markov model. Trajectories of pause times (**a**) observed and (**b**) simulated over the course of 150 pulse cycles*

if a long or short pause will occur in the next cycle. The pause length is then randomly selected from the appropriate long or short pause time distribution. The cycle lengths and pause times for an example simulation are shown in Fig. 7 with the corresponding data taken from an actual medusa.

It is important to note that pulsing dynamics exhibit a large amount of variation between individuals and within a single individual. It is likely that the frequency and duration of long pause times is influenced by the lighting conditions, nutrient availability, time of day, and temperature. The data provided here is by no means a complete description of the variation in pause lengths and pulsing dynamics.

3.3. Numerical Simulations. Numerical parameters in the immersed boundary simulations are chosen to approximate the actual organism and are listed in Table 2. In all cases, the contraction time is set to 0.6 s, the first pause time is set to 0.13 s, and the expansion time is set to 0.7 s. To compare the effect of the changes in the length of the second pause, simulations are performed with no pause and a constant pause time of 2 s. The discrete time Markov chain (DTMC) is then used as an input into the simulations to randomly generate the lengths of the pauses between the contraction and the subsequent expansion.

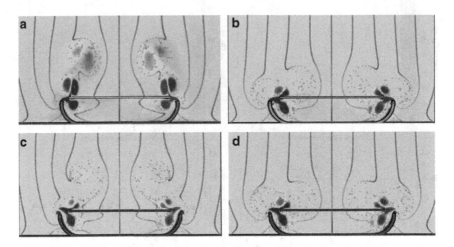

FIG. 8. *Vorticity plots of the flow fields at the end of the expansion phase after four pulse cycles for pulsing kinematics with (a) no pause between the expansion and subsequent contraction, (b) a constant pause time of 2 s, (c) and (d) pause times generated by the Markov model. Tracer particles are denoted by open circles and move with the local fluid velocity. The tracers were initially placed in vertical lines across the fluid domain*

Vorticity plots of the flow fields at the end of the expansion phase after four pulse cycles are shown in Fig. 8. The flow fields are shown for the simulations with no pause, a constant 2 s pause, and two simulations with pause times generated by the DTMC. Tracer particles are denoted by open circles and move with the local fluid velocity. The tracers are initially placed in vertical lines across the fluid domain. The warmer colors correspond to regions of positive vorticity, while the cooler colors indicate negative vorticity. Note in all cases that the fluid is driven by the motion of the bell and pushed through the permeable layer that represents the oral arms. In the case without a second pause, the fluid is advected upwards and away from the jellyfish. In the case of the constant pause time, the strong mixing occurs around the porous layer representing the oral arms. When the pause times are dictated by the DTMC, the pattern alternates between advection up and away from the jellyfish and mixing over the oral arms during the relaxation phase. An examination of the location of the tracers shows the history of mixing in the fluid.

Vorticity plots of the flow fields at the end of the expansion phase of each pulse cycle for pulsing kinematics generated by the DTMC are shown in Fig. 9. The corresponding times of each snapshot are shown in each panel. Tracer particles are initially placed in vertical lines across the fluid domain and move with the local fluid velocity. The locations of the tracer particles over time show the history of mixing in the fluid.

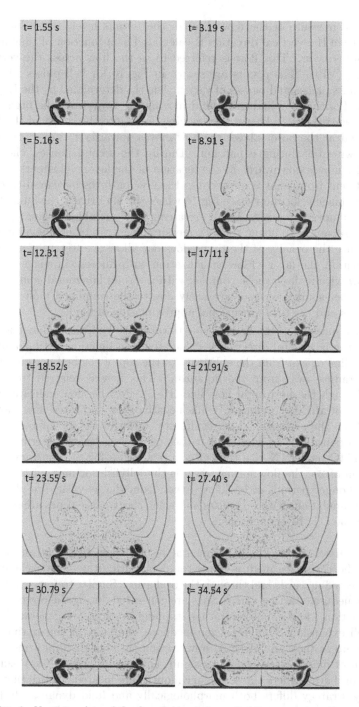

FIG. 9. *Vorticity plots of the flow fields at the end of the expansion phase of each pulse cycle for pulsing kinematics generated by the Markov model. The corresponding time that each snapshot was taken is shown in each panel. Tracer particles are denoted by open circles and move with the local fluid velocity. The tracers were initially placed in vertical lines across the fluid domain*

As a measure of the flow through the porous layer, the volumetric flow rate in the vertical direction is calculated as a function of time along the porous layer. This flow is referred to as the flow across the vertical flow line (VFL). The results are shown in Fig. 10 for bell pulses with no pause, a 2 s pause, and pauses of varying lengths determined by the DTMC model. Positive flow indicates fluid moving up away from the organism, while negative flow indicates flow moving down into the cavity of the bell. Note that at the porous layer the flow moves upwards during contraction and downwards during expansion. During the second pauses, the flow typically moves downwards at the location of the layer. This pattern is a bit different from what is observed in the flow patterns measured through particle image velocimetry (see Fig. 5) where the majority of the fluid above the layer moves upwards. There are a couple of possible explanations for this discrepancy. To begin, the PIV measurements are taken above the oral arms since the opacity of the tissue does not allow one to resolve the flow fields within the layer. Flow visualization with fluorescein suggests that the flow within the layer is complex with a strong horizontal component. Some other sources of error are likely due to the simplified model of the oral arms and the two-dimensional approximation of fluid.

As a measure of the flow moving towards the bell, the volumetric flow rate in the horizontal direction is calculated as a function of time along a vertical line drawn from the substrate to the top of the bell margin about half a radius away from the organism. This flow is referred to as the flow through the horizontal flow line (HFL). Positive flow indicates fluid moving toward the organism, while negative flow indicates flow moving away from the organism. The results are shown in Fig. 11 for bell pulses with no pause, a 2 s pause, and pauses of varying length determined by the DTMC model. Notice that after a few cycles for the constant pause duration of 2 s, the flow moves towards the jellyfish for most of the pulse cycle. Longer periods of backflow (away from the jellyfish) are observed for the simulations with no pauses and when short pause times are selected during the DTMC simulations.

4. Discussion. The experimental results show that fluid moves towards the *Cassiopea* medusa along the floor during the pulse cycle and is then ejected upwards and through the oral arms. Similar to free-swimming oblate medusa, starting vortex rings are generated during the power stroke (contraction), and counter-rotating stopping vortices are generated during bell expansion. Unlike pelagic medusae, the pairs of vortex rings are broken up as they are advected upward and through the oral arms. This pattern of fluid motion generates significant mixing over the oral surface of the oral arms where the secondary mouths are located. Notice that this feeding strategy differs both morphologically and fluid dynamically from *Aurelia* where starting and stopping vortices are stationed over the tentacles located at the edges of the bell. Given the pattern of fluid motion

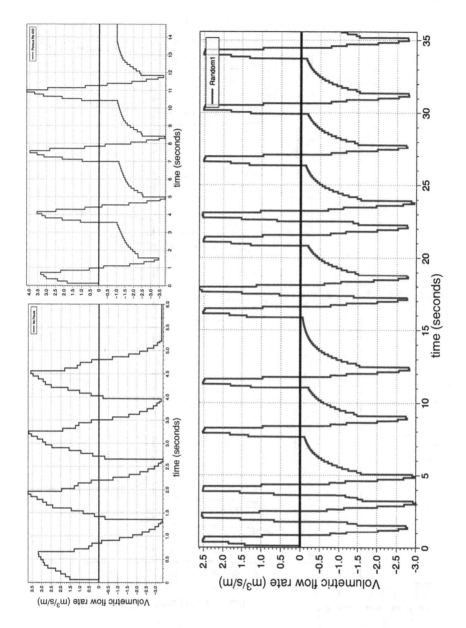

FIG. 10. *Volumetric flow rates along the VFL describing vertical flow moving through the porous layer region for (a) simulations with no pause between expansion and the subsequent contraction, (b) simulations with a constant 2 s pause between expansion and contraction, and (c) simulation with pause times generated by the Markov model. These plots indicate the normalized vertical flow from through the region where the porous structure (if present) is defined. Positive flow indicates fluid moving up away from the structure, while negative flow indicates flow moving down into the cavity of the structure*

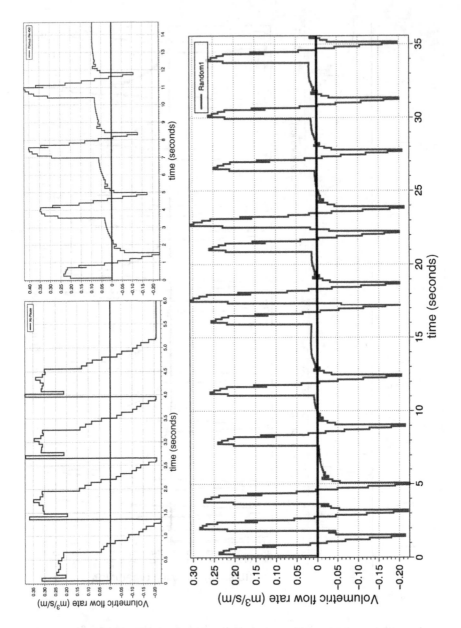

FIG. 11. *Volumetric flow rates along the HFL describing horizontal flow moving towards the bell for (a) simulations with no pause between expansion and the subsequent contraction, (b) simulations with a constant 2 s pause between expansion and contraction, and (c) simulation with pause times generated by the Markov model. These plots indicate the horizontal flow from the left of the domain toward the model organism. Positive flow indicates fluid moving toward the structure, while negative flow indicates flow moving away from the structure*

TABLE 1
Experimentally determined values for the Markov model

Individual	# cycles	Long pause (s)	Short pause (s)	P_{sl}	P_{ls}
1	146	2.59 ± 0.79	0.20 ± 0.11	0.55	0.62
2	247	1.56 ± 0.88	0.20 ± 0.17	0.17	0.82
3	240	2.17 ± 1.12	0.16 ± 0.16	0.20	0.90
4	359	2.17 ± 1.29	0.15 ± 0.13	0.013	0.60
5	88	1.15 ± 0.62	0.46 ± 0.14	0.35	0.70

TABLE 2
Parameter values for the numerical simulations

Parameter	Units	Value
Density (ρ)	kg/m^3	998
Body length (L)	m	0.00002
Porosity (λ)	m$^2/(N \cdot s)$	0.0508
Contraction time	s	0.6
1st pause	s	0.13
Expansion time	s	0.7
Fluid domain	m $\times$ m	0.232×0.232

towards the bell, *Cassiopea* likely feeds on organisms that live either near or within the substrate. Moreover, it is possible that the pulsations of the bell act as a suction pump drawing nutrient-rich waters towards the oral arms and projecting nutrient-poor waters up and away from the medusa. This pattern also suggests that *Cassiopea* may be an 'ecosystem engineer' capable of harnessing sediment-locked nutrients.

The numerical results suggest that the duration of the pause times between bell expansion and the subsequent contraction have significant implications for the resulting patterns of fluid mixing. For example, pauses might allow for the water that is brought into the bell to be sampled for a longer period of time. Pulsing cycles without pauses might be used for swimming and to move fluid up and away from the animal. Pause times generated by the DTMC suggest that an effective strategy for these benthic jellyfish under certain conditions might be to alternate between sampling phases and advective phases via long and short pause times. Furthermore, the porous structure of the jellyfish oral arms can significantly alter bulk flow properties around the organism [15]. The results also suggest that the role of secondary structures such as the oral arms should be taken into account when designing mathematical models of some jellyfish.

Acknowledgements. The authors would like to thank the organizers for putting together the workshop and for handling the publication of the associated papers. This work was supported by a BWF CASI Award-1005782.01 and NSF grant DMS-1022802 awarded to Miller.

REFERENCES

[1] Arai MN (1997) A functional biology of scyphozoa. Chapman and Hall, London

[2] Arai MN (2001) Pelagic coelenterates and eutrophication: a review. Hydrobiologia 451:69–87

[3] Bigelow RP (1900) The anatomy and development of *Cassiopeia xamachana*. Boston Soc Nat Hist Mem 5:191–236

[4] Brown GO (2002) Henry Darcy and the making of a law. Water Resour Res Volume 38, Issue 7, pp. 11–1.

[5] Colin SP, Costello JH (2002) Morphology, swimming performance, and propulsive mode of six co-occurring hydromedusae. J Exp Biol 205:427–437

[6] Costello JH, Colin SP (1994) Morphology, fluid motion and predation by the scyphomedusa Aurelia aurita. Mar Biol 121:327–334

[7] Costello JH, Colin SP, Dabiri JO (2008) Medusan morphospace: phylogenetic constraints, biomechanical solutions, and ecological consequences. Invertebr Biol 127:265–290

[8] Daniel TL (1983) Mechanics and energetics of medusan jet propulsion. Can J Zool 61:1406–1420

[9] Daniel TL (1984) Unsteady aspects of aquatic locomotion. Am Zool 24:121–134

[10] Dabiri JO, Colin SP, Costello JH, Gharib M (2005) Flow patterns generated by oblate medusan jellyfish: field measurements and laboratory analyses. J Exp Biol 208:1257–1265

[11] Dabiri JO, Colin SP, Costello JH (2006) Fast-swimming jellyfish exploit velar kinematics to form an optimal vortex wake. J Exp Biol 209:2025–2033

[12] Dabiri JO, Colin SP, Costello JH (2007) Morphological diversity of medusan lineages is constrained by animal-fluid interactions. J Exp Biol 210:1868–1873

[13] Fauci JJ, Peskin CS (1988) A computational model of aquatic animal locomotion. J Comput Phys 77:85–108

[14] Fogelson AL (1984) A mathematical model and numerical method for studying platelet adhesion and aggregation during blood clotting. J Comput Phys 56:111–134

[15] Hamlet C, Santhanakrishnan A, Miller LA (2011) A numerical study of the effects of bell pulsation dynamics and oral arms on the exchange currents generated by the upside-down jellyfish Cassiopea spp. J Exp Biol 214:1911–1921

[16] Hayward RT (2007) Modeling experiments on pacemaker interactions in scyphomedusae. Master's thesis, University of North Carolina at Wilmington, Department of Biology and Marine Biology

[17] Heller HC, Sadava DE, Orians GH (2006) Life, the science of biology. W.H. Freeman, New York

[18] Herschlag G, Miller LA (2011) Reynolds number limits for jet propulsion: a numerical study of simplified jellyfish. J Theor Biol 285:84–95

[19] Jung E, Peskin CS (2001) Two-dimensional simulations of valveless pumping using the immersed boundary method. SIAM J Sci Comput 23:19–45

[20] Kim Y, Peskin CS (2006) 2-D Parachute simulation by the immersed boundary method. SIAM J Sci Comput 28:2294–2312

[21] Larson RJ (1991) Diet, prey selection and daily ration of Stomolophus meleagris, a filter- feeding scyphomedusa from the NE Gulf of Mexico. Estuar Coast Shelf Sci 32:511–525

[22] Lim S, Peskin CS (2004) Simulations of the whirling instability by the immersed boundary method. SIAM J Sci Comput 25:2066–2083

[23] Lipinski D, Mohseni K (2009) Flow structures and fluid transport for the hydromedusae Sarsia tubulosa and Aequorea victoria. J Exp Biol 212:2436–2447

[24] McHenry MJ, Jed J (2003) The ontogenetic scaling of hydrodynamics and swimming performance in jellyfish *Aurelia aurita*. J Exp Biol 206:4125–4137

[25] Miller LA, Peskin CS (2004) When vortices stick: an aerodynamic transition in tiny insect flight. J Exp Biol 207:3073–3088

[26] Miller LA, Peskin CS (2009) Flexible clap and fling in tiny insect flight. J Exp Biol 212:3076-3090

[27] Omoto C, Dillon RH, Fauci LJ, Yang X (2007) Fluid dynamic models of flagellar and ciliary beating. Ann N Y Acad Sci 1101:494-505

[28] Peskin CS (2002) The immersed boundary method. Acta Numer 11:479-517

[29] Peskin CS, McQueen DM (1996) Fluid dynamics of the heart and its valves. In: Othmer HG, Adler FR, Lewis MA, Dallon JC (eds) Case studies in mathematical modeling: ecology, physiology, and cell biology, 2nd edn. Prentice-Hall, New Jersey

[30] Peskin CS, Kramer PR, Atzberger PJ (2008) On the foundations of the stochastic immersed boundary method. Comput Method Appl Mech Eng 197:2232-2249

[31] Sahin M, Mohseni K, Colin SP (2009) The numerical comparison of flow patterns and propulsive performances for the hydromedusae *Sarsia tubulosa* and *Aequorea victoria*. J Exp Biol 212:2656-2667

[32] Santhanakrishnan A, Hamlet C, Dollinger M, Colin S, Miller LA Flow structure and transport characteristics of feeding and exchange currents generated by upside-down Cassiopea jellyfish. J Exp Biol. 215: 2369-2381

[33] Satterlie RA (2002) Neuronal control of swimming in jellyfish: a comparative story. Canad J Zool 80:1654-1669

[34] Stockie JM (2009) Modelling and simulation of porous immersed boundaries. Comput Struct 87:701-709

[35] Templeman MA, Kingsford MJ (2010) Trace element accumulation in Cassiopea sp. (Scyphozoa) from urban marine environments in Australia. Mar Environ Res 69:63-72

[36] Todd BD, Thornhill DJ, Fitt WK (2006) Patterns of inorganic phosphate uptake in Cassiopea xamachana: a bioindicator species. Mar Poll Bull. 52:515-521

[37] Verde EA, McCloskey LR (1998) Production, respiration, and photophysiology of the mangrove jellyfish *Cassiopea xamachana* symbiotic with zooxanthellae: effect of jellyfish size and season. Mar Ecol Prog Ser 168:147-162

[38] Welsh DT, Dunn RJK, Meziane T (2009) Oxygen and nutrient dynamics of the upside down jellyfish (*Cassiopea* sp.) and its influence on benthic nutrient exchanges and primary production. Hydrobiologia 635:351-362

KINETIC MODELS FOR BIOLOGICALLY ACTIVE SUSPENSIONS

DAVID SAINTILLAN(✉)*

Abstract. Biologically active suspensions, such as suspensions of swimming microorganisms, exhibit fascinating dynamics including large-scale collective motions and pattern formation, complex chaotic flows with good mixing properties, enhanced passive tracer diffusion, among others. There has been much recent interest in modeling and understanding these effects, which often result from long-ranged fluid-mediated interactions between swimming particles. This paper provides a general introduction to a number of recent investigations on these systems based on a continuum mean-field description of hydrodynamic interactions. A basic kinetic model is presented in detail, and an overview of its applications to the analysis of coherent motions and pattern formation, chemotactic interactions, and the effective rheology in active suspensions, is given.

Key words. Suspensions, microorganisms, kinetic theory, collective dynamics

AMS(MOS) subject classifications. 35Q35, 35Q92, 76D07, 76T20, 76Z99

1. Introduction. Microorganisms are present in every part of the biosphere, ranging from harmful and beneficial bacteria in our bodies to phytoplankton in the oceans. They play a central role in many biological and ecological phenomena, among which pathogenic infection, digestion, reproduction, CO_2 capture and mixing in the oceans, and they are also at the base of the marine food web. Understanding their behavior, motility, dynamics, and interactions, is therefore a central step in the modeling of these various phenomena.

Much previous work in this field has focused on the hydrodynamics of single swimming microorganisms [27], which exhibit interesting and unusual strategies for locomotion in environments where viscous effects dominate and inertia is negligible. In this regime of low Reynolds numbers, the disturbance flows generated by moving particles (such as swimming microorganisms) decay very slowly with the distance from the particle center, thereby resulting in strong particle–particle hydrodynamic interactions in suspensions of many swimmers. These interactions in turn are known to result in a variety of complex and fascinating phenomena that have been reported in experiments, including: enhanced passive tracer diffusion [26, 28, 30, 49] and swimming speeds [13], large-scale chaotic flows with unsteady jets and vortices [11, 13, 29, 45], emergence of density fluctuations and patterns [11, 13], etc. Direct numerical simulations of these systems have also been performed, using various models and levels of approximation, including: simple dumbbell models [19, 20], boundary integral

*Department of Mechanical Science and Engineering, University of Illinois at Urbana-Champaign, Urbana, IL 61801, USA, dstn@illinois.edu.

S. Childress et al. (eds.), *Natural Locomotion in Fluids and on Surfaces*, IMA 155, DOI 10.1007/978-1-4614-3997-4_4, © Springer Science+Business Media New York 2012

simulations [25], slender-body models [40, 43], and Stokesian dynamics simulations [24]. These simulations are often successful at capturing the qualitative features observed experimentally, and provide a wealth of useful information on the details of interactions and on the structure of the suspensions. Yet, they sometimes fail at elucidating the fundamental mechanisms leading to collective motion.

A different and complementary approach consists of developing continuum equations to capture the dynamics of various field variables such as swimmer concentration and orientation. These models are typically variants of existing kinetic theories for passive suspensions, liquid crystals, or polymer solutions, which all share similarities with active suspensions. The first notable model of this kind was proposed by Aditi Simha and Ramaswamy in a seminal paper [1], in which they adapted equations for the dynamics of liquid crystals, coupled to the Navier-Stokes equations for the fluid flow, to study the stability of aligned suspensions of active particles. A number of similar models have been developed since then [4, 6, 31, 32, 48], which have been applied to investigate collective effects in concentrated active suspensions. These models, however, often include ad hoc terms to account for near-field steric interactions, so that they are not always appropriate to study the sole effect of hydrodynamic interactions.

Another simpler kinetic model was developed recently by Saintillan and Shelley [41, 42], and is the focus of this paper.[1] The model is based on the use of a probability distribution function $\Psi(\mathbf{x}, \mathbf{p}, t)$ of finding a particle at position $\mathbf{x}$ with orientation $\mathbf{p}$ in the suspension (here $\mathbf{p}$ is a unit vector pointing in the direction of swimming). A conservation equation is written for the distribution function, with fluxes that depend on the local fluid velocity. This fluid velocity is in turn obtained by solving the Stokes equations with a coarse-grained effective stress tensor capturing the effect of the swimming particles on the flow. These basic equations can then either be analyzed theoretically (for instance in a stability analysis) or integrated numerically in simulations.

Here, we review this basic kinetic model and some of its applications. We briefly discuss single-particle hydrodynamics and derive an expression for the effective stress tensor induced by a collection of particles in Sect. 2. The governing equations for the kinetic theory are exposed in Sect. 3. We then describe their application to the study of instabilities and coherent motions in active suspensions in Sect. 4, chemotaxis in thin bacterial films in Sect. 5, and the effective rheology of active suspensions in Sect. 6. We conclude and discuss directions for future work in Sect. 7.

2. Single-Particle Hydrodynamics and Coarse-Graining. A large body of work exists on the analysis and modeling of propulsion mechanisms for microorganisms and on single-organism hydrodynamics,

---

[1]Note that a very similar model was also proposed independently and around the same time by Subramanian and Koch [47].

e.g. [27]. Here, we only review a few basic features that we will use to construct an expression for the mean-field stress tensor generated by a collection of swimmers. In nature, numerous swimming mechanisms exist at low Reynolds numbers, which all rely on non-reciprocal shape deformations as prescribed by Purcell's famous scallop theorem [34]. Most microorganisms make use of flexible appendages named flagella, which are actuated in a non-reciprocal fashion, thereby exerting a net thrust on the surrounding fluid. This is the case of many types of bacteria such as the common *Escherichia coli* and *Bacillus subtilis*, which use a bundle of flagella for propulsion, and of some types of microphytes such as *Chlamydomonas reinhardtii*, which beats two flagella in a breaststroke-like fashion (Fig. 1).

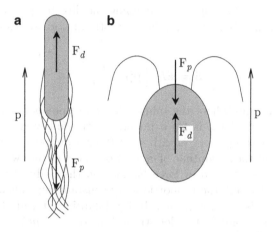

Fig. 1. *Propulsion of two model swimming microorganisms: (a) a pusher (such as a bacterium) exerts a propulsive force near its tail, whereas (b) a puller (such as a microalga) exerts a thrust near its head*

While the resulting propulsive force $\mathbf{F}_p$ will in general be time-dependent, we will assume here for simplicity that it is steady: its value may be interpreted as a time average over one beat cycle (an approximation that is not necessarily easy to justify as unsteady effects may also have an impact on hydrodynamic interactions). If gravitational effects can be neglected, i.e. if the microorganism and the fluid have nearly matching densities, the swimmer is force-free and must therefore exert an equal and opposite drag force $\mathbf{F}_d = -\mathbf{F}_p$ on the fluid: this drag force is likely to be exerted mostly by those parts of the body that do not contribute to propulsion (i.e. the cell body for bacteria and microalgae). Because the application points of $\mathbf{F}_p$ and $\mathbf{F}_d$ differ by a distance l (of the order of the organism size), the net leading-order effect on the surrounding fluid is that of a force dipole, whose sign may depend on the mechanism for swimming. In the case of a bacterium (Fig. 1a), the propulsive force is exerted near the rear of the particle, and such a swimmer will be called a *pusher*. Conversely, an alga swimming the breaststroke (Fig. 1b) exerts a thrust near its front, and will be called a *puller*.

The force dipole exerted by a swimmer can be characterized by the so-called stresslet $\mathbf{S}$, which is a second-order tensor defined as the symmetric first moment of the two forces:

$$\mathbf{S} = -\sum_i \left[\frac{1}{2}(\mathbf{x}_i\mathbf{F}_i + \mathbf{F}_i\mathbf{x}_i) - \frac{1}{3}(\mathbf{x}_i \cdot \mathbf{F}_i)\mathbf{I} \right], \tag{1}$$

where the sum is over the two forces $\mathbf{F}_p$ and $\mathbf{F}_d$. In Eq. 1, $\mathbf{x}_i$ is the point of application of force $\mathbf{F}_i$, and the last term on the right-hand side involving the idem tensor $\mathbf{I}$ is added to make $\mathbf{S}$ traceless. In the case of the two swimmers illustrated in Fig. 1, and defining the director $\mathbf{p}$ as a unit vector pointing in the direction of swimming, it is straightforward to simplify this expression to:

$$\mathbf{S} = \pm Fl \left(\mathbf{pp} - \frac{\mathbf{I}}{3} \right), \tag{2}$$

with $F = |\mathbf{F}_p|$, and where the minus sign corresponds to the case of a pusher and the plus sign is for a puller. In the following, we introduce the dipole strength $\sigma_0 = \pm Fl$, with $\sigma_0 < 0$ for a pusher and $\sigma_0 > 0$ for a puller. Note that the magnitude of σ_0 is also related to the swimming speed U_0 of the particle. Indeed, a force balance on the body of the organism yields $F \propto \mu U_0 l$ where the proportionality constant depends on the exact shape, which leads to $\sigma_0 \propto \mu U_0 l^2$. In the following, it will be convenient to define a dimensionless stresslet strength as $\alpha = \sigma_0/\mu U_0 l^2$, which is an $O(1)$ constant of the same sign as σ_0.

Of course, the description of Fig. 1 in terms of two equal and opposite point forces is simplistic, and in reality the microorganism exerts a distribution of stresses over the entire surface of its body. The definition of the stresslet Eq. 1 is then easily generalized as

$$\mathbf{S} = -\int_S \left[\frac{1}{2}(\mathbf{x}\mathbf{f} + \mathbf{f}\mathbf{x}) - \frac{1}{3}(\mathbf{x} \cdot \mathbf{f})\mathbf{I} \right] dS, \tag{3}$$

where the integral is over the surface of body, and $\mathbf{f}(\mathbf{x})$ is the traction (force per unit area) at any point $\mathbf{x}$ on the body. For an axisymmetric microorganism, this expression must also simplify to

$$\mathbf{S} = \sigma_0 \left(\mathbf{pp} - \frac{\mathbf{I}}{3} \right), \tag{4}$$

where the value of σ_0 will depend on the details of the traction distribution. This approach provides a more general and rigorous definition of pushers and pullers than that provided above: a pusher can be defined as a self-propelled particle for which $\sigma_0 < 0$ in Eq. 4, whereas a puller is a particle

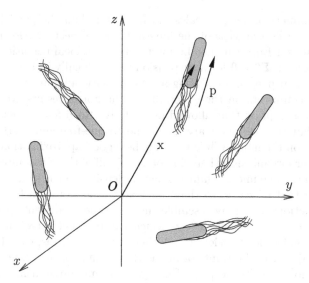

FIG. 2. *Active suspension: the configuration of the particles at time t is modeled in terms of a distribution function $\Psi(\mathbf{x}, \mathbf{p}, t)$ of the center-of-mass $\mathbf{x}$ and director $\mathbf{p}$ of the swimmers*

for which $\sigma_0 > 0$. The case $\sigma_0 = 0$, which corresponds to a zero net force dipole, is unlikely to occur in nature as any small fore-aft asymmetry will result in $\sigma_0 \neq 0$, but it may occur for some types of artificial swimmers. This case will not be considered further, as the kinetic model we introduce below is based on a non-zero stresslet.

Having obtained the stresslet $\mathbf{S}$, it is then possible to determine the effective extra stress induced by a collection of swimmers as a volume average. This classic result, sometimes known as the Kirkwood formula, was originally derived to model the stress in polymer solutions [12]. It was adapted to the case of suspensions by Batchelor [7], who expressed the extra stress tensor $\mathbf{\Sigma}^p$ for a suspension of torque-free particles in a volume V as a volume average of the stresslets on all the particles (with index α):

$$\mathbf{\Sigma}^p = \frac{1}{V} \sum_\alpha \mathbf{S}_\alpha. \tag{5}$$

If we model the configuration of a suspension at time t in terms of a probability distribution function $\Psi(\mathbf{x}, \mathbf{p}, t)$ of finding a particle with center-of-mass $\mathbf{x}$ and director $\mathbf{p}$ (see Fig. 2), the particle extra stress at point $\mathbf{x}$ is then more readily expressed as

$$\mathbf{\Sigma}^p(\mathbf{x}, t) = \int_\Omega \Psi(\mathbf{x}, \mathbf{p}, t) \mathbf{S} \, d\mathbf{p} = \sigma_0 \int_\Omega \Psi(\mathbf{x}, \mathbf{p}, t) \left(\mathbf{p}\mathbf{p} - \frac{\mathbf{I}}{3} \right) d\mathbf{p}, \tag{6}$$

where Ω denotes the unit sphere. The expression (6) for the coarse-grained particle stress tensor is valid for describing the flow generated by a suspen-

sion of swimmers on length scales much greater than the particle dimensions, and will be the basis of the kinetic model of Sect. 3. Also note that
the stress tensor Eq. 6 can also be interpreted as a local nematic order parameter: indeed, $\mathbf{\Sigma}^p = \mathbf{0}$ for a suspension that is locally isotropic, whereas
$\mathbf{\Sigma}^p \neq \mathbf{0}$ for a suspension exhibiting a local nematic alignment.

A few comments on the generality of the above results are in order.
The main assumption that allowed us to derive Eq. 6 for the active stress
tensor is that the swimmers are force- and torque-free and exert a steady
force dipole on the fluid. While this may be a good approximation for some
types of microorganisms, others do not satisfy all of these assumptions [39].
In particular, some microorganisms have a density that is significantly different from that of water, so that they exert a net force on the fluid and
have interactions that may resemble those between sedimenting particles:
this is the case of *Volvox carteri*, which exerts a net force on the fluid
and has a negligible stresslet [14]. Other microorganisms, called gyrotac-
tic, are subject to a buoyant torque as their center of mass and center of
buoyancy do not coincide [33]. Finally, some microorganisms, including
Chlamydomonas reinhardtii, drive oscillatory flows in time as they swim,
that sometimes even result in a reversal of the stresslet over the course of
one swimming stroke [16]. For simplicity, all of these effects are neglected
here, though we realize that they should be incorporated into more detailed
theories.

3. Basic Kinetic Model. The basic model of interest here was first
introduced by Saintillan and Shelley [41, 42] and is based on an evolution
equation for the distribution function $\Psi(\mathbf{x}, \mathbf{p}, t)$ defined in Sect. 2, coupled
to an equation for the fluid motion. By conservation of particles, Ψ must
indeed satisfy a Smoluchowski Equation [12]:

$$\frac{\partial \Psi}{\partial t} + \nabla_x \cdot (\dot{\mathbf{x}} \Psi) + \nabla_p \cdot (\dot{\mathbf{p}} \Psi) = 0, \tag{7}$$

where ∇_p is the gradient on the unit sphere Ω. Ψ is also normalized as

$$\frac{1}{V} \int_V \int_\Omega \Psi(\mathbf{x}, \mathbf{p}, t) \, d\mathbf{p} \, d\mathbf{x} = n, \tag{8}$$

where V is the volume of interest and n is the number density (number of
particles per unit volume). The solution of Eq. 7 requires knowledge of the
center-of-mass and rotational flux velocities $\dot{\mathbf{x}}$ and $\dot{\mathbf{p}}$, which describe the
dynamics of a given swimmer. In a dilute suspension, these can be modeled
as

$$\dot{\mathbf{x}} = U_0 \mathbf{p} + \mathbf{u}(\mathbf{x}) - D \nabla_x (\ln \Psi), \tag{9}$$

$$\dot{\mathbf{p}} = (\mathbf{I} - \mathbf{pp}) \cdot \nabla_x \mathbf{u} \cdot \mathbf{p} - d \nabla_p (\ln \Psi). \tag{10}$$

Specifically, the center-of-mass velocity of a particle is modeled as the sum
of its swimming velocity $U_0 \mathbf{p}$, which is assumed to be unchanged by interactions, and of the local fluid velocity $\mathbf{u}(\mathbf{x})$, which may result from an

external flow or from hydrodynamic interactions. Similarly, the particle rotational velocity is modeled using Jeffery's Equation [10] in terms of the velocity gradient $\nabla_x \mathbf{u}$. Both flux velocities also account for diffusion, with isotropic diffusivities D and d which are assumed to be independent of $\mathbf{x}$ and $\mathbf{p}$. These may model hydrodynamic fluctuations in the suspension [40], or thermal fluctuations if the swimmers are small enough to be affected by Brownian motion (generally not the case for microorganisms).

To close the equations, a model for the fluid velocity $\mathbf{u}$ appearing in Eqs. 9 and 10 is needed. Here we consider the situation where there is no external flow, in which case $\mathbf{u}$ is simply the velocity driven by the swimming particles themselves. As we argued in Sect. 2, swimming particles (in most cases) exert force dipoles on the surrounding fluid, which can be captured in a mean-field description using the active stress tensor of Eq. 6. More precisely, the flow field driven by the distribution of dipoles on all the particles satisfies the momentum and continuity equations:

$$ -\mu \nabla_x^2 \mathbf{u} + \nabla_x q = \nabla_x \cdot \boldsymbol{\Sigma}^p, \quad \nabla_x \cdot \mathbf{u} = 0, \tag{11} $$

where $\boldsymbol{\Sigma}^p$ is obtained in terms of Ψ using Eq. 6. Equation 7, together with Eqs. 9–11, form a closed system that may in principle be integrated in time for the distribution function Ψ and fluid velocity $\mathbf{u}$ in the suspension, given an initial condition.

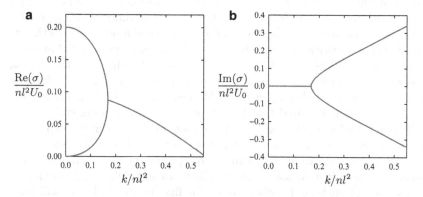

FIG. 3. (a) Real and (b) imaginary parts of the complex growth rate σ (normalized by $nl^2 U_0$), as functions of the wavenumber k in a suspension of pushers with $\alpha = -1$ and $D = d = 0$ (Adapted with permission from [42])

4. Instabilities and Coherent Motions.

The first study of interest that we describe here concerns the evolution of an initially isotropic and uniform suspension of swimmers [22, 41, 42]. This situation can first be analyzed as a stability problem: given a small perturbation in such a system, under which conditions will this perturbation grow or decay? This question was addressed by Saintillan and Shelley [41, 42], who considered a plane-wave perturbation at wavenumber $\mathbf{k}$:

$$\Psi(\mathbf{x}, \mathbf{p}, t) = \frac{n}{4\pi} \left[1 + \epsilon \widetilde{\Psi}(\mathbf{p}, \mathbf{k}) \exp(i\mathbf{k} \cdot \mathbf{x} + \sigma t) \right]. \qquad (12)$$

By substituting Eq. 12 into the kinetic equations of Sect. 3, linearizing to order ϵ, and neglecting rotational diffusion, it is possible to reduce the equations to an eigenvalue problem for the active particle stress tensor [41]:

$$\widetilde{\boldsymbol{\Sigma}}^p(\mathbf{k}) = \boldsymbol{\Pi}(\mathbf{k}, \sigma) : \widetilde{\boldsymbol{\Sigma}}^p(\mathbf{k}), \qquad (13)$$

where

$$\widetilde{\boldsymbol{\Sigma}}^p(\mathbf{k}) = \int_\Omega \widetilde{\Psi}(\mathbf{p}, \mathbf{k}) \left(\mathbf{pp} - \frac{\mathbf{I}}{3} \right) d\mathbf{p}, \qquad (14)$$

and the operator $\boldsymbol{\Pi}(\mathbf{k}, \sigma)$ is a fourth-order tensor. A dispersion relation for this eigenvalue problem can be obtained as [22, 41, 42]

$$-\frac{3\alpha}{4} \int_0^\pi \frac{\cos^2 \theta \sin^3 \theta}{\sigma + k^2 D + ik \cos \theta} d\theta = 1. \qquad (15)$$

This equation can be solved numerically for the complex growth rate σ in terms of the wavenumber $k = |\mathbf{k}|$, and such a solution is shown in Fig. 3 for a suspension of pushers ($\alpha = -1$) in the absence of diffusion ($D = d = 0$). The main conclusion of this study is the existence of a positive growth rate $\mathrm{Re}(\sigma)$ below a critical wavenumber $k_c \approx 0.55 \, nl^2$ in suspensions of pushers. The case of pullers ($\alpha > 0$) is simply obtained by changing the sign of $\mathrm{Re}(\sigma)$ and is therefore characterized by a negative growth rate. Beyond the critical wavenumber k_c, a more detailed analysis by Hohenegger and Shelley [22] demonstrated that both types of suspensions are stable when diffusion is included. Also note that the numerical factor in the expression for k_c depends on diffusion, and decreases as either D or d increases.

This critical wavenumber can be interpreted as corresponding to the smallest linear system size L above which an instability will occur in a suspension of pushers. In other words, the fluctuations in Eq. 12 will grow if the system size L satisfies:

$$\frac{2\pi}{L} < 0.55 \, nl^2, \qquad (16)$$

or equivalently,

$$\left(\frac{L}{l} \right) \times nl^3 > \frac{2\pi}{0.55}, \qquad (17)$$

i.e. when the product of the system size L (normalized by the characteristic dimension l of the swimmers) by the effective volume fraction nl^3 exceeds

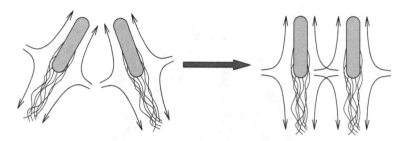

FIG. 4. *Hydrodynamic interaction of two pushers. The arrows illustrate the disturbance flow field driven by the force dipole exerted on the fluid, in the reference frame of the particle. When two pushers come together, they tend to align as a result of these flows*

a given threshold. It is important to realize that this linear instability pertains to the active stress tensor Eq. 14, which as we mentioned earlier can be viewed as a nematic order parameter: the instability will therefore result in a local nematic alignment of the particles, but the linear eigenmodes can be shown not to be associated with spatial concentration fluctuations [22, 42]. This local alignment of pushers can be understood qualitatively by simply considering the disturbance flow induced by the force dipole exerted by a pusher on the fluid (Fig. 4): as two pushers come together, it is easy to see that their disturbance flows will tend to align them, whereas the disturbance flow driven by a puller is of opposite direction. This effect was also observed in direct numerical simulations [40].

Further insight can be gained into the effects of nonlinearities and into the long-time dynamics in the suspensions by calculating numerical solutions of the kinetic equations, starting from a weakly perturbed homogeneous and isotropic distribution function (Fig. 5). Such simulations were performed in two dimensions by Saintillan and Shelley [41, 42], and in three dimensions by Alizadeh Pahlavan and Saintillan [2]. These simulations confirm the stability of puller suspensions, and the instability criterion (17) for pusher suspensions. In unstable suspensions, local nematic alignment of the particles is observed in agreement with the prediction of the linear analysis, but this alignment is followed by the growth of density fluctuations on the scale of the system. These fluctuations eventually saturate as a result of diffusion, and undergo complex time dynamics in which dense sheet-like structures form and break up repeatedly in time, see Fig. 5a. This growth of concentration fluctuations, which is a nonlinear effect, can be explained as a result of the swimming of the particles, which causes them to aggregate in regions of negative divergence of the mean director field [42]. The dynamics in the unstable suspensions are complex and chaotic, and are characterized by large-scale flows with jets and vortices, efficient fluid

a No imposed flow

b Weak imposed shear flow

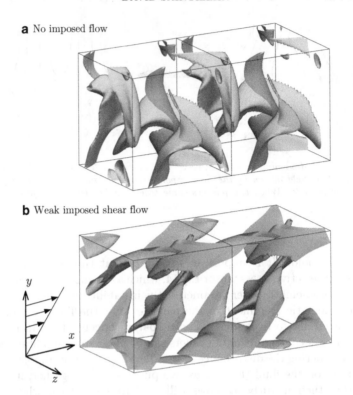

FIG. 5. *Pattern formation in suspensions of swimming microorganisms: concentration iso-surfaces at $c = 1.5$, (**a**) in a quiescent suspension (no imposed flow), and (**b**) when a weak external shear flow is imposed (Adapted with permission from [2])*

mixing, enhanced swimming speeds, and correlated dynamics on length scales of the order of the system size, in good qualitative agreement with experimental observations.

Note that the prediction of Eq. 17 was also recently tested in direct particle simulations using a slender-body model [43]. In these simulations, where individual particles are tracked and interact hydrodynamically, a transition from uncorrelated to correlated motions is also observed at a critical volume fraction (for a fixed system size) that matches the prediction of Eq. 17 within a factor of 2. The transition manifests itself in many different ways, and for instance affects velocity correlation lengths and times, swimming velocities, density fluctuations, passive tracer mixing rates and diffusivities, among others.

While the analysis and simulations described above considered swimmer dynamics and interactions in a quiescent fluid, microorganisms in nature often evolve in complex flow environments, e.g. in the oceans. It is therefore interesting and important to understand how an external flow may affect these instabilities and dynamics. Such a study is also important to understand the effective rheology of microorganism suspensions, as we discuss further in Sect. 6. The case of a simple shear flow was recently analyzed by Alizadeh Pahlavan and Saintillan [2], using both a stability theory and continuum simulations. The main finding of this study is that an external shear flow tends to stabilize the suspensions by controlling the orientation of the particles. A sample simulation in a weak shear flow is illustrated in Fig. 5b, and shows alignment of the density patterns with a 45° axis with respect to the flow direction. As shear rate increases, the instabilities can be shown to become weaker and are eventually suppressed.

5. Chemotaxis. The previous analysis, in which boundaries and external fields were entirely neglected, provides useful insight into the dynamics in active suspensions, but is highly idealistic and may be difficult to recreate in a laboratory experiment. Swimming microorganisms indeed often interact with boundaries and chemical cues. Here, we describe how the kinetic model of Sect. 3 can be amended to include some of these effects, and we specifically discuss the modeling of a thin active suspension film surrounded on both sides by an oxygen bath [3]. This study is motivated by the many experiments on active suspensions that have been performed in stabilized liquid films [16, 45, 49], and more specifically by the recent investigation of Sokolov et al. [46]. In this latter study, the dynamics in a free-standing thin film containing a suspension of *Bacillus subtilis* were observed, and demonstrated a transition from quasi-two-dimensional collective motion to three-dimensional chaotic behavior as the film thickness was increased. Migration of the bacteria towards the boundaries where the strongest concentration in oxygen occurs was also reported. The existence of this transition is not too surprising in light of Eq. 17, which predicts instabilities only above a critical system size, but a more accurate model should include coupling with the oxygen field and interactions with the free surfaces.

The first modification consists in coupling the equations of motion to the dynamics of the oxygen field, whose concentration we denote by $s(\mathbf{x}, t)$. It obeys an advection-diffusion equation:

$$\frac{\partial s}{\partial t} + \mathbf{u}(\mathbf{x}) \cdot \nabla_x s - d_0 \nabla_x^2 s = -\kappa s(\mathbf{x}, t) c(\mathbf{x}, t), \qquad (18)$$

which expresses transport of the oxygen by the disturbance flow $\mathbf{u}(\mathbf{x})$ driven by the microorganisms, and diffusion with constant diffusivity d_0. The last term on the right-hand side of Eq. 18 models consumption of oxygen by the swimmers as a second-order reaction. If we were to model the release of a chemical cue by the swimmers (as would arise in a simulation of quorum sensing [36]), this term may be replaced by $+\kappa c(\mathbf{x}, t)$.

Secondly, the effect of the oxygen concentration on the microorganism dynamics must also be modeled. This coupling is more subtle, and here we mention two different approaches:

- *Gradient-detecting model*: this model, which is the simplest of the two but also the least realistic, assumes that the swimmers are able to detect the local oxygen gradient and adjust their orientation to swim towards the regions of high oxygen concentration. This is achieved by adding an extra deterministic torque that aligns particles with the oxygen gradient in Eq. 10 for the rotational velocity, which becomes:

$$\dot{\mathbf{p}} = (\mathbf{I} - \mathbf{pp}) \cdot [\nabla_x \mathbf{u} \cdot \mathbf{p} + \chi \nabla_x s] - d\nabla_p(\ln \Psi). \tag{19}$$

- *Run-and-tumble model*: In reality, bacteria for instance are unable to sense local concentration gradients, but instead use a stochastic process of random orientation changes (so-called 'tumbling' events) whose characteristic frequency depends on the local oxygen concentration. The net effect of this biased random walk is a migration towards the regions of high oxygen concentration [9]. As explained in detail by Bearon and Pedley [8], this effect may be captured by a modification of the conservation equation (7) as follows:

$$\frac{\partial \Psi}{\partial t} + \nabla_x \cdot (\dot{\mathbf{x}} \Psi) + \nabla_p \cdot (\dot{\mathbf{p}} \Psi) = -\lambda \Psi + \frac{1}{4\pi} \int_\Omega \lambda \Psi(\mathbf{x}, \mathbf{p}', t) \, d\mathbf{p}', \tag{20}$$

where the quantity $\lambda(\mathbf{x}, \mathbf{p}, t)$ is the stopping rate and is related to the probability for a bacterium to undergo a tumbling event over a fixed time interval. This stopping rate depends on the oxygen field sampled by the bacterium as it swims, and is modeled as

$$\lambda(\mathbf{x}, \mathbf{p}, t) = \lambda_0 \left(1 - \xi \frac{Ds}{Dt} \right), \tag{21}$$

where

$$\frac{Ds}{Dt} = \frac{\partial s}{\partial t} + [U_0 \mathbf{p} + \mathbf{u}(\mathbf{x})] \cdot \nabla_x s \tag{22}$$

is akin to a material derivative, and denotes the rate of change of $s(\mathbf{x}, t)$ sampled by a swimmer along its trajectory.

Thirdly, non-periodic boundary conditions need to be implemented to account for the free surfaces of the liquid film. A natural boundary condition for the disturbance velocity field is zero shear stress (although the no-slip boundary condition may be more appropriate if surfactants are present). Boundary conditions for the distribution function Ψ are slightly more subtle and must express the inability of the swimmers to cross the boundaries. This can be achieved by letting the normal component of the center-of-mass flux velocity (9) vanish at the free surface: $\mathbf{n} \cdot \dot{\mathbf{x}} = 0$, where $\mathbf{n}$ is the unit normal at the boundary. A slightly weaker condition consists in prescribing zero net concentration flux, and is expressed as

$$\mathbf{n} \cdot \int_\Omega \dot{\mathbf{x}} \, d\mathbf{p} = 0. \tag{23}$$

a   Swimmer concentration   **b**   Oxygen concentration

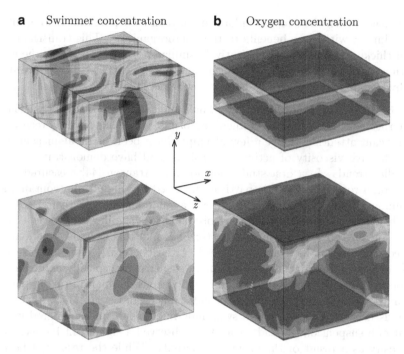

FIG. 6. *Simulations of chemotaxis performed using the run-and-tumble model of Eq. 20. The domains are doubly-periodic in the x and z directions, with interfaces in the y-direction. Two film thicknesses are shown: (a) $L_y = 20(nl^2)^{-1}$, and (b) $L_y = 50(nl^2)^{-1}$. The left column shows the local swimmer concentration, whereas the right panel shows the oxygen concentration*

The zero-shear-stress boundary condition, together with Eq. 23, can be implemented in simulations using a reflection method, and details of the numerical algorithm are forthcoming [3].

Two typical simulations are illustrated in Fig. 6 for two different film thicknesses: (a) $L_y = 20(nl^2)^{-1}$, and (b) $L_y = 50(nl^2)^{-1}$. These simulations were performed using the run-and-tumble model of Eq. 20, although the gradient detecting method is found to qualitatively produce very similar results. The figure shows both swimmer concentration and oxygen concentration. In thin films (Fig. 6a), the dynamics are found to be quasi two dimensional: the swimming microorganisms still organize into dense patterns that form and break up repeatedly in time, but these patterns are nearly uniform in the y-direction, and all the dynamics take place in the x-z plane. As the film thickness increases (Fig. 6b), a transition occurs to three-dimensional chaotic behavior: the density patterns are no longer uniform in the y-direction, and the dynamics near both interfaces become uncorrelated on average. This also leads to the emergence of three-dimensional flows, which drive more fluctuations in the oxygen field, with the formation of oxygen plumes that penetrate into the bulk of the film.

This has in turn the effect of enhancing oxygen transport and mixing into the film [3], with clear benefits to the microorganisms. This transition as film thickness increases is qualitatively similar to that reported in the experiments of Sokolov et al. [46], suggesting that the effect is likely a result of hydrodynamic interactions between swimmers.

6. Effective Rheology. As a final application, we discuss the effective rheology of suspensions of microorganisms, which has recently received significant attention. Only a few attempts have been made at measuring the effective viscosity of active suspensions, and have demonstrated very peculiar trends. In a first study, Sokolov and Aranson [44] measured the drag on a rotating magnetic particle immersed in a liquid film containing swimming bacteria, and used it to infer a value for the effective viscosity of the suspension. The value they obtained was significantly lower than that for pure solvent, by up to a factor of 7, and this decrease was found to correlate with the swimming speed of the bacteria. This decrease in viscosity is quite unusual, as particulate suspensions typically exhibit enhanced viscosities owing to the additional viscous dissipation taking place near the particle surfaces. In a second study, Rafaï et al. [35] measured the viscosity of a suspension of swimming microalgae, and compared it to that of a suspension of dead algae: they observed a significant increase in viscosity as a result of the swimming activity. While the results of both studies seem to contradict each other, this discrepancy is easily resolved by realizing that bacteria are pushers whereas microalgae are pullers.

A number of models [15, 17, 18] and numerical simulations [23] have been proposed to address this problem. Here, we briefly discuss the analysis of Saintillan [37, 38], which uses a model very similar to that of Sect. 3. In a dilute suspension, hydrodynamic interactions between microorganisms can be neglected to a first approximation, and particle positions become uncorrelated. In this limit, the configuration of a spatially homogeneous suspension is entirely captured by an orientation distribution $\Psi(\mathbf{p}, t)$, which satisfies a special case of Eq. 7:

$$\frac{\partial \Psi}{\partial t} + \nabla_p \cdot (\dot{\mathbf{p}} \Psi) = 0. \tag{24}$$

If an external linear flow with constant velocity gradient $\mathbf{A}$ is applied, the angular flux velocity $\dot{\mathbf{p}}$ captures the rotation and alignment of the swimmers in the flow:

$$\dot{\mathbf{p}} = (\mathbf{I} - \mathbf{pp}) \cdot \mathbf{A} \cdot \mathbf{p} - d\nabla_p(\ln \Psi). \tag{25}$$

Steady-state solutions $\Psi(\mathbf{p})$ of Eqs. 24–25 can be obtained analytically for irrotational flows [38], and numerically for other types of linear flows such as a simple shear flow [37]. Once $\Psi(\mathbf{p})$ is known, it can be used to calculate the effective particle stress tensor $\boldsymbol{\Sigma}^p$ in the suspension as a configurational average of the force dipoles (or stresslets) on the swimmers according to

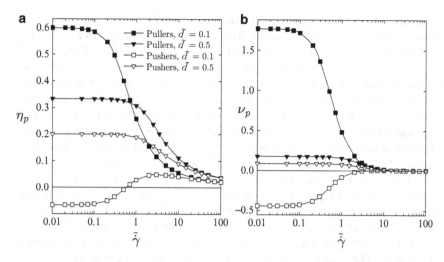

FIG. 7. *(a) Effective viscosity η_p, and (b) first normal stress difference coefficient ν_p, as functions of dimensionless shear rate $\tilde{\dot{\gamma}}$ in suspensions of pushers and pullers in a simple shear flow (Adapted with permission from [38])*

Eq. 5. Note that for particles in an external flow, an additional stresslet must be included, which arises from the inability of the particles to stretch under flow. This flow-induced stresslet, which is added to the permanent stresslet Eq. 4, can be shown to be of the form [21]:

$$\mathbf{S} = C(\mathbf{pp} : \mathbf{A}) \left(\mathbf{pp} - \frac{\mathbf{I}}{3} \right), \tag{26}$$

where the constant C depends on the shape of the particle. Having determined the extra stress tensor, quantities such as the effective viscosity η_p, and first and second normal stress difference coefficients ν_p and κ_p, can be inferred as functions of flow strength as

$$\eta_p = \frac{\Sigma^p_{xy}}{\dot{\gamma}}, \quad \nu_p = \frac{\Sigma^p_{xx} - \Sigma^p_{yy}}{\dot{\gamma}^2}, \quad \kappa_p = \frac{\Sigma^p_{yy} - \Sigma^p_{zz}}{\dot{\gamma}^2}, \tag{27}$$

where $\dot{\gamma}$ denotes the deformation rate and is obtained as: $\dot{\gamma} = (\mathbf{A} : \mathbf{A})^{1/2}$. A more detailed discussion of the model and its underlying assumptions can be found in [37, 38].

Results for the effective viscosity η_p and first normal stress difference coefficient ν_p in a simple shear flow are shown in Fig. 7. In this figure, we have defined a dimensionless flow strength as $\tilde{\dot{\gamma}} = \dot{\gamma} t_c$ and a dimensionless rotary diffusivity as $\tilde{d} = d t_c$, where the time scale t_c is defined as

$$t_c = \frac{\pi \mu l^3}{6|\sigma_0| \ln(2/\varepsilon)}. \tag{28}$$

Here, ε is the inverse aspect ratio of the microswimmer, which is assumed to be an axisymmetric slender body. Suspensions of pullers (full symbols) exhibit a shear-thinning behavior and a positive first normal stress difference coefficient, much like suspensions of passive rodlike particles. As the swimming activity becomes stronger (i.e. as the dimensionless diffusivity $\tilde{d}$ decreases), both η_p and ν_p are found to increase. The trends are reversed in suspensions of pushers (open symbols): activity causes a decrease in effective viscosity at low shear rates, which can even result in a negative value of η_p at low values of $\tilde{d}$, corresponding to a strong level of activity. Swimming activity also results in a sign change in ν_p at low flow rates. These unusual findings are consistent with experimental observations that reported an increase in viscosity in suspensions of microalgae [35] but a decrease in suspensions of motile bacteria [44]. Similar effects have been predicted in other types of flows as well, including uniaxial extensional and compressional flows and planar extensional flow [37]. All of these previous studies have only considered dilute suspensions and have neglected particle-particle interactions; such interactions may be accounted for by including an external shear flow in the kinetic model of Sect. 3, see Fig. 5b [2].

7. Conclusions and Outlook. We have presented a simple kinetic model for suspensions of self-propelled particles such as swimming microorganisms based on a continuum description of the particle phase, that was first proposed by Saintillan and Shelley [41, 42]. This model, which is simply based on a conservation equation for the distribution of particles, coupled to the Stokes equations for the fluid motion driven by the force dipoles on the particles, was developed with the aim of understanding the effects of hydrodynamic interactions in these systems. The model was applied to study the emergence of collective motion in bacterial suspensions: as we demonstrated, such collective motion is predicted to arise when the product of the system size by the volume fraction exceeds a given threshold, in agreement with results from recent numerical simulations [43]. Other more complex applications also include the modeling of chemotaxis in external chemical fields, and the effective rheology of active suspensions.

Quite naturally, these models are only useful and valid inasmuch as they faithfully capture and explain phenomena observed in physical or biological systems. Suspensions of microorganisms are extremely complex systems, in which effects such as swimming noise, steric interactions, chemical cues, gravity, temperature variations, surfactants, among others, may all play a role in the observed dynamics. The kinetic model presented in this work neglected most of these effects, with the aim of isolating the contribution of mean-field hydrodynamic interactions. While this approach is useful from a fundamental and theoretical standpoint, attempts at quantitatively capturing the dynamics in bacterial suspensions will likely require improvements of our model to capture some of these more complex effects. Of ongoing interest to us are:

- *Steric interactions*: Experiments on bacterial suspensions are often performed at high concentrations, where excluded volume interactions between swimmers become important [11, 45]. In fact, the emergence of collective motion discussed in Sect. 4 often occurs in the semi-dilute to concentrated regime, in which the mean-field description of interactions may no longer be appropriate. While including direct particle-particle contacts and interactions is fairly natural in direct particle simulations [40, 43], it is far from trivial in continuum models. One possible approach is the inclusion of an interaction potential that causes particles to align with their neighbors as a result of steric interactions, in the manner of the classic model of Doi and Edwards for the isotropic-nematic transition in passive rod suspensions [12]. This additional interaction potential significantly complicates the analysis of the model, but numerical progress is still possible.

- *Confinement*: Most theoretical calculations and simulations described herein assumed periodic boundary conditions, both for convenience and as a means to model the dynamics in bulk suspensions away from boundaries. The effects of boundaries and of confinement, however, are non-trivial in real systems, as we discussed briefly in Sect. 5. The boundary condition Eq. 23 of zero net concentration flux, while physically justified, does not include any details of the particle interactions with the boundary. Such details are again difficult to include in a continuum model, but may become significant in highly confined systems.

- *Unsteady swimming actuation*: Finally, as we mentioned briefly in Sect. 3, swimming microorganisms do not exert a steady force dipole on the fluid around them, but rather perform repeated swimming cycles during which the flow field around them fluctuates periodically [16]. These fluctuations are likely to add noise to the dynamics, but their precise effects remain unknown. A previous model on a related system suggests that the effects of these fluctuations on coherent motions and ordering may be significant [5]. Amending the present kinetic model to account for unsteady swimming is not straightforward, but will be attempted in future work.

Acknowledgments. I am indebted to Michael Shelley, with whom the models described herein were developed, and I also thank my graduate students A. Alizadeh Pahlavan and B. Ezhilan at the University of Illinois for their contributions to this work. I am grateful to Christel Hohenegger and Enkeleida Lushi for useful conversations. I started thinking about kinetic models for active suspensions during a summer workshop at the Aspen Center for Physics, whose hospitality and support are gratefully acknowledged. Financial support for this work was provided in part by the National Science Foundation under Grant No. DMS-0920931, and computing resources were provided by the National Center for Supercomputing Applications (NCSA) under Teragrid Grant No. TG-CTS100007.

REFERENCES

[1] Aditi Simha R, Ramaswamy S (2002) Hydrodynamic fluctuations and instabilities in ordered suspensions of self-propelled particles. Phys Rev Lett 89:058101

[2] Alizadeh Pahlavan A, Saintillan D (2011) Instability regimes in flowing suspensions of swimming micro-organisms. Phys Fluids 23:011901

[3] Ezhilan B, Alizadeh Pahlavan A, Saintillan D (2012) Chaotic dynamics and oxygen transport in thin films of aerotactic bacteria. (under consideration)

[4] Aranson IS, Sokolov A, Kessler JO, Goldstein RE (2007) Model for dynamical coherence in thin films of self-propelled micro-organisms. Phys Rev E 75:040901

[5] Bartolo D, Lauga E (2010) Shaking-induced motility in suspensions of soft active particles. Phys Rev E 81:026312

[6] Baskaran A, Marchetti MC (2009) Statistical mechanics and hydrodynamics of bacterial suspensions. Proc Natl Acad Sci USA 106:15567

[7] Batchelor GK (1970) The stress system in a suspension of force-free particles. J Fluid Mech 41:419–440

[8] Bearon RN, Pedley TJ (2000) Modelling run-and-tumble chemotaxis in a shear flow. Bull Math Biol 62:775–791

[9] Berg HC (1983) Random walks in biology. Princeton University Press, Princeton

[10] Bretherton FP (1962) The motion of rigid particles in a shear flow at low Reynolds number. J Fluid Mech 14:284

[11] Cisneros LH, Cortez R, Dombrowski C, Goldstein RE, Kessler JO (2007) Fluid dynamics of self-propelled micro-organisms, from individuals to concentrated populations. Exp Fluid 43:737

[12] Doi M, Edwards SF (1986) The theory of polymer dynamics. Oxford University Press, New York

[13] Dombrowski C, Cisneros L, Chatkaew S, Goldstein RE, Kessler JO (2004) Self-concentration and large-scale coherence in bacterial dynamics. Phys Rev Lett 93:098103

[14] Drescher K, Goldstein RE, Michel N, Polin M, Tuval I (2010) Direct measurement of the flow field around swimming microorganisms. Phys Rev Lett 105:168101

[15] Giomi L, Liverpool TB, Marchetti MC (2010) Sheared active fluids: thickening, thinning, and vanishing viscosity. Phys Rev E 81:051908

[16] Guasto JS, Johnson KA, Gollub JP (2010) Oscillatory flows induced by microorganisms swimming in two-dimensions. Phys Rev Lett 105:168102

[17] Haines BM, Sokolov A, Aranson IS, Berlyand L, Karpeev DA (2009) Three-dimensional model for the effective viscosity of bacterial suspensions. Phys Rev E 80:041922

[18] Hatwalne Y, Ramaswamy S, Rao M, Simha RA (2004) Rheology of active-particle suspensions. Phys Rev Lett 92:118101

[19] Hernández-Ortiz JP, Stoltz CG, Graham MD (2005) Transport and collective dynamics in suspensions of confined self-propelled particles. Phys Rev Lett 95:204501

[20] Hernández-Ortiz JP, Underhill PT, Graham MD (2007) Dynamics of confined suspensions of swimming particles. J Phys Condens Matter 21:204107

[21] Hinch EJ, Leal LG (1976) Constitute equations in suspension mechanics. Part 2. Approximate forms for a suspension of rigid particles affected by Brownian rotations. J Fluid Mech 76:187–208

[22] Hohenegger C, Shelley MJ (2009) Stability of active suspensions. Phys Rev E 81:046311

[23] Ishikawa T, Pedley TJ (2007) The rheology of a semi-dilute suspension of swimming model micro-organisms. J Fluid Mech 588:399–435

[24] Ishikawa T, Pedley TJ (2007) Diffusion of swimming model microorganisms in a semi-dilute suspension. J Fluid Mech 588:437

[25] Kanevsky A, Shelley MJ, Tornberg A-K (2010) Modeling simple locomotors in Stokes flow. J Comput Phys 229:958

[26] Kim MJ, Breuer KS (2004) Enhanced diffusion due to motile bacteria. Phys Fluids 16:L78

[27] Lauga E, Powers TR (2009) The hydrodynamics of swimming microorganisms. Rep Prog Phys 72:1–36

[28] Leptos KC, Guasto JS, Gollub JP, Pesci AI, Goldstein RE (2009) Dynamics of enhanced tracer diffusion in suspensions of swimming eukaryotic microorganisms. Phys Rev Lett 103:198103

[29] Mendelson NH, Bourque A, Wilkening K, Anderson KR, Watkins JC (1999) Organized cell swimming motions in Bacilus subtilis colonies: patterns of short-lived whirls and jets. J Bacteriol 181:600

[30] Miño G, Mallouk TE, Darnige T, Hoyos M, Dauchet J, Dunstan J, Soto R, Wang Y, Rousselet A, Clement E (2011) Enhanced diffusion due to active swimmers at a solid surface. Phys Rev Lett 106:048102

[31] Mishra S, Baskaran A, Marchetti MC (2010) Fluctuations and pattern formation in self-propelled particles. Phys Rev E 81:061916

[32] Pedley TJ (2010) Instability of uniform micro-organism suspensions revisited. J Fluid Mech 647:335

[33] Pedley TJ, Kessler JO (1992) Hydrodynamic phenomena in suspensions of swimming microorganisms. Annu Rev Fluid Mech 24:313358

[34] Purcell EM (1977) Life at low reynolds number. Am J Phys 45:3–11

[35] Rafaï S, Jibuti L, Peyla P (2010) Viscosity of microswimmer suspensions. Phys Rev Lett 104:098102

[36] Redfield RJ (2002) Is quorum sensing a side effect of diffusion sensing? Trends Microbiol 19:365

[37] Saintillan D (2010) Extensional rheology of active suspensions. Phys Rev E 81:056307

[38] Saintillan D (2010) The dilute rheology of swimming suspensions: a simple kinetic model. Exp Mech 50:1275–1281

[39] Saintillan D (2010) A quantitative look into microorganism hydrodynamics. Physics 3:84

[40] Saintillan D, Shelley MJ (2007) Orientational order and instabilities in suspensions of self-locomoting rods. Phys Rev Lett 99:058102

[41] Saintillan D, Shelley MJ (2008) Instabilities and pattern formation in active particle suspensions: kinetic theory and continuum simulations. Phys Rev Lett 100:178103

[42] Saintillan D, Shelley MJ (2008) Instabilities, pattern formation, and mixing in active suspensions. Phys Fluids 20:123304

[43] Saintillan D, Shelley MJ (2012) Emergence of coherent structures and large-scale flows in motile suspensions. J R Soc Interface 9:571

[44] Sokolov A, Aranson IS (2009) Reduction of viscosity in suspension of swimming bacteria. Phys Rev Lett 103:148101

[45] Sokolov A, Aranson IS, Kessler JO, Goldstein RE (2007) Concentration dependence of the collective dynamics of swimming bacteria. Phys Rev Lett 98:158102

[46] Sokolov A, Golstein RE, Feldchtein FI, Aranson IS (2009) Enhanced mixing and spatial instability in concentrated bacterial suspensions. Phys Rev E 80:031903

[47] Subramanian G, Koch DL (2009) Critical bacterial concentration for the onset of collective swimming. J Fluid Mech 632:359

[48] Wolgemuth CW (2008) Collective swimming and the dynamics of bacterial turbulence. Biophys J 95:1564–1574

[49] Wu X-L, Libchaber A (2000) Particle diffusion in a quasi two-dimensional bacterial bath. Phys Rev Lett 84:30173020

INDIVIDUAL TO COLLECTIVE DYNAMICS OF SWIMMING BACTERIA

LUIS H. CISNEROS(✉)[†], SUJOY GANGULY[‡],
RAYMOND E. GOLDSTEIN[†§], AND JOHN O. KESSLER[¶]

Abstract. Spatial order and fast collective coherent dynamics of populations of the swimming bacteria *Bacillus subtilis* emerges from local interactions and from flows generated by the organisms' locomotion. The transition from dilute, to intermediate, to high concentrations of cells is analyzed and presented as probability density functions for swimming velocity. The low concentration phase, which exhibits swimming speeds characteristic of individual bacteria, arrives at the anomalously high speed phase,the ZoomingBioNematic (ZBN), via an intermediate phase that exhibits surprisingly low mean speeds. We show that these low speeds at intermediate concentrations are due to transitional speeds that occur after collisions of the organisms, while the flagella that propel the bacteria re-form into a "bundle". Measurement of individual and collective velocities, as well as correlation of speeds with alignment of velocity directions, within and adjacent to coherent patches, were found by Particle Imaging Velocimetry (PIV). The significance for mixing of the ZBN dynamic is demonstrated.

Key words. Bacteria dynamics, Bacillus subtilis, swimming, micro-organisms, self propelled particles, collisions, Onsager, liquid crystal, mixing, phase transition, flocking, schooling

AMS(MOS) subject classifications. 00Bxx

1. Introduction. The locomotion of self-propelled individuals, such as ants, bacteria, single algal cells, birds and others may often be characterized by probability distribution functions that describe their velocities. When these statistics describe individuals, singly or at low concentration, means and higher moments may depend on distribution of ages, sizes, response to directional spatial cues, and intrinsic stochastic processes. At increased concentrations, individuals per area or per volume, interaction between individuals may cause changes in the distribution of velocities. Depending on the spatial scale of the action, and on the characteristics of the embedding medium, these interactions may come from visual or sonic cues, as in the case of birds or fish, from fluid-dynamic fields associated with individuals' locomotion, from collisions, and from steric interactions with neighboring organisms. Thus, when the concentration of organisms is increased so that within some domains they are closely adjacent, taking up most of the space available, then they can only move together coherently,

[†]PSOC, Arizona State University, Tempe, AZ 85287, USA, cisluis@asu.edu.

[‡]DAMPT, The University of Cambridge, Cambridge, UK, S.Ganguly@damtp.cam.ac.uk.

[§]DAMPT, The University of Cambridge, Cambridge, UK, R.E.Goldstein@damtp.cam.ac.uk.

[¶]Department of Physics, University of Arizona, Tucson, AZ 85721, USA, kessler@physics.arizona.edu. Work supported by DOE-W-31-109-ENG-38.

S. Childress et al. (eds.), *Natural Locomotion in Fluids and on Surfaces*, IMA 155, DOI 10.1007/978-1-4614-3997-4_5,
© Springer Science+Business Media New York 2012

at the same speed and in the same direction. In the case of swimming organisms, the collective speeds depend on the hydrodynamics generated by individual organism's propulsion apparatus. They are strongly modified by proximity, via group-based hydrodynamic interactions, collective propulsion and drag, all in the context of a variety of accidentally occurring geometries of particular domains. The progression from behavior of individuals to the dynamics of condensed populations is somewhat analogous to the transition from energy levels of individual atoms to the electronic band structure of crystals.

The experiments described here, and their interpretation, follow the phenomena that arise with increase of the concentration of the motile soil bacteria *Bacillus subtilis*. Although our experiments were performed in an in vitro laboratory setting, the fact that these bacteria swim up gradients of dissolved oxygen, self-concentrate in sloping environments [5, 11], and multiply rapidly in "rich" media implies that our results are likely to be applicable to natural conditions that prevail in humid soils.

Our observations follow populations of *B subtilis* from low concentration to nearly "close packed". In the highest states of concentration the bacteria form domains of co-directionally oriented and swimming individuals, moving together at nearly the same velocity. These domains are typically separated by relatively disordered populations, as discussed below. A population of bacteria that exhibits these regions of spontaneous alignment, is a phase, in the sense of materials science. By analogy with the nematic liquid crystal phase, this phase consisting of parallel rod-shaped bacteria rather than rod-shaped molecules, is called "ZoomingBioNematic" (ZBN). Unlike molecular nematics, the ZBN alignment is polar. The constituent cells swim co-directionally. The lifetime of any specific domain is of order seconds, due to collisions with other domains and instabilities of coherence that are currently being investigated. Collapse is followed by birth of new domains, moving in directions that appear random. The velocity of particular domains can be greater than the velocity of individual swimmers. That fact, together with the succession of random directions of domain movement, and apparently random choice of constituent occupants, implies that this coherent, collective behavior is efficient in mixing solutes, suspended inert particles, and indeed the members of the bacterial population. We have demonstrated that inference by following synthetic particles along path lines that follow the measured bacterial trajectories (Fig. 4).

The managed transition from low to high concentration of bacteria revealed an anomalous result: The measured means of the velocity of locomotion, following swimming bacteria, using Particle Imaging Velocimetry (PIV), was substantially slower at intermediate concentrations than for individuals at low concentration or for the members of the coherent domains. We have related this phenomenon to the slow swimming intervals that characterize the re-acceleration of individuals that have collided.

The locomotion of domains at speeds greater than the speed of the individuals comprising them is less well understood. It was shown [3] that coherent patches of bacteria may move more quickly than individual swimmers, driven only by cells attached at their periphery exerting on the surrounding water the hydrodynamic force normally used for swimming as single organisms. This model is incomplete. It ignores the flow of water through self-propelled domains which are, in effect, self-propelled porous media. Work on these models is also in progress.

2. Materials and Methods. *Bacillus subtilis* are rod-shaped soil bacteria. A typical bacterium is four micrometers long and about $0.8\,\mu m$ wide. Typical mean swimming speeds are $25\,\mu m/s$. Propulsion is due to six or more flagella, attached to rotating motors embedded at nonspecific locations within the bacterial cell wall. During forward locomotion the flagella form a bundle of co-rotating helices. The cell bodies are not polar: The bundle may form at either end of the rod. *B subtilis* require oxygen. They swim up gradients of oxygen concentration. They are approximately 10% denser than water. These two properties play a major role in self-concentration of these organisms, in the laboratory, and presumably in nature [5, 11].

Experiments were carried out with the strain 1085B of *Bacillus subtilis*. Bacterial spores were added to 10 ml of sterile Terrific Broth (TB) [10] at room temperature for 18 h, 1/2 ml of which was mixed in equal parts with glycerol and stocked at $-20°C$. Experimental samples were prepared by adding 1 ml of the stock to 10 ml of fresh TB in a petri dish for 18 h, then 1 ml of the resulting bacterial suspension was re-inoculated in 50 ml of TB and incubated in a shaker bath (37°C, 100 rpm) for 4 h. Prior to each experiment, 1 ml of the cell culture was centrifuged for 2 min at 4,000 g. The concentrated bacteria sediment was extracted and re-suspended in a controlled amount of medium.

Square chambers 5 mm side and 0.1 mm deep were used to image the bacteria suspension under an inverted microscope (Nikon Diaphot 300). Each chamber was enclosed in a petri dish with water reservoirs providing a humid environment avoiding evaporation flows. Digital videos were captured using a high speed camera (Phantom V5.1, Vision Research) at 100 frames per second. Such videos were processed to remove the background and increase the contrast. A commercial Particle Imaging Velocimetry system (PIV, Dantec Dynamics) was used to estimate cell velocities and generate quantifiable data. Also, Particle Tracking Velocimetry (PTV) of individual bacteria in dilute conditions was used to characterize the free swimming phase. Such samples were prepared by re-suspending the spun down bacteria pellets, diluting to 1/100th of the original concentration and placing in micro chambers for imaging. Thereafter, multiple cell trajectories were measured from the digital videos using a PTV system written in MATLAB [9].

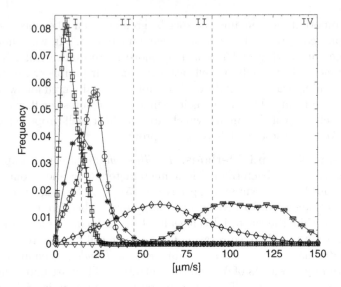

FIG. 1. *Probability distribution of speeds for different case studies: (○) free swimming cells, (□) semi-dilute, (◇) ZBN phase. And for the ZBN case, filtered data according to the measure Φ: (▽) highly organized regions with (0.98 < Φ < 1); (*) isotropic regions with (−0.15 < Φ < 0.15). Regions of the velocity range: (I) Jammed, (II) Free, (III) Collective, (IV) Super fast*

3. Results.

3.1. Dilute Populations: Individual Swimmers. Instantaneous speeds can be directly calculated from the trajectories generated by PTV on a dilute bacterial suspension (see Fig. 2a). Interactions in this sample are very rare and for the most part, bacteria move completely independently from each other. The corresponding probability distribution of speeds for the free swimming cells is shown in Fig. 1. As is evident, the most likely speed occurs around 25 µm per second. The distribution appears approximately Maxwellian. Cells moving faster than 35 µm per second are quite rare while low speeds occur with significant frequency.

3.2. Semi-dilute System: Jamming. Semi-dilute samples were prepared as indicated in Sect. 2 and filmed at 100 frames per second. PIV analysis of these digital movies return a measurement of the velocity field of the concentrated bacterial suspension. Taking the average distribution of speeds over all realizations (frames) we obtain the curve shown in Fig. 1. As we can see, this distribution is shifted to the left, giving typical speeds that are only of about 5 µm/s and quickly decaying for speeds well below the typical free swimming speeds. The inferred cause of this finding is discussed in Sect. 4.1.

3.3. Concentrated Suspension: Fast Collective Swimming. A dramatically different regime occurs with further increase of the

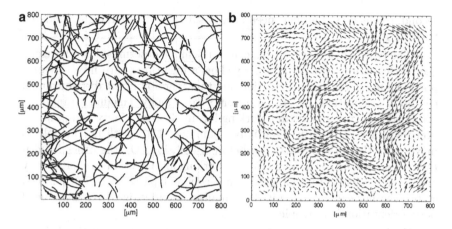

FIG. 2. *(a) Tracks of bacteria trajectories generated with PTV (b) Example of a typical PIV frame of bacteria in the ZBN mode*

concentration of bacteria. This regime, the ZBN, is again analyzed using PIV (Fig. 2b) to generate a probability distribution of speed (Fig. 1). The distribution now centers at 60 μm/s with a dispersion that ranges from below free swimming speeds to greater than 100 μm/s, several times greater than free swimming speed! The correlation of speeds with angular alignment of adjacent swimming trajectories is discussed in Sect. 4.2

4. Discussion.

4.1. Swimming Speeds at Intermediate Concentration. Unexpectedly the distributions of swimming speeds during the transition from low to high concentrations of organisms is not monotonic as a function of concentration. The relatively low average speeds at intermediate concentrations can be ascribed to cell-cell collisions that cause dissolution of the bundle of flagella that propel the organisms. During its re-formation the bacteria swim at speeds ranging from zero to normal.

The bacteria are propelled by helical flagella which are rotated by motors which are rigidly attached to the cells bodies. From six to ten helical flagella emerge from each cell. The diameter of a flagellum is approximately 20 nm. Locomotion is achieved when a cell's flagella, all having the same helix pitch and all rotating at the same rate, form a tight rotating bundle. In *B. subtilis*, the bundle is located at the posterior, relative to the direction of swimming. When a swimming bacterium collides with another, or with an obstacle, its bundle may disperse and the orientation of its body axis may change. Subsequently, the bundle re-forms and the bacterium accelerates to its normal cruising speed. After a collision it takes about 1 s of acceleration to return to full speed [2]. That is presumably the time required for the bundle to reconstitute, i.e. for reaching normal propulsion efficiency. Thus, if each organism collides and re-orients every fraction of a second, the mean swimming speed is necessarily less than along unhindered trajectories.

The typical distance between collisions can be determined by Onsager's geometrical arguments [8, 12]. They show a mean free path inversely proportional to bacterial concentration and to a function of the distribution of cell body orientations [4]. In an automotive analogy, imagine a randomly arranged collection of cars in a parking lot, in close proximity and pointing every which way. This is essentially a jammed state, with no car able to move more than a short distance before a collision. On the other hand if the cars are aligned as in a functioning parking lot, they can circulate freely and even accelerate to high speeds in a coherent pattern. In the jammed state, whether cars or bacteria, the angular distribution of velocities or of body orientation is isotropic.

For large concentrations, Onsager's passive rods, thermally jostled, exhibit a transition to the nematic liquid crystal phase. Similarly, due again to steric repulsion, the axes of rod-shaped active bacteria must align locally to a nematic state. Boundaries separate adjacent domains of different orientations. Alignment of body axes is necessary for coherent movement of domains. Transition to collective polarity of locomotion is required also. It has been shown [2] that bacteria that meet a rigid boundary stop, reconstitute their bundles of flagella at the original front end of the cell, causing them then to swim in the direction opposite to the original, all without turning their bodies. Thus when a bacterium aligned in a domain collides with another, oppositely directed one, one of them presumably flips its bundle of flagella, thereby reversing its direction of swimming. Some form of quorum sensing, based on this process and on hydrodynamics, is presumably responsible for domains of alignment becoming a coherent phalanx.

Angular coherence and directional coherence imply long mean free paths and the attainment of at least terminal velocity for each cell. But that cannot yet be the whole story. For limited durations, the velocities in any given phalanx must be at least approximately uniform in order to maintain the required concentration. Furthermore, the measurements reported here show that the mean speed in the ZBN phase of phalanxes is greater than the mean speed of individuals.

4.2. Speed and Coherence. To show that the exceptionally high speeds are associated with local order, the directional alignment of velocity vectors is evaluated by means of a local order parameter, or alignment measure [3, 4]. This measure is a scalar field Φ_R defined by the local average $\langle \cos \theta \rangle_R$ of the scalar product of adjacent unit velocity vectors over a small region:

$$\Phi_R(i, j, t) = \frac{1}{N_R} \sum_{B_R} \frac{\mathbf{v}_{ij}(t) \cdot \mathbf{v}_{lm}(t)}{|\mathbf{v}_{ij}(t)||\mathbf{v}_{lm}(t)|}, \tag{1}$$

where $\mathbf{v}_{ij}(t)$ is the PIV velocity field and B_R is a region of radius R centered in (i, j) and with N_R elements. If $\Phi_R \sim 1$ the vectors inside B_R are nearly parallel, corresponding to coherent motion. Values close to zero indicate strong misalignment, or isotropic orientations, while negative values imply locally opposing streamlines. Figure 3a shows an example of this field for a particular time frame in a ZBN sample. Near perfect alignment is indicated by dark gray, less perfect alignments are shown by a sequence of gray tones ending at white. Figure 3b shows the spatial distribution of speeds for the same PIV frame. The highest speeds are shown in black; lower speeds are grey, lowest in white. The incisive fact is the strong correlation of the highest speeds with the best alignments, the lowest with misalignments. This qualitative correlation is consistent over all time frames and all ZBN cases that have been analyzed. It demonstrates that fast domains consist of co-directionally swimming bacteria. For quantitative correlation, the alignment measure Φ_R was used as a localized filter of speeds: The data used to calculate speed distribution functions can be restricted to localities with particular degrees of alignment. In this way, the distribution of speeds in highly organized regions ($0.95 < \Phi_R < 1$, dark gray in Fig. 3a), is shown in triangles ($\bigtriangledown$) in Fig. 1. And the distribution for very disorganized regions ($-0.15 < \Phi_R < 0.15$ is shown in asterisks ($*$) in Fig. 1. Note that the distribution indicated with diamonds ($\Diamond$) for the full data set covers the entire spectrum of speeds in the ZBN, whereas the ones for partial data, ($\bigtriangledown$) and ($*$), cover only two narrow cases; therefore they do not add up to the curve indicated by the ($\Diamond$) symbols. The two distinct levels of organization manifest clearly distinct behaviors in terms of the typical speeds observed, supporting the conclusion that alignment corresponds to fast motion while disorganized orientations correspond to slow motions. The inter-domain spaces exhibit low alignment associated with low speeds, due partly to collisions and jamming, as shown in Sect. 3.2, and partly characteristic of the swimming velocity of individual organisms, as shown in Sect. 3.1.

4.3. Mixing and Transport. The dynamics of the ZBN, rapid self-ordering into domains, followed by their disappearance and succeeded by re-formation of coherent regions that move in apparently random directions suggest that these concentrated populations of bacteria are efficient mixers of themselves and solutes in their embedding liquid (Fig. 4). When these bacteria are located in a deep environment that can deliver the oxygen required by them from only one direction which is far from the bulk of the population, measured in diffusion/consumption time distance units, the rapid, energy requiring dynamics continue for many hours [5]. That fact implies excellent transport and distributive mixing of oxygen from the free surface. The locomotion-driven dynamics would otherwise have quickly ceased as previously shown [7].

For the present case, taking 5×10^9 bacteria/ml, and a consumption rate of $\sim 10^6$ oxygen molecules per second per organism [1], and allowing

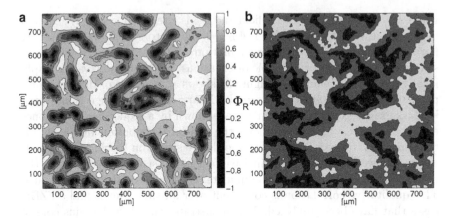

FIG. 3. *(a) Contour levels for the orientation order parameter Φ_R in a snapshot of a ZBN sample (with $R = 18:91\,\mu m$). (b) for the same frame, contour levels for values of speeds in four intervals: I (black) $v < 15\mu/s$, II (dark gray) $15\,\mu m/s < v < 45\,\mu m/s$, III (light gray) $45\mu m/s < v < 90\mu m/s$, and IV (white) $v > 90$ $\mu m/s$. Taking Fig. 1 into consideration, these intervals correspond to regions of: jamming (I), free motion (II), typical collective motion (III) and super fast motion (IV) respectively*

saturated concentration of oxygen of $\sim 10^{17}$ molecules/ml, all the oxygen in the fluid medium would be consumed in ~ 20 s! After this time if there were no mixing within the suspension and transport from the surface of it all the bacterial dynamics would have ceased, Other suspended molecules, as well as the cells themselves, are transported and mixed within the ZBN. The interaction with its environment of a concentrated population of bacterial cells, the communication of the organisms with each other, and the dispersal of metabolites and wastes are changed qualitatively as a function of population concentration.

The experiments reported here provided the opportunity for actually demonstrating the required folding and stretching that had previously only been inferred. So, does the measured velocity field of the ZBN generate the stretching and folding implied by these earlier results?

Using PIV data, bacterial velocity fields are generated for each time frame. Having the fields quantified for each time frame, artificial tracers can be set to evolve and illustrate the fluid dynamics. In other words, each member of a set of initial points that constitute a square grid, as shown in Fig. 4a, are translated according to the velocity field given by PIV data on each time step. Figure 4b shows the locations of the points after only 200 frames (corresponding to 2 s of real time data). The mixing efficiency of this system is evident.

5. Conclusions. Analysis of the distributions of velocities of the bacteria *Bacillus subtilis* as a function of the concentration of these microorganisms showed a decreased mean velocity at concentrations intermediate

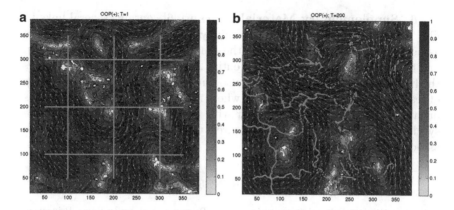

FIG. 4. *Computer generated points starting on a square grid at t = 0, left figure, are followed along their local velocity vectors which are generated by the collective dynamics and measured by PIV. The right figure shows the location of these grid points at t = 2 s. Gray scale levels indicate* Φ_R *values as in Fig. 3a*

between dilute and very concentrated. These relatively low velocities were quantitatively attributable to collisions among the organisms, followed by velocities lower than during free swimming, due to the time required for reassembly of the bundle of propelling flagella.

The mean collective velocities in the regime of self-assembled coherent domains were found to be higher than mean velocities of individual cells swimming at low concentration. These anomalously high velocities were shown to be well correlated locally with excellent local alignment of trajectories, as measured by PIV. A mechanism believed to be responsible [3] but not yet proven, for the anomalously high velocities in the coherent patches was mentioned in the introduction. The low velocity tail of the probability distribution of velocities was ascribed to the disordered regions that are present simultaneously with the ordered domains, interspersed between them.

The velocity field measured by PIV provided time lines following velocity vectors. These time lines, with initial positions on a square grid, were followed for 2 s of real time observation of the collective dynamics. They were derived from the live data (i.e. not a computer simulation). They exhibited the stretching and folding that underlies efficient mixing and transport of solutes as well as of the driving occupants. The bacteria supply the energy for the dynamics; their evidently fairly weak coherent interactions supply the intermittency. That transport and mixing is vastly enhanced by the dynamics of the ZBN was inferred previously, but never before was it demonstrated explicitly by measurements on an actual population of bacteria. We believe this demonstration of enhanced autonomous mixing and transport will provide new insights into the interactions of bacterial populations with their environment.

6. Supplementary Materials. The url [6] contains several links to film strips illustrating (1) the PIV analysis of speed in colors that are associated with the magnitude of the scalar alignment parameter (Φ_R), and (2) the mixing sequence for the entire 200 frames of which Fig. 4 shows the first and last frames. There are also film strips showing the trajectories of real passive markers in the ZBN, where real means actual large particles or latex spheres suspended together with the bacteria, as opposed to the synthetic markers that follow measured time lines as in Fig. 4 and the associated film strip.

REFERENCES

[1] Berg HC (1993) Random walks in biology. Princeton University Press, Princeton, NJ, USA

[2] Cisneros LH, Dombrowski C, Goldstein RE, Kessler JO (2006) Reversal of Bacteria locomotion at an obstacle. Phys Rev E 73:030901(R).

[3] Cisneros LH, Cortez R, Dombrowski C, Goldstein RE, Kessler JO (2007) Fluid dynamics of self propelled microorganisms, from individuals to concentrated populations. Exp Fluids 43:737

[4] Cisneros LH, Kessler JO, Ganguly S, Goldstein RE (2011) Dynamics of swimming bacteria: transition to directional order at high concentration. Phys Rev E 83:061907

[5] Dombrowski C, Cisneros LH, Chatkaew S, Goldstein RE, Kessler JO (2004) Self-concentration and large-scale coherence in bacterial dynamics. Phys Rev Lett 93:098103

[6] http://www.physics.arizona.edu/~kessler/IMA/

[7] Kessler JO, Wojciechowski MF (1997) Collective behavior and dynamics of swimming bacteria. In: Shapiro JA, Dworkin M (eds) Bacteria as multicellular organisms. Oxford University Press, Oxford, pp 417–450

[8] Onsager L (1949) The effects of shape on the interaction of colloidal particles. Ann N Y Acad Sci 51:627

[9] PTV system written in MATLAB based original source code by Nicholas Darnton and Jacob D. Jaffe (Rowland Institute at Harvard University 2003) with modifications by one of the authors (LHC).

[10] Terrific Broth (TB): 48.2 g Ezmix Terrific Broth (Sigma) and 8 ml glycerol in sufficient water to make 1 liter of medium

[11] Tuval I, Cisneros LH, Dombrowski C, Wolgemuth CW, Kessler JO, Goldstein RE (2005) Bacterial swimming and oxygen transport near contact lines. PNAS 102:2277

[12] Vroege G, Lekkerkerker H (1992) Phase transition in lyotropic colloidal and polymer liquid crystals. Rep Prog Phys 55:1241

DYNAMICS, CONTROL, AND STABILIZATION OF TURNING FLIGHT IN FRUIT FLIES

LEIF RISTROPH(✉)*, ATTILA J. BERGOU†, GORDON J. BERMAN‡,
JOHN GUCKENHEIMER§, Z. JANE WANG¶, AND ITAI COHEN‖

Abstract. Complex behaviors of flying insects require interactions among sensory-neural systems, wing actuation biomechanics, and flapping-wing aerodynamics. Here, we review our recent progress in understanding these layers for maneuvering and stabilization flight of fruit flies. Our approach combines kinematic data from flying insects and aerodynamic simulations to distill reduced-order mathematical models of flight dynamics, wing actuation mechanisms, and control and stabilization strategies. Our central findings include: (1) During in-flight turns, fruit flies generate torque by subtly modulating wing angle of attack, in effect paddling to push off the air; (2) These motions are generated by biasing the orientation of a biomechanical brake that tends to resist rotation of the wing; (3) A simple and fast sensory-neural feedback scheme determines this wing actuation and thus the paddling motions needed for stabilization of flight heading against external disturbances. These studies illustrate a powerful approach for studying the integration of sensory-neural feedback, actuation, and aerodynamic strategies used by flying insects.

Key words. Insect flight, aerodynamics, flight dynamics, control, stability

AMS(MOS) subject classifications. 37N25, 76Z10, 92B05, 70E99

1. Introduction. The flight of insects is a beautiful example of an organism's complex interaction with its physical environment. Consider, for example, a fly's evasive dodge of an approaching swatter. The insect must orchestrate a cascade of events that starts with the visual system perceiving information that is then processed and transmitted through neural circuits. Next, muscle actions are triggered that induce changes to the insect's wing motions, and these motions interact with fluid flows to generate aerodynamic forces. As another example, even the simple task of flying straight requires similarly complex events in order to overcome unexpected disturbances and suppress intrinsic instabilities. Here, we review our recent progress in dissecting the many layers that comprise maneuvering and stabilization in the flight of the fruit fly, *D. melanogaster* [1–3]. Our emphasis is on aspects of flight at the interface of biology and physics, and we seek to understand how physical effects both constrain and simplify biological strategies.

---

*Department of Physics, Cornell University, Ithaca, NY 14853, USA, lgr24@cornell.edu. The authors thank the NSF for support.

†Department of Engineering, Brown University, Providence, RI 55555, USA

‡Lewis-Sigler Institute for Integrative Genomics, Princeton University, Princeton, NJ 08544, USA

§Department of Mathematics, Cornell University, Ithaca, NY 14853, USA

¶Departments of Mechanical and Aerospace Engineering and Physics, Cornell University, Ithaca, NY 14853, USA

‖Department of Physics, Cornell University, Ithaca, NY 14853, USA

S. Childress et al. (eds.), *Natural Locomotion in Fluids and on Surfaces*, IMA 155, DOI 10.1007/978-1-4614-3997-4_6,
© Springer Science+Business Media New York 2012

Over the last 40 years, turning flight in insects has emerged as an archetype of complex animal behaviors [4–12]. When searching for food, flies exhibit a stereotyped exploratory behavior in which straight flight paths are separated by rapid turns called saccades. Fruit flies turn when triggered by specific visual stimuli, and a typical saccade through 90° is completed in 50 ms or about 10 wing-beats [13]. Given that a blink of an eye is about 250 ms, these maneuvers are quite impressive. Is it difficult for a fly to perform a saccade? Physically, one might address this question by comparing the torque needed to turn its body with the torque exerted simply to keep the body aloft. The scale of the turning torque is given by the body moment of inertia times angular acceleration: $I\alpha \approx (10^{-13} \text{ kg/m}^2)(90°/(50 \text{ ms})^2) \approx 10^{-10}$ Nm. To hold its milligram body up during hovering, the millimeter-scale wings exert torques of about $Mgr \approx (10^{-6} \text{ kg})(10 \text{ m/s}^2)(1 \text{ mm}) \approx 10^{-8}$ Nm. This simple estimate shows that the torque needed to turn is only a few percent of the torque produced during hovering. Thus, counter-intuitively, even these extreme flight maneuvers are achieved with little additional effort.

However, what appears to be effortless in terms of torque exertion is difficult in nearly all other respects. For example, the changes in wing motions needed to induce such a maneuver are also expected to be a few percent [9], which amounts to adjustments in wing orientation on the scale of a few degrees! What modulations to wing motions do insects actually make, and how small are these changes? Of course, such minuscule adjustments demand precise muscular actuation [14]. How do muscle actions lead to subtle modulations of wing motions? Further, the time-scales involved in such maneuvers are so fast that the turn is often complete within the visual system reaction time [7]. How are these wing motions orchestrated if the insect is effectively blind during the maneuver? Armed with knowledge of the force scales, one can now also appreciate the difficulty of simply maintaining straight flight. Air is a messy environment, and small external torque or noise in the flight motor will knock the insect off its intended path. How do fruit flies resist unwanted body rotations and keep on-course?

In this work, we take a tour through some of these aspects of turning behavior of fruit flies. We show that analyzing motions of flying insects reveals a remarkable amount of information about sensory, neural, muscular, and aerodynamic processes. First, we review recent developments in techniques for motion tracking of flying insects and in modeling aerodynamic forces on flapping wings. We apply these tools to reveal how subtle adjustments to wing motions drive turning maneuvers. Next, we go a level deeper to examine how the wing motions themselves emerge from the interaction of muscular actuation, aerodynamic forces, and biomechanics of the wing hinge. Finally, we examine the fruit fly's "auto-pilot", a sensory-neural scheme that uses feedback to maintain body orientation during straight flight. By combining results from each of these studies, we demonstrate

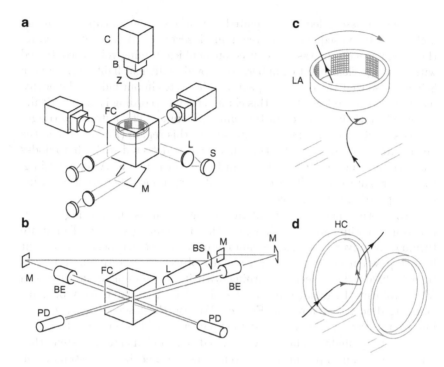

FIG. 1. *Schematics of the experimental set-ups [1–3]. (**a**) Three orthogonal high-speed cameras capture the flight of insects within a clear flight chamber. Each camera is back-lit by a single bright light-emitting diode focused with a lens. (**b**) An automatic trigger consists of crossed laser beams to detect the presence of an insect in the filming volume and initiate recording. (**c**) A computer-controlled array of lights presents insects with rotating striped patterns and reliably generates turning maneuvers. (**d**) Helmholtz coils generate a field that applies a torque to a magnetic pin glued to the back of an insect, thus disrupting its flight*

that complex flight behaviors can be understood in terms of the integration of reduced-order mathematical models.

2. Experimental Methods: Videography, Behavioral Stimulation, and Motion Tracking. Given that adjustments to wing motions are expected to be subtle, our experimental emphasis is on developing precision techniques for gathering large quantities of flight data. Here, we present three experimental advances needed to address flight maneuverability and stability. First, we show how to automate the high-speed video capture process in order to obtain many flight sequences. Second, we elicit specific behaviors by presenting insects with visual stimuli and by mechanically perturbing their flight. Third, we outline our algorithm for automatically extracting wing and body motion data from flight videos.

We have assembled an automated, versatile system for capturing many high-speed video sequences of free-flying insects [1]. As shown in Fig. 1a, three high-speed cameras are focused on a cubical filming volume contained within a large clear flight chamber. To provide sufficient light, each camera is back-lit so that the insects appear as silhouettes in our videos. Typically, many insects are released into this chamber, and recording is automatically triggered by their presence in the filming volume. The automatic trigger consists of two laser beams that intersect in this region (Fig. 1). When the beams are simultaneously broken, this event is detected with photodiodes and an electronic circuit initiates recording of the cameras. After recording, the cameras automatically become available to film another event, allowing us to capture many sequences.

To capture many movies of rare events such as turning flight, it is necessary to stimulate the behavior within the filming volume. To initiate turning maneuvers, we take advantage of a well-known behavior of fruit flies [15]: when presented with a moving object or pattern, these insects tend to fixate the object by turning with it. We assembled an arena in which rotating light patterns can be played on a circular array of light-emitting diodes, as shown in Fig. 1c. We use the laser trigger signal to initiate rotation of a light-dark striped pattern, yielding saccadic turns [3].

To study flight stability, we devised a complementary system that imposes mechanical perturbations to insects, causing them to "stumble" in flight [2]. We first glue tiny ferromagnetic pins to insects' backs and image their flight using the set-up described above. In Fig. 1d, we show that as a fly crosses the filming volume, the trigger used to initiate recording also activates a pair of magnetic Helmholtz coils that generate a brief magnetic field. Here, the field and pin are both oriented horizontally, so the resulting magnetic torque on the pin reorients the yaw, or heading, of the insect. This experiment can be thought of as an experimental simulation of a disruptive gust of wind. With this technique, however, we have control over the magnitude, direction, and duration of the perturbing torque.

We use Hull Resolution Motion Tracking (HRMT) to extract the wing and body motions from flight videos [1]. We developed this algorithm using computer vision techniques for estimating the shape of a 3D object from its silhouettes (Fig. 2a–c). For each image, the algorithm first reconstructs the maximal volume 3D shape that is consistent with the three 2D shadows captured by the cameras. We find that this shape is sufficiently close to the real insect's shape to allow extraction of coordinates. As shown in Fig. 2d, portions of the hull that correspond to the body, right wing, and left wing are then "dissected" by applying a clustering algorithm that identifies groups of nearby points. Then, we apply a variety of geometric techniques to determine the center of mass position and angular orientation of each group (Fig. 2e, f). For example, using principal components analysis reveals the long axis of each cluster, which identifies the yaw and pitch angles of the body and the stroke and deviation angles for each wing. The wing chord

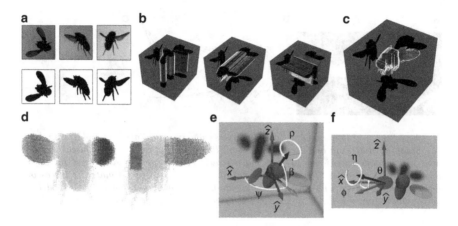

FIG. 2. *Hull Reconstruction Motion Tracking (HRMT) [1].* **(a)** *Silhouette information is obtained for each frame of a high-speed video.* **(b)** *To construct the visual hull of the insect, each silhouette is extruded into the third dimension.* **(c)** *The intersection of these extrusions forms the hull, the maximal volume 3D shape that is consistent with the three 2D shadows.* **(d)** *A clustering algorithm 'dissects' the insect into portions corresponding to the body, right wing, and left wing.* **(e)** *The body yaw, pitch, roll, and center-of-mass coordinates are extracted by applying geometrical and statistical measures to the body cluster.* **(f)** *Similar procedures yield the stroke, deviation, and pitch angles that describe the orientation of each wing*

is determined as the diagonal of the wing points, yielding its pitch angle. In the end, the positions and angles of each component are determined for all images, yielding 18 coordinates for each frame of the movie. These data can then be used in further studies, for example, in simulations that predict the aerodynamic forces generated.

3. Simulation Methods: Aerodynamics and Dynamics. Experimental studies of insect flight have been accompanied by theoretical efforts aimed at identifying flapping-wing aerodynamic mechanisms [16–18]. Early studies focused on comparing the forces generated by flapping and flipping insect wings to quasi-steady estimates appropriate for translating wings at fixed orientation [19, 20]. Quasi-steady aerodynamic models approximate the instantaneous fluid forces using a mathematical form that depends on the state variables of the wing, for example, its orientation and velocity. This technique is rapid, tractable, and intuitive, and quasi-steady calculations were able to account for the forces needed to sustain simple flight modes such as level forward motion [19]. However, the general applicability of these early approaches was controversial since the models were not able to account for all experimental observations [20].

These discrepancies inspired researchers to use dynamically-scaled flapping wing models to investigate aerodynamic mechanisms more closely [21–24]. This experimental approach prescribes flapping motions while

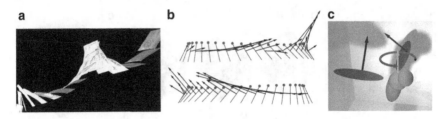

FIG. 3.   *Flapping-wing aerodynamics at intermediate Reynolds number.*
(a) Studying fluttering and tumbling plates allows one to extract fluid forces associ-
ated with complex motions [32]. A quasi-steady model approximates fluid forces in
terms of state variables. (b) Such a model can then be used to estimate the forces on
flapping wings [30]. (c) A computational simulation couples a quasi-steady model of
aerodynamic forces with the linked rigid-body mechanics of the insect body and wings

measuring fluid forces and visualizing flow structures. These experiments
have been used in conjunction with simulations to reveal the influence of
unconventional aerodynamic effects, and the most important effect is elevated flight forces that result from the presence of a leading-edge vortex
(LEV). The enhanced lift can be achieved either by steadily revolving a
wing about a root thereby stabilizing the attached LEV, as is the case for
a spinning maple seed [25], or by flipping the wing before the vortex has had
enough time to shed [18]. Both mechanisms are expected to be involved
in the revolving and flapping motions of insect wings. Mechanized wing
experiments have also revealed that other effects associated with wing rotation and wake interference may significantly influence aerodynamic forces
[21, 24]. While these techniques are useful for elucidating basic flapping-
wing aerodynamics, it has remained challenging to integrate this approach
with flight measurements from actual insects [9]. For example, until recently [26], such experiments did not couple the body motion to the forces
produced by the wings, an effect that is crucial for understanding the dynamics of free-flight maneuvers [12].

Finally, with the advent of more powerful computers, it has become
possible to use computational fluid dynamics (CFD) simulations to numerically solve the Navier-Stokes equations and determine the flight forces and
flow structures associated with flapping wings [27–29]. These techniques
have been instrumental in showing that flapping flight can be more efficient
than fixed-wing flight [30] and in determining the role of wing flexibility
[31]. While the CFD approach is promising, it remains computationally-
intensive thus prohibiting its use for studies that require the analysis of
many wing-strokes. For example, understanding the control of a given
flight mode requires many instances of the maneuver each consisting of
many wing-strokes [3], and such a statistical analysis remains a challenge
for the CFD method.

Our computational approach borrows elements from these different
techniques, enabling rapid calculation of flight forces via a quasi-steady

model that is modified to incorporate unconventional mechanisms (Fig. 3). Specifically, we use a quasi-steady model whose form is distilled from experiments conducted on the fluttering and tumbling of plates [32, 33]. We include the effect of the LEV by using enhanced lift and drag coefficients determined by mechanized wing experiments [24], and the effect of wing flipping is modeled by including a rotational lift term that couples rotational and translational velocities [32, 34]. Forces are computed for 2D blades and integration over the span of the wing yields an estimate of the 3D force. When compared with CFD calculations and experiments, these quasi-steady estimates have been shown to account for the average forces with an accuracy of about 90% [33, 34].

In order to understand how the forces generated by the wing motions lead to changes in the body position and orientation, we combine the quasi-steady aerodynamic calculations with a rigid-body dynamic solver [3, 35]. By solving the Newton-Euler dynamical equations for the coupled wings-body system, we are able to prescribe wing motions and determine the resulting body motion. Alternatively, the torque at the wing base can be prescribed and the resulting wing and body motions are then computed. These techniques have become increasingly important for elucidating how body motions alter the wing motions relative to the air and hence alter the aerodynamic forces associated with different maneuvers [36–41]. In addition, it has recently become possible to study aspects of free-flight using CFD flow solvers coupled to rigid-body dynamics solvers [42, 43]. We use our quasi-steady implementation in conjunction with the experimental data to analyze the flight dynamics, wing actuation, and sensory-neural control of flying insects.

4. Maneuvering Dynamics. To determine the aerodynamic basis of turns, we develop a systematic procedure for distilling physical mechanisms from kinematic data [3]. We first use our LED arena to elicit turning events, and then reconstruct the maneuver using our HRMT algorithm to extract wing and body kinematics. The results of this procedure for an extreme turn through 120° performed in 80 ms are shown in Fig. 4. To determine the wing kinematics that lead to turning, we use a phase-averaging algorithm [3] to collapse the data for wing-strokes during the turn. We then examine the measured kinematics in search of differences in the right and left wing motions. Perhaps not surprisingly, during the turn asymmetries appear in all three Euler angles, as shown in Fig. 4c–e. To assess which of these changes is aerodynamically important, we form various symmetrized versions of the wing kinematics and play these modified motions in simulation. For example, to assess the importance of the observed changes in stroke angle, we form kinematics that consist of the measured stroke but with the deviation and pitch angles symmetrized to be the mean of those measured for the right and left wings. In this case, the simulated insect fails to turn, indicating that the differences in stroke

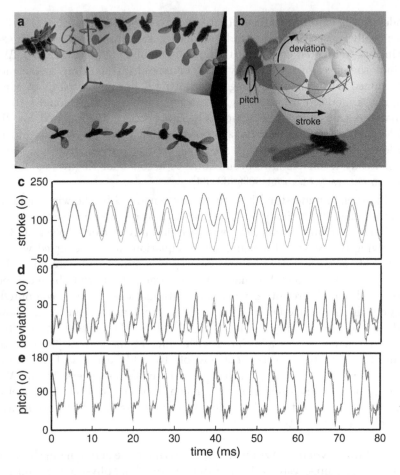

FIG. 4. *Body and wing motions during a turning maneuver [3]. (a) The wing and body configuration are captured by the three high-speed cameras, and selected snapshots are displayed on the side panels. We use a motion tracking algorithm to extract the wing and body posture for each frame, and these data are used to render the model insect. The insect is initially hovering on the right, and performs a rightward turn through about 120° while drifting to the left in this image. (b) Representation of wing motion. Each wing sweeps a path along a globe centered about its root on the body. (c–e) Wing orientation angles throughout the maneuver. Stroke angle is measured in the horizontal plane, deviation angle is the vertical excursion, and pitch angle is measured between the wing chord and the horizontal plane*

angle are not crucial to generating yaw torque. Surprisingly, this procedure reveals that small changes in wing pitch are responsible for about 90% of the yaw torque generated.

For turning maneuvers, the changes in wing pitch correspond to rowing or paddling motions of the wings (Fig. 5a). To turn to the right, the right wing pushes off the air at a high angle of attack on the forward sweep and

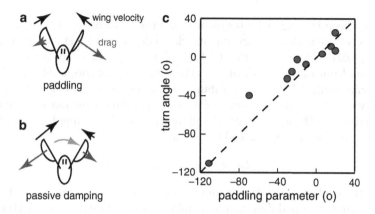

FIG. 5. *Minimal model of turning dynamics [2, 3].* (**a**) *Paddling drives turns. In this top-view schematic, a fly turns rightward by inclining its wings at different angles, thus generating differential drag (red arrows) on its wings.* (**b**) *Passive damping resists body rotations. As the wings beat during body rotation, their airspeeds are modified and thus induce a net resistive drag.* (**c**) *Testing the model. The model predicts that the body turn angle is equal to the paddling parameter, a quantity related to the wing motions. The model is consistent with the measurements of wing and body motions for ten maneuvers. See text for model details*

slices through the air at a low attack angle for the backward sweep. This paddling motion generates torque through differential drag. In addition to this driving mechanism, we find that the insect must overcome a resistive aerodynamic torque during its rotation. Here, we build on recent findings that rotations are passively damped during flapping flight [12, 44], and we idealize the damping mechanism in Fig. 5b. As the insect rotates, its wings encounter different airspeeds and thus set up resistive aerodynamic forces that oppose the rotation.

These observations can be quantified using a minimal model that incorporates the relevant aerodynamic torques into the Euler equation for yaw rotations,

$$I\ddot{\psi} = N_{\mathrm{aero}}, \tag{1}$$

where I is the yaw moment of inertia of the insect body and N_{aero} is the net aerodynamic torque on the insect. The drag on each wing depends on its attack angle α through the drag coefficient, $C_{\mathrm{D}}(\alpha)$, times the square of its speed relative to air. To analyze turning flight by paddling, we consider the case in which the right and left wing angles of attack are different, and each wing beats with average angular speed ω relative to the body. For an insect body rotating at angular velocity $\dot{\psi}$, the stroke-averaged net aerodynamic torque is found by summing each wing's contribution:

$$\begin{aligned} N_{\mathrm{aero}} &\sim -C_{\mathrm{D}}(\alpha_{\mathrm{L}}) \cdot (\omega + \dot{\psi})^2 + C_{\mathrm{D}}(\alpha_{\mathrm{R}}) \cdot (\omega - \dot{\psi})^2 \\ &\approx -C_{\mathrm{D}}(\alpha_0) \cdot 4\omega \cdot \dot{\psi} + 2C_{\mathrm{D}}'(\alpha_0) \cdot \omega^2 \cdot \Delta\alpha. \end{aligned} \tag{2}$$

Here, we keep leading-order terms in $\dot{\psi}$ and take advantage of the linearity of the coefficient dependence on attack angle [45]: $C_D(\alpha) \approx C_D(\alpha_0) + C_D'(\alpha_0) \cdot \Delta\alpha$, where $\alpha_0 = 45°$, $C_D'(\alpha_0)$ is the slope, and $\Delta\alpha$ is the mean deviation from α_0 in attack angle. The aerodynamic torque of Eq. 2 has two components. The first is a damping torque proportional to the yaw velocity $\dot{\psi}$. The second is a torque due to the paddling wing motions that is proportional to the angle of attack difference $\Delta\alpha$. Combining Eqs. 1 and 2, we arrive at the yaw dynamical equation

$$I\ddot{\psi} = -\beta\dot{\psi} + \gamma \cdot \Delta\alpha. \tag{3}$$

where the coefficients β and γ depend on aerodynamic properties of the wings. Thus, the paddling torque combines with damping and inertia to generate the body rotational dynamics.

To provide a test of the aerodynamic model, we derive a prediction for the body turn angle based on the wing motions. Integrating Eq. 3 over the entire turn yields

$$I \cdot \Delta\dot{\psi} = -\beta\Delta\psi + \gamma \int dt(\Delta\alpha), \tag{4}$$

where $\Delta\psi$ and $\Delta\dot{\psi}$ indicate the net change in each quantity. Once the maneuver is complete, the change in yaw velocity $\Delta\dot{\psi} = 0$ so that the left hand side of Eq. 4 is eliminated. Solving the remaining portion for the body turn angle yields

$$\Delta\psi = \frac{\gamma}{\beta} \int dt(\Delta\alpha) = \omega \int dt(\Delta\alpha), \tag{5}$$

where ω is mean angular speed of the wings relative to the body. Thus, the model predicts that the stronger the paddling and the longer such motions are applied, the greater the body rotates.

To test this prediction, we use our videography apparatus to capture many instances of saccadic turns. We then use our motion tracking algorithm to extract the complete wing and body kinematics for ten such sequences, and use these data to distill the paddling parameter, $\omega \int dt(\Delta\alpha)$, and the body turn angle, $\Delta\psi$. In Fig. 5c, we plot the body turn angle against the paddling parameter and find that the model captures the overall trend in the relationship. This indicates that paddling and damping are the key physical factors in the dynamics of turning maneuvers. Recently, work on a dynamically-scaled robot has confirmed many of the aerodynamic findings presented here [26].

5. Wing Actuation and Dynamics. At a level deeper, we next investigate how the paddling wing motions themselves arise. Our procedure for investigating wing actuation uses experimental measurements and aerodynamic simulation to extract the torque exerted by the insect on

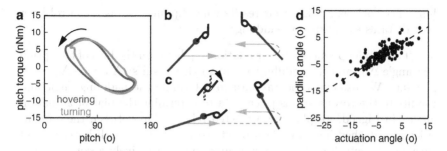

FIG. 6. *A biomechanical model of wing actuation [3].* (a) *Stress–strain relation for wing pitch. The pitch torque that the insect exerts is plotted against the pitch angle itself. During both hovering and turning flight, the torque exerted by the insect primarily acts like a viscoelastic brake that resists flipping due to inertial and aerodynamic forces. During turning, the mean pitch angle is biased to a pitch angle greater than 90°.* (b–c) *Biomechanical interpretation. The wing hinge or musculature acts like a damped torsional spring that is biased to generate paddling motions.* (d) *The degree of paddling varies with the bias applied. The paddling angle is approximately one-half the bias angle (dashed line)*

its wing [3]. The pitch rotational acceleration is dictated by aerodynamic torque on the wing, the torque exerted by the insect, and the torque associated with driving the wing about an axis above its center of mass. Applying the quasi-steady aerodynamic model to the measured kinematics, we solve for the torque exerted by the insect to pitch the wing. As an analogy to a stress-strain curve or work-loop for a material, this torque can be plotted against the pitch angle itself, as in Fig. 6a. For the symmetric wing motions in hovering and the asymmetric paddling motions, this relation forms a loop which is traversed in a counter-clockwise sense. This direction indicates that work is being done by the fluid to pitch the wing. Even though the insect is actively applying torque to sweep the wings back-and-forth, the wing rotations arise passively.

The difference between wing pitching during hovering and during a turn corresponds to a shift in the loop toward greater values of pitch (Fig. 6a). As discussed above, greater pitch corresponds to paddling motions. To physically interpret the actuation scheme, we note that the derived stress-strain relation is similar to that of a damped torsional spring. The general negative correlation indicates elastic or spring-like behavior, and the open loop indicates damping or viscous-like dissipation. The torque exerted is then well-approximated by $\tau = -\kappa \cdot (\eta - \eta_0) - C \cdot \dot{\eta}$, where η is the wing pitch angle, κ is the torsional spring constant, η_0 is the rest angle of the spring, and C defines the degree of damping. For hovering, $\Delta\eta_0 = \eta_0 - 90° \approx 0$ so that the wings pitch symmetrically back and forth, as in Fig. 6b. For turning, the spring rest angle is biased such that $\Delta\eta_0 \neq 0$ and asymmetric paddling motions result, as shown in Fig. 6c. These results suggest a biomechanical interpretation in which the wing musculature and hinge act both as a brake that resists wing flipping and as a control lever that dictates the degree of paddling.

This model predicts a linear relationship between the bias of the spring rest angle $\Delta\eta_0$ and the paddling angle $\Delta\alpha$ that results: $\Delta\alpha = \mu \cdot \Delta\eta_0$ with $\mu = 0.6$. We use experimental data to determine the paddling angle and use fits to the torque versus pitch data to determine the bias or actuation angle. In Fig. 6d, we plot the predicted relationship between these quantities as a dashed line, and data points corresponds to both hovering and turning wing-strokes. The theory is able to quantitatively account for the relationship between actuation and resulting wing motion.

Physically, this actuation model can be interpreted as a transmission system that converts flapping motions to flapping-plus-flipping motions. It is an active-passive hybrid system. It is passive in the sense that the insect need not invest power directly to the pitching degree-of-freedom but instead harvests power from the flapping motions. It is active in the sense that pitching motions can be modulated in a simple way to drive maneuvers. The transmission system has other nice properties. First, it eliminates the need to actuate on the time-scale of a single wing-beat, a convenience and perhaps necessity for such animals in which the wing-beat period approaches the fastest neural time-scales. Though the wing motions themselves change within a wing-beat, the actuation need only be applied on the scale of an entire maneuver, say 10–20 wing-beats. Second, the system takes a relatively large actuation to a finer change in wing motion. The data of Fig. 6d show that if the base of the wing is biased by 10° then paddling motions of about 6° result. This down-gearing may be important considering the precision required for such a sensitive dependence of torque on wing motion modulation.

6. Auto-Stabilization and Sensory-Neural Control.

More frequently than performing a turn, fruit flies simply fly forward and maintain their current orientation. However, slight perturbations, such as gusts of wind, will knock them off course. How do these insects maintain their flight course? To explore this issue, we use the magnetic perturbation apparatus to apply torques to free-flying insects [2]. A reconstruction of a typical experiment is shown in Fig. 7a. Remarkably, these insects react to perturbations by quickly recovering their flight heading. In Fig. 7b we show overhead views of this recovery and in (c) we plot the yaw dynamics for this sequence. This example is typical of responses to moderate perturbations in that it is accurate, usually to within a few degrees, and it is fast, usually complete in under 15 wing-beats. To drive this recovery, these insects use paddling wing motions, just as during visually-elicited turning maneuvers. In Fig. 7d, we quantify these motions by plotting the time-course of the difference between right and left wing attack angles.

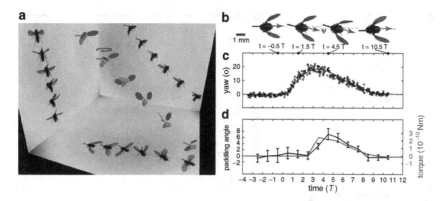

FIG. 7. *Flight stabilization in the fruit fly [2].* (**a**) *Reconstruction of a recovery maneuver. The flight of the insect is perturbed by a magnetic torque (red arrow) that is applied for one wing-beat.* (**b**) *Top-view stills show the fast and accurate recovery of yaw orientation.* (**c**) *Yaw dynamics for an insect perturbed by a magnetic torque (red stripe).* (**d**) *Recovery is driven by paddling wing motions which are quantified by the time-course of difference in right and left wing angles of attack. Blue curves in (c) and (d) represent results from a feedback control model*

These wing motions must be selected in order to generate accurate recovery of body yaw, and our measurement of both body and wing motions provides a window into this control strategy. A simple strategy would involve the use of sensory measurements of body orientation in order to determine the necessary wing response. A minimal control model [46] that guarantees perfect correction requires that the exerted torque contain a term proportional to the yaw angle ψ. However, we find that this so-called proportional (P) controller fails to account for the fast recovery time observed in the flight data. By adding a term that is proportional to the yaw angular velocity $\dot{\psi}$, we arrive at a good match to the yaw data, as shown by the model fit (blue curve) shown in Fig. 7c. This model is a proportional-derivative (PD) scheme [46], and the corrective paddling torque $N_{\text{fly}} = \gamma \cdot \Delta\alpha$ can be written as:

$$N_{\text{fly}}(t) = K_{\text{P}}\psi(t - \Delta t) + K_{\text{D}}\dot{\psi}(t - \Delta t). \tag{6}$$

Here, K_{P} and K_{D} are gain constants and Δt is the response delay time that we measure to be about three wing-beat periods. This loop delay may reflect both neural latency and inertia of the sensors and motor. In Fig. 7d, we overlay the torque, N_{fly}, predicted by Eq. 6 on the measured $\Delta\alpha$ data and find a strong agreement between the model and experiment.

These findings suggest that these insects sense their body motion and use this information to determine the appropriate response. In fact, flies are equipped with a pair of small vibrating organs called halteres that act as gyroscopic sensors [47]. Anatomical, mechanical, and behavioral evidence indicates that the halteres serve as detectors of body angular

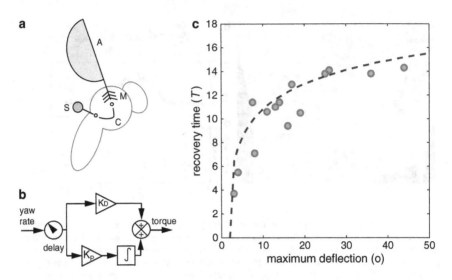

FIG. 8. *Feedback control model of yaw stabilization response [2]. (**a**) Stabilization involves the mechanosensory halteres (S) detecting undesired body rotations, neural circuits (C) processing this information, to drive muscular (M) response and thus aerodynamic (A) torque due to paddling wing motions. (**b**) The neural controller can be modeled by a yaw and yaw-rate output. (**c**) The control model accounts for the recovery time taken by the insects in response to perturbations of different strengths*

velocity [47, 48]. These findings suggest that these insects drive their corrective response using an auto-stabilizing feedback loop in which the sensed angular velocity serves as the input to the flight controller. As diagrammed in Fig. 8a, the velocity is sensed by the halteres (S), processed by a neural controller (C), and transmitted by the flight motor (M) into specific wing motions that generate aerodynamic torque (A). In the control diagram of Fig. 8b, the loop is triggered when an external torque induces a yaw velocity that is sensed and processed by neural circuitry to determine the paddling torque N_{fly}.

To test the control model, we use it to predict the total recovery time, Δt_{rec}, as a function of perturbation strength. As shown by the dashed blue curve in Fig. 8c, the model predicts that this time rises sharply and then plateaus for increasingly strong imposed deflections. The experimentally measured recovery times confirm this trend, indicating that the control model captures general features of stabilization behavior in fruit flies.

7. Synthesis and Implications. Collectively, these investigations outline the structure of maneuvering and stabilization in the flight of insects. We have put together experimental and analytical techniques that quantify the nature of these flight behaviors and reveal some of the solutions that insects have evolved in interacting with their aerial environment. Having dissected the many levels involved, the next step is to put these elements back together. Here, we will outline this synthesis for the specific case of yaw stabilization in fruit flies.

A description of yaw stabilization must include the roles of several key elements: the mechanosensory halteres, neural circuits, wing muscles and hinge, motion of the wings, aerodynamic forces, and the resulting body rotational dynamics. The interaction of these elements forms a closed loop of information flow. At each step in the loop, a given process can be viewed as a transfer function that converts one quantity into another. In the context of our simplest reduced-order models, these transfer functions are linear operators, and the entire feedback loop can be written as a system of linear differential equations that can be summarized by

$$I\ddot{\psi} = -\beta\dot{\psi} + N_{\text{fly}} + N_{\text{ext}} \tag{7}$$

where N_{ext} is the externally-imposed torque and

$$N_{\text{fly}}(t) = \gamma \cdot \Delta\alpha(t) = \gamma\mu \cdot \Delta\eta_0(t) = K_{\text{P}}\psi(t - \Delta t) + K_{\text{D}}\dot{\psi}(t - \Delta t) \tag{8}$$

is the response torque. Here, we have written the fly's response torque in several equivalent ways to emphasize the roles of the different systems that generate it. At the level of the wing motions, the torque is proportional to the paddling angle $\Delta\alpha$. At the actuation level, the torque is also proportional to the bias angle $\Delta\eta_0$. At the sensory and neural levels, the response is proportional to the time-delayed yaw plus yaw-rate.

This study of yaw control should provide a template for future studies aimed at dissecting insect flight behaviors. For the problem of fruit fly flight, the next steps entail devising similar reduced-order models for pitch and roll stabilization. Experiments that apply sequential perturbations could be used to understand the coordination of different degrees-of-freedom and may reveal non-linear interactions in the corrective response. Measurements aimed at probing other sensory modalities, such as vision, can be used to elucidate additional feedback circuits [49, 50]. Moreover, experiments that present simultaneous and conflicting stimuli could reveal the interactions between these circuits.

Overall, the combination of carefully-designed experiments and numerical simulations that capture the key physics promise to reveal much about the integration of subsystems in structuring complex flight behaviors. More generally, the organization of processes involved in controlling flight should be shared by a broad array of animal behaviors. As such, we envision that the framework we have outlined here will also provide a useful strategy for unraveling these problems.

Acknowledgements. The work reviewed here was performed by the authors along with K. Coumes and G. Ristroph. We thank the IMA and the organizing committee for putting together a stimulating conference. This paper benefitted from useful discussions with Andy Ruina and Noah Cowan as well as with IMA visitors Russ Tedrake, John Roberts, David Lentink, and Mimi Koehl.

REFERENCES

[1] Ristroph L, Berman GJ, Bergou AJ, Wang ZJ, Cohen I (2009) Automated hull reconstruction motion tracking (HRMT) applied to sideways maneuvers of free-flying insects. J Exp Biol 212:1324–1335

[2] Ristroph L, Bergou AJ, Ristroph G, Coumes K, Berman GJ, Guckenheimer J, Wang ZJ, Cohen I (2010) Discovering the flight autostabilizer of fruit flies by inducing aerial stumbles. PNAS 107:4820–4824

[3] Bergou AJ, Ristroph L, Guckenheimer J, Wang ZJ, Cohen I (2010) Fruit flies modulate passive wing pitching to generate in-flight turns. Phys Rev Lett 104:148101

[4] Collett TS, Land MF (1975) Visual control of flight behavior in the hoverfly, Syritta pipiens L. J Comp Physiol A 99:1–66

[5] Reichardt W, Poggio T (1976) Visual control of orientation behavior in the fly. Q Rev Biophys 9:311–375

[6] Mayer M, Vogtmann K, Bausenwein B, Wolf R, Heisenberg M (1988) Flight control during 'free yaw turns' in Drosophila melanogaster. J Comp Physiol A 163:389–399

[7] Heisenberg M, Wolf R (1993) The sensory-motor link in motion-dependent flight control of flies. In: Visual motion and its role in the stabilization of gaze. Elsevier, Amsterdam, pp 265–283

[8] Heide G, Goetz KG (1996) Optomotor control of course and altitude in Drosophila is correlated with distinct activities of at least three pairs of steering muscles. J Exp Biol 199:1711–1726

[9] Fry SN, Sayaman R, Dickinson MH (2003) The aerodynamics of free-flight maneuvers of Drosophila. Science 300:495–498

[10] Dickinson MH (2005) The initiation and control of rapid flight maneuvers in fruit flies. Integr Comp Biol 45:274–281

[11] Bender JA, Dickinson MH (2006) Visual stimulation of saccades in magnetically tethered Drosophila. J Exp Biol 209:3170–3182

[12] Hedrick TL, Cheng B, Deng X (2009) Wingbeat time and the scaling of passive rotational damping in flapping flight. Science 324:252–255

[13] Tammero LF, Dickinson MH (2002) The influence of visual landscape on the free flight behavior of the fruit fly Drosophila melanogaster. J Exp Biol 205:327–343

[14] Dickinson MH, Tu MS (1997) The function of Dipteran flight muscle. Comp Biochem Physiol A 116:223–238

[15] Reiser MB, Dickinson MH (2008) A modular display system for insect behavioral response. J Neurosci Methods 167:127–139

[16] Sane SP (2003) The aerodynamics of insect flight. J Exp Biol 206:4191–4208

[17] Lehmann F-O (2004) The mechanisms of lift enhancement in insect flight. Naturwissenschaften 91:101–122

[18] Wang ZJ (2005) Dissecting insect flight. Annu Rev Fluid Mech 37:183–210

[19] Jensen M (1956) Biology and physics of locust flight. III. The aerodynamics of locust flight. Philos Trans R Soc Ser B 239:511–552

[20] Ellington CP (1984) The aerodynamics of hovering insect flight. I. The quasi-steady analysis. Philos Trans R Soc Ser B 305:1–15

[21] Bennett L (1970) Insect flight: lift and the rate of change of incidence. Science 167:177–179

[22] Dickinson MH, Lehmann F-O, Goetz KG (1993) The active control of wing rotation by Drosophila. J Exp Biol 182:173–189

[23] Ellington CP, van den Berg C, Willmott AP, Thomas ALR (1996) Leading-edge vortices in insect flight. Nature 384:626–630

[24] Dickinson MH, Lehmann F-O, Sane S (1999) Wing rotation and the aerodynamic basis of insect flight. Science 284:1954–1960

[25] Lentink D, Dickinson MH (2009) Rotational accelerations stabilize leading edge vortices on revolving fly wings. J Exp Biol 212:2705–2719

[26] Dickson WB, Polidoro P, Tanner MM, Dickinson MH (2010) A linear systems analysis of the yaw dynamics of a dynamically scaled insect model. J Exp Biol 213:3047–3061

[27] Sun M, Tang J (2002) Unsteady aerodynamic force generation by a model fruit fly wing in flapping motion. J Exp Biol 205:55–70

[28] Ramamurti R, Sandberg WC (2002) A three-dimensional computational study of the aerodynamic mechanisms of insect flight. J Exp Biol 205:1507–1518

[29] Wang ZJ, Birch J, Dickinson MH (2004) Unsteady forces in hovering flight: computation vs experiments. J Exp Biol 207:449

[30] Pesavento U, Wang ZJ (2009) Flapping wing flight can save aerodynamic power compared to steady flight. Phys Rev Lett 103:118102

[31] Young J, Walker SM, Bomphrey RJ, Taylor GK, Thomas ALR (2009) Details of wing design and deformation enhance aerodynamic function and flight efficiency. Science 325:1549–1552

[32] Pesavento U, Wang ZJ (2004) Falling paper: Navier-stokes solutions, model of fluid forces, and center of mass elevation. Phys Rev Lett 93:144501

[33] Andersen A, Pesavento U, Wang ZJ (2005) Unsteady aerodynamics of fluttering and tumbling plates. J Fluid Mech 541:65–90

[34] Sane SP, Dickinson MH (2002) The aerodynamic effects of wing rotation and a revised quasi-steady model of flapping flight. J Exp Biol 205:1087–1096

[35] Featherstone R, Orin D (2000) Robot dynamics: equations and algorithms. In: IEEE international conference robotics & automation, San Francisco, pp 826–834

[36] Deng X, Schenato L, Wu WC, Sastry SS (2006) Flapping flight for biomimetic insects: part I – system modeling. IEEE Trans Robot 22:776–788

[37] Deng X, Schenato L, Sastry SS (2006) Flapping flight for biomimetic insects: part II – flight control design. IEEE Trans Robot 22:789–803

[38] Hedrick TL, Daniel TL (2006) Flight control in the hawkmoth Manduca sexta: the inverse problem of hovering. J Exp Biol 209:3114–3130

[39] Dickson WB, Straw AD, Dickinson MH (2008) Integrative model of Drosophila flight. AIAA J 46:2150–2164

[40] Faruque I, Humbert JS (2010) Dipteran insect flight dynamics. Part 1: longitudinal motion about hover. J Theor Biol 264:538–552

[41] Faruque I, Humbert JS (2010) Dipteran insect flight dynamics. Part 2: lateral-directional motion about hover. J Theor Biol 265:306–313

[42] Sun M, Wu JH (2003) Aerodynamic force generation and power requirements in forwar flight in a fruit fly with modeled wing motion. J Exp Biol 206:3065–3083

[43] Gao N, Aono H, Liu H (2011) Perturbation analysis of 6DoF flight dynamics and passive dynamic stability of hovering fruit fly Drosophila melanogaster. J Theor Biol 270:98–111

[44] Hesselberg T, Lehmann F-O (2007) Turning behavior depends on frictional damping in the fruit fly Drosophila. J Exp Biol 210:4319–4334

[45] Dickinson MH, Lehmann F-O, Sane S (1999) Wing rotation and the aerodynamic basis of insect flight. Science 284:1954–1960

[46] Bechhoefer J (2005) Feedback for physicists: a tutorial essay on control. Rev Mod Phys 77:783–836

[47] Pringle JWS (1948) The gyroscopic mechanism of the halteres of Diptera. Philos Trans R Soc Lond B 233:347–384

[48] Dickinson MH (1999) Haltere-mediated equilibrium reflexes of the fruit fly, Drosophila melanogaster. Philos Trans R Soc Lond B 354:903–916

[49] Heide G (1983) Neural mechanisms of flight control in Diptera. In: BIONA report 2, Fischer, Stuttgart, pp 35–52

[50] Taylor GK, Krapp HG (2007) Sensory systems and flight stability: what do insects measure and why? Adv Insect Physiol 34:231–316

GEOMETRIC MECHANICS, DYNAMICS, AND CONTROL OF FISHLIKE SWIMMING IN A PLANAR IDEAL FLUID

SCOTT DAVID KELLY(✉)*, PARTHESH PUJARI†, AND HAILONG XIONG‡

Abstract. We summarize the geometric treatment of locomotion in an ideal fluid in the absence of vorticity and link this work to a planar model incorporating localized vortex shedding evocative of vortex shedding from the caudal fin of a swimming fish. We present simulations of open-loop and closed-loop navigation and energy-harvesting by a Joukowski foil with variable camber shedding discrete vorticity from its trailing tip.

Key words. Geometric mechanics, vortex dynamics, locomotion

AMS(MOS) subject classifications. Primary 70H33, 76B47, 93C10

1. Introduction. In this paper, we present an idealized model for planar fishlike swimming and explore the dynamics and control of this model. The model represents the synthesis of two lines of research pertaining to ideal hydrodynamics, one addressing the locomotion of free deformable bodies in the absence of vorticity and the other addressing the interaction of free rigid bodies with systems of discrete vorticity. The paper begins with a survey of concepts arising in this earlier work, focused in particular on the realization of locomotion problems in terms of analytical mechanics and nonlinear control systems on differentiable manifolds. The latter sections of the paper focus on the computational study of idealized fishlike swimming with an eye toward simple feedback control design for navigation and energy harvesting.

2. The Geometric View of Locomotion. A mathematical framework for modeling the locomotion of a deformable body can be derived from the geometry of the space in which the variables parameterizing the problem reside. Implicit in the following discussion are the smoothness assumptions needed to justify our regarding this space and various subspaces thereof as differentiable manifolds. The material that follows has been addressed in mathematical detail elsewhere [4]; we endeavor to summarize the essential concepts with only as much detail as is needed for practical calculations.

2.1. Configuration Space as a Principal Fiber Bundle. We refer to a time-parameterized curve in a manifold as a *trajectory* and the image of this curve – in other words, the sequential locus of points it

---

*Department of Mechanical Engineering and Engineering Science, University of North Carolina at Charlotte, Charlotte, NC 28205, USA scott@kellyfish.net.

†Department of Mechanical Engineering and Engineering Science, University of North Carolina at Charlotte, Charlotte, NC 28205, USA

‡Quantitative Risk Management, Inc., Chicago, IL 60602, USA. The work of all three authors was supported in part by NSF grant CMMI 04-49319.

S. Childress et al. (eds.), *Natural Locomotion in Fluids and on Surfaces*, IMA 155, DOI 10.1007/978-1-4614-3997-4_7,
© Springer Science+Business Media New York 2012

comprises, without regard to time parametrization – as a *path*. We regard the instantaneous shape of a deformable body – distinct from its position or orientation relative to a frame of reference affixed to its environment – as corresponding to a point in a *shape manifold*. We assume that a body changing shape for the sake of locomotion has direct authority over its trajectory in its shape manifold; we forego analysis of the forces or moments needed to assert this authority. Our focus will be the manner in which prescribed changes in shape generate locomotion. We refer to a periodic change in a body's shape, corresponding to a closed path in the shape manifold, as a *gait*. The variability in a body's shape may be infinite-dimensional, but the concrete examples we describe will focus on bodies with finitely many internal degrees of freedom, and for simplicity's sake we will assume a finite-dimensional shape manifold for the remainder of the discussion. We use the symbol M to denote a body's shape manifold and the symbol r to denote a point in M corresponding to a particular shape.

As a rigid body moves through ambient space, its position and orientation may be specified relative to a reference configuration, perhaps corresponding to the body's position and orientation at some initial time. In the case of a deformable body, such a comparison requires that a method be specified for affixing a preferred frame of reference to the body – henceforth the *body frame* – in each of its different shapes. With such a method in place, we may identify each position and orientation of a deformable body with an element of a Lie group G of rigid motions relative to the body's reference configuration. In general, this group is the Euclidean group $SE(3)$ comprising displacements and rotations in three spatial dimensions, but certain locomotion problems require only strict Lie subgroups of $SE(3)$ to be considered. We will describe problems in planar locomotion for which $G = SE(2)$ and problems in rectilinear locomotion for which $G = (\mathbb{R}, +)$. In every case, we use the symbol g to denote an arbitrary element in the relevant group G, and the symbol e to denote the identity element associated with the reference configuration.

The locomotion of a deforming body may thus be described in terms of a trajectory in $M \times G$, resulting from dynamics influenced by the prespecified projection of this trajectory into M alone. It may be the case that the path in M is sufficient to determine the corresponding path in $M \times G$, or it may not. It may be the case that coordinates on $M \times G$, together with their time derivatives, furnish sufficiently many variables to model the governing dynamics, or it may not. In particular, when a deforming body swims through a fluid, additional variables are generally needed to represent the dynamics of the fluid. Only under special circumstances is the evolution of a fluid determined uniquely by the motion of a body through it, obviating – or formally eliminating via dynamic *reduction* [9] – additional variables to represent the fluid's internal degrees of freedom. Sect. 2.3 examines the locomotion of deforming bodies through irrotational ideal flows and through Stokes flows, representing precisely these special circumstances.

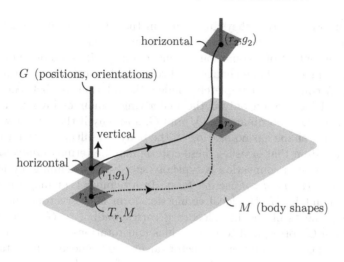

FIG. 1. *A principal fiber bundle $M \times G$. The base manifold M is shown with two dimensions and the group G with one. A connection defines a horizontal subspace of each tangent space $T_{(r,g)}(M \times G)$, isomorphic to the tangent space $T_r M$ and complementary to the vertical subspace of $T_{(r,g)}(M \times G)$. The curve in M passing through r_1 has a unique horizontal lift passing through (r_1, g_1)*

Sect. 3 summarizes a model in which a finite collection of point vortices is used to represent additional degrees of freedom in the fluid surrounding a planar swimmer.

The product $M \times G$ may be regarded as a fiber bundle over M. Each base point corresponds to a particular body shape; the points comprising the fiber over this base point represent the various ways of situating and orienting the body in its environment. Since G is a Lie group, the bundle over M is a *principal bundle* [7], accommodating the constructions defined below.

2.2. Principal Connections and Driftless Locomotion. A *connection* on the principal fiber bundle $M \times G$ specifies a manner in which a path in M passing through the point r may be identified with a unique path in $M \times G$ passing through a given point in the fiber over r. A connection corresponds to a splitting of the tangent space at each point in $M \times G$ into *horizontal* and *vertical* subspaces. The horizontal subspace at each point in the fiber over r is isomorphic to $T_r M$. The vertical subspace has dimension equal to the dimension of G; identifying $T_{(r,g)}(M \times G)$ with $T_r M \times T_g G$ we regard the vertical subspace as comprising vectors in $T_r M \times T_g G$ of the form $(0, \cdot)$. See Fig. 1.

A trajectory in $M \times G$ is said to be horizontal relative to a given connection if every vector tangent to the trajectory is horizontal. A *horizontal lift* of a trajectory in M is a horizontal trajectory in $M \times G$ that projects onto the trajectory in M. If two trajectories in M correspond to the same path – in other words, they differ only in their time-parameterizations – then they will lift to the same set of paths in $M \times G$.

In the context of locomotion, a connection specifies a unique mapping between sequential shape changes and sequential translations and reorientations. A connection completely models the relationship between shape change and locomotion only if the underlying dynamics involve no variables other than coordinates on M and G, and only if the rate at which a body changes shape has no bearing on the body's resulting motion through space. A system that satisfies these criteria may be termed *kinematic*.

In practice, a connection is typically specified by defining a one-form on M called the *local connection form*. In a manner that may vary from point to point in M, the local connection form maps vectors tangent to M into elements of the Lie algebra $\mathfrak{g}$ corresponding to G. Recall that elements in G correspond to translations and rotations of a body relative to its reference configuration. A vector tangent to G encodes instantaneous rates of change in position and orientation. Left-translating such a vector to the tangent space at the identity in G – in other words, into the Lie algebra $\mathfrak{g}$ – corresponds to mapping these instantaneous rates of change into the body frame.

If $A : TM \to \mathfrak{g}$ denotes the local connection form, then the requirement that a system evolve along horizontal trajectories is equivalent to the requirement that

$$\dot{g} + T_e L_g \left[A(r)\dot{r} \right] = 0. \tag{1}$$

This first term in this equation represents the rate of change of a body's position and orientation, and may be regarded as a column vector with height equal to the dimension of G. Within the square braces in the second term, the local connection form maps the rate of change of the body's shape into $\mathfrak{g}$ in a manner that depends on the shape r. Elements of the Lie algebra also correspond to column vectors with height equal to the dimension of G. The Lie algebra element $A(r)\dot{r}$ is mapped into the space in which $\dot{g}$ resides – the space tangent to G at the point g – through multiplication by the square Jacobian matrix $T_e L_g$ corresponding to left translation in G.

A common variation of (1) is obtained by prepending $(T_e L_g)^{-1}$ to each term, translating the objects they represent into $\mathfrak{g}$. We will invoke the symbol ξ for the *body velocity* $(T_e L_g)^{-1}\dot{g}$ in Sects. 2.4 and 3.2.

The significance of (1) is that it separates the local connection form $A(r)$, which involves only shape variables, from aspects of the equations of motion that depend only on the group G. The equations governing the motion of any body relative to a spatially fixed frame of reference inherit significant complexity, in particular, when the body frame is able to rotate relative to the spatial frame. This complexity is generic to the choice of spatial frame, however, and is completely encoded in the matrix representing $T_e L_g$. The distinctions between two different kinematic systems for locomotion in the same ambient space – in other words, involving the same

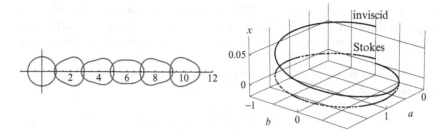

FIG. 2. *Driftless squirming locomotion. The panel on the left shows several snap-shots of the locomotion of a massless cylinder with time-varying radius given by (2), with $(a, b) = (1 - \cos t, \sin t)$ and $\epsilon = 1/10$, in an ideal flow. The snapshots depict the configuration of the cylinder at two-hundred-unit intervals in time. The panel on the right depicts the displacement of the cylinder throughout a single cycle of this gait in ideal flow and in Stokes flow. The panel on the right corresponds to Fig. 1, with the (a, b) plane fibered by copies of the real line corresponding to the group $(\mathbb{R}, +)$ of trans-lations in the x direction. The two curves correspond to horizontal lifts determined by different principal connections*

group G – are encoded entirely in the differences between the two local connection forms.

Note that (1) may be interpreted as a nonlinear control system in control-affine form, with the components of the column vector $\dot{r}$ viewed as control inputs. This system is *driftless* in the sense that no dynamics persist when the controls are equal to zero. In other words, (1) describes the locomotion of a body which instantly ceases to move if it ceases to deform. The controllability of (1), which depends only on the local connection form and the Lie bracket operation on $\mathfrak{g}$, is examined in [5].

2.3. Squirming Circles. We illustrate the constructions of Sects. 2.1 and 2.2 with a pair of examples involving rectilinear swimming. Consider an evacuated circular cylinder with unit radius, initially centered about the origin in its cross-sectional plane, surrounded by a stationary incompressible fluid without no outer boundary. If the cylinder's profile is perturbed from circular in a time-varying way, it's possible for the cylinder to accelerate from rest.

Specifically, suppose the radius of the cylinder to be given, in body-fixed polar coordinates, by

$$r(t) = 1 + \epsilon \left(a(t) \cos 2\theta + b(t) \cos 3\theta \right). \tag{2}$$

If ϵ is a small parameter, we can use perturbation theory to determine the horizontal motion of the cylinder resulting from variations in the shape parameters a and b.

If the fluid is inviscid, the cylinder will translate such that

$$\dot{x} + [-\epsilon^2 b \quad \epsilon^2 a] \begin{bmatrix} \dot{a} \\ \dot{b} \end{bmatrix} + O\left(\epsilon^3\right) = 0. \tag{3}$$

Compare this equation to (1). The group G comprises only translations in the x direction, and the Jacobian $T_e L_g$ is the one-by-one identity matrix. The local connection form is represented (to order ϵ^2) by the matrix $[-\epsilon^2 b \quad \epsilon^2 a]$.

If Stokes flow is assumed instead of ideal flow, the cylinder will translate such that

$$\dot{x} + \begin{bmatrix} \dfrac{\epsilon^2 b}{4} & \dfrac{\epsilon^2 a}{2} \end{bmatrix} \begin{bmatrix} \dot{a} \\ \dot{b} \end{bmatrix} + O\left(\epsilon^3\right) = 0. \tag{4}$$

Fig. 2 illustrates the difference between swimming in ideal flow and swimming in Stokes flow with a gait that favors the former. In either case, the displacement of the cylinder after a single cyclic deformation may be regarded as the *geometric phase* or *holonomy* associated with the corresponding closed path in the shape manifold with coordinates a and b.

The local connection form in (3) encodes the fact that the *Kelvin impulse* [8] in the system – effectively the total momentum – must remain zero even after the cylinder begins to deform, requiring that certain deformations be accompanied by translation. The conservation of impulse in the horizontal direction is equivalent, via *Noether's theorem* [9], to a one-dimensional symmetry manifest in the invariance of the system's kinetic energy under horizontal translations of the spatial frame of reference. Formally, we derive the local connection form – following a procedure outlined in [4] – by constructing the *momentum map* associated with this symmetry. In the context of the inviscid swimmer in Fig. 2, the momentum map corresponds to the component of impulse in the horizontal direction, which is related to the left-hand side of (3) by a nonzero multiplicative factor.[1] The impulse, initially zero, is conserved if and only if the left-hand side of (3) equals zero.

This procedure applies to inviscid swimming in three dimensions as well, involving rotation as well as translation. In the most general case, the symmetry group is all of $SE(3)$, and the momentum map has six components corresponding to linear and angular impulse. The local connection form in this case – taking values in $\mathfrak{se}(3)$ – has six components as well, specifying the linear and angular velocity of the swimmer relative to the body frame such that all components of impulse remain zero.

Significantly, the same procedure is used to derive the local connection form governing locomotion in three-dimensional Stokes flow.[2] In this case, the relevant symmetry is the invariance not of the kinetic energy but of the rate of energy dissipation – the *Rayleigh dissipation function* – under translations and rotations of the spatial reference frame. Specifying that the corresponding momentum map remain zero is equivalent to specifying that the net force and moment on a swimmer remain zero in the low

---

[1]Specifically, the swimmer's effective mass in the horizontal direction.
[2]Planar Stokes flow requires a modified treatment due to *Stokes' paradox*[1].

Reynolds number limit, consistent with the negation of inertial phenomena by viscous effects.

2.4. Extensions to Systems with Drift. In certain cases, the notion of a connection on the bundle $M \times G$ over M may assist in the mathematical description of locomotion even when trajectories in M do not lift uniquely. Suppose, for instance, that a system is governed by Lagrange's equations subject to viscous dissipation, and that both the Lagrangian – generalizing the role played by kinetic energy in Sect. 2.3 – and the dissipation function are invariant under transformations in G. In such a case, a deforming body may accumulate momentum through cyclic deformation, translating or rotating to different degrees during successive cycles. The position and orientation of the body evolve dynamically in a manner coupled to the dynamics of the corresponding components of momentum.

If A_{mech} and A_{Stokes} denote the local connection forms derived from the Lagrangian and the dissipation function, respectively, then the dynamics of the system may be expressed in the form

$$\begin{aligned}
\dot{g} &= T_e L_g \left(-A_{\text{mech}} \dot{r} + I^{-1} p \right) \\
\dot{p} &= V \left(A_{\text{Stokes}} - A_{\text{mech}} \right) \dot{r} + V I^{-1} p + \operatorname{ad}_\xi^* p.
\end{aligned} \tag{5}$$

Here p, a vector in the space $\mathfrak{g}^*$ dual to $\mathfrak{g}$, comprises the components of the momentum map derived from the Lagrangian – in other words, the linear and angular momentum or impulse – expressed in the body frame. If the momentum map is conserved, its representation in the body frame will change as the body frame does; the final term in the second line tracks this change. The *local locked inertia tensor* I expresses the body's shape-dependent inertia, and V the body's shape-dependent directional viscous resistance, in the body frame.

Observe that if $V = 0$ and if $p = 0$ initially – in other words, if all dissipation is removed from the system and the system is initially at rest – then (5) simplifies to (1) with A_{mech} as the local connection form. The significance of inertial effects relative to viscous effects is represented by IV^{-1}. Prepending this tensor term-by-term to the second line in (5) and letting $IV^{-1} \to 0$, we recover (1) with A_{Stokes} as the local connection form.[3]

Fig. 3 depicts a system described by equations of the form (5). A simple walker with a uniformly dense elliptical body makes contact with the ground at four points, each at the end of a rigidly pivoting leg. The angles ϕ and ψ specify the shape of the walker as viewed from above. An additional parameter, representing the walker's ability to lift its legs in opposite pairs, specifies the distribution of the walker's weight on its four feet. Each foot is assumed to experience isotropic viscous resistance to sliding along the ground; the effective damping coefficients c_1 and c_2 vary linearly with the

---

[3]Compare the role of of IV^{-1} here to that of the Reynolds number in the Navier-Stokes equations, which become the equations for creeping flow as $Re \to 0$ [3].

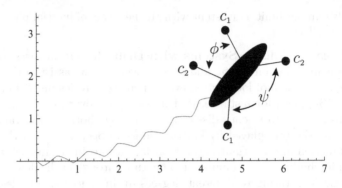

FIG. 3. *A quadrupedal walker governed by* (5), *accelerating from rest. The gait shown corresponds to sinusoidal oscillations in* ϕ *and* ψ *that are exactly out of phase –* ψ *varies with a larger amplitude – overlaid with cosinusoidal oscillations in the balance between the effective viscous drag coefficients* c_1 *and* c_2

normal forces applied to the ground by the feet. The path of the walker's center of mass is shown as it executes a shuffling gait defined by phased variations in leg position whereby no leg is ever completely unloaded.

A set of equations akin to (5), but representing the case in which inertial dynamics interact with nonholonomic constraints rather than dissipative forces, is developed in [2] in terms of the *nonholonomic momentum map* and *nonholonomic connection*.

3. Propulsion via Localized Discrete Vortex Shedding. Neither of the idealized models described in Sect. 2.3 captures the essential physics of fishlike swimming. The deficiency of both models is apparent in their time-reversibility. The undulation of a single rigid appendage will accelerate a macroscopic body from rest in water, yet the models from Sect. 2.3 prohibit the net displacement or reorientation of a swimmer as a result of cyclic variation in only one shape variable.[4]

Fishlike propulsion hinges on the interplay of inertial and viscous phenomena in a manner that also eludes the model from Sect. 2.4, which accommodates only those dynamic variables that directly represent the state of a self-propelling body. When a body moves through water, viscosity causes the water to adhere to the surface of the body, generating shear forces that might plausibly be modeled using the machinery of Sect. 2.4. Depending on the size and geometry of the body and its motion, however, the inertia of the water may prevent the flow from remaining attached at all points on the body. When a fluid flow separates from a body, vorticity is shed into the fluid and the body experiences a force reflecting the resulting momentum exchange. The vortical wake of a fishlike swimmer exhibits

---

[4]In the context of Stokes flow, this is the famous *scallop theorem* [13].

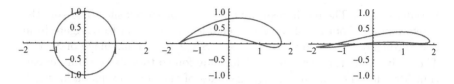

FIG. 4. *A circle in the complex plane and its images under Joukowski (center) and von Mises (right) transformations, both of the form* (6)

dynamics of its own, coupled to those of the swimmer itself but requiring independent accounting.

For the remainder of the paper, we focus on a model for planar swimming that seeks to encompass the central role played by vortex shedding in fishlike locomotion while sacrificing as little as possible of the mathematical structure present in the idealized models of Sect. 2. In fact, the geometric mechanics underpinning this model – a model involving the interaction of a free deforming body with a system of point vortices, with the number of vortices varying in a manner coupled to the time-varying circulation around the body – exhibits a richness beyond the scope of the present paper. We summarize features linking the model to its predecessors from Sect. 2, then proceed to a discussion of its dynamics. The model is detailed in [18].

3.1. Geometry and Modeling Assumptions. Fig. 4 depicts a circle with radius $\varrho = 1$ in the complex ζ plane, centered about the origin, and its images under a pair of transformations of the form

$$z = (\zeta + \zeta_c) + \frac{a_1}{(\zeta + \zeta_c)} + \cdots + \frac{a_n}{(\zeta + \zeta_c)^n}. \tag{6}$$

Such transformations can be used to realize a variety of contours reminiscent of the profiles of practical airfoils [17]. Regarded as coordinates on a shape manifold, the radius ϱ and the real and imaginary parts of ζ_c and $a_1, \ldots, a_n$ may be varied periodically with time to specify the undulations of a planar swimmer.[5]

In particular, the first two panels in Fig. 4 depict the *Joukowski transformation* obtained from (6) when $n = 1$ and a_1 is a positive real number. If ϱ is fixed, the imaginary part of ζ_c may be varied to realize variations in the camber of the resulting profile, while complementary variations in the real part of ζ_c and in a_1 preserve the area enclosed within. In the present paper, we focus on the self-propulsion of a Joukowski foil with a single internal degree of freedom corresponding to camber.

In the absence of vortex shedding, variations in a single shape parameter cannot accelerate a body through an ideal fluid if the system is

---

[5]Our analysis relies on the use of conformal maps in the manner described in [11], prohibiting the arbitrary assignment of values to the parameters in (6). The conformal nature of the transformations used in our simulations is addressed in [18].

initially at rest. The mechanism we introduce for vortex shedding from the Joukowski foil from Fig. 4, however, enables the free foil to navigate from rest throughout the plane, so that $G = SE(2)$ in the language of Sect. 2. The body frame is affixed to the deforming foil so that its coordinate axes coincide with the real and imaginary axes of the complex plane in which the foil's shape is obtained from the Joukowski map.

Adopting the perspective of [16] and significant subsequent literature, we focus on vortex shedding from the sharp trailing tip of the foil – modeling the caudal fin of a fish viewed in cross-section – as a mechanism for propulsion. We assume the fluid surrounding the foil to be ideal, and confine the representation of viscous phenomenology to the time-periodic enforcement of a *Kutta condition* [3] at this single point.

The Kutta condition compels smooth flow separation from the foil's trailing tip by requiring that the preimage of the tip in the ζ plane correspond to a stagnation point for the preimage of the flow. This condition can be enforced continuously through the continuous introduction of vorticity to the fluid, but we amend the flow only at regular instants in time by introducing discrete vortices near the foil's tip. Given the freedom to choose both the location and the strength of a newly introduced vortex, we can enforce the Kutta condition in more than one way. The simulations depicted below rely on a method adapted from [15] whereby the location of each new vortex is specified before its strength, and shed vortices need not have the same strength. A comparison of alternate methods appears in [18]. Once shed, each vortex in our model remains constant in strength.

New vortices are introduced to the fluid in a manner that respects two conservation laws, each of which can be traced to a symmetry of the fluid-foil system. Consistent with *Kelvin's circulation theorem* [3], the circulation around the foil changes discretely with the shedding of each new vortex to ensure that the circulation around any contour enclosing the foil and its wake remain zero. Changes in the circulation around the foil may be attributed to the introduction of *image vortices* [11] within the foil. The relationship between Kelvin's theorem and the invariance of kinetic energy under fluid particle relabeling is discussed in [10].

Consistent with the conservation of total impulse in the system, the linear and angular momenta of the foil must also change discretely with the introduction of each new vortex. Note that changes in the foil's velocity affect the fluid velocity, requiring the simultaneous application of the constraints imposed on shed vortex strength and placement by the Kutta condition and by impulse conservation.

3.2. Hamiltonian Structure. The interaction of a free solid body with a finite collection of point vortices in an infinite ideal fluid is governed by a system of ordinary differential equations exhibiting a non-canonical Hamiltonian structure. This is demonstrated in [14] for the case in which the body is rigid. We summarize the extension in [18] to the case in which the body deforms as the image of a circle with radius ϱ, centered about the origin in the complex ζ plane, under a time-varying transformation $z = F(\zeta)$. We assume $z = F(\zeta)$ to be parameterized by coordinates r_j on the appropriate shape manifold. Between vortex-shedding events, the swimming foil in our model interacts with ambient vortices – whether previously shed by the foil or present initially – according to these equations.

The phase space in question is the product of $\mathfrak{g}^*$ – elements p in which encode the linear and angular impulse in the system – and the space of coordinate pairs (x_i, y_i) specifying the positions of the vortices relative to the body frame. If n denotes the total number of vortices at a given time and γ_i the strength of the ith vortex, we may define a Hamiltonian of the form

$$H(p, x_1, y_1, \ldots, x_n, y_n) = H_0 - 2\pi H_1 \tag{7}$$

relative to which the fluid impulse evolves according to the *Lie-Poisson equation*

$$\dot{p} = \mathrm{ad}^*_{\delta H/\delta p}\, p$$

and the positions of the vortices evolve according to the equations

$$-2\pi\gamma_i \dot{x}_i = \frac{\partial H}{\partial y_i}, \qquad -2\pi\gamma_i \dot{y}_i = -\frac{\partial H}{\partial x_i}.$$

The significance of the Lie-Poisson equation in Hamiltonian mechanics is detailed in [9].

While expressible in terms of the phase variables named above, the first term in the Hamiltonian (7) is most easily understood as

$$H_0 = \frac{1}{2}\xi^T I \xi,$$

where ξ and I are the body velocity vector and locked inertia tensor from Sect. 2. The second term in the Hamiltonian is proportional to

$$H_1 = \sum_i \gamma_i \sum_j \dot{r}_j \psi_j(x_i, y_i) - \frac{1}{2}\sum_i \gamma_i^2 \left(\log\left|\zeta_i\bar{\zeta}_i - \varrho^2\right| + \log\left|F'(\zeta_i)\right|\right)$$
$$+ \frac{1}{2}\sum_i \sum_{i\neq j} \gamma_i\gamma_j \left(\log\left|\zeta_i - \zeta_j\right| - \log\left|\zeta_k\bar{\zeta}_j - \varrho^2\right|\right),$$

where ζ_i denotes the preimage of the location of the ith vortex in the ζ plane and $\dot{r}_j\psi_j$ is the component of the stream function in the ζ plane due to variations in the shape parameter r_j.

3.3. Steady Swimming. Vortex shedding from the trailing tip of a free Joukowski foil serves to rectify oscillations in the foil's camber to generate locomotion. Fig. 5 depicts snapshots from simulations of three different

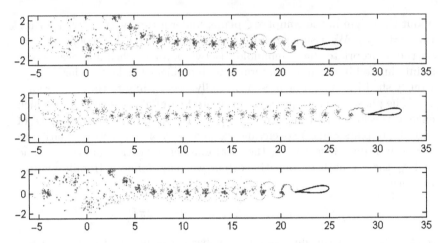

FIG. 5. *Three different swimming gaits for the Joukowski foil. As a function of time, the rate of change of the camber parameter $r(t)$ is a sine wave in the top panel, a square wave in the middle panel, and a triangle wave in the bottom panel*

swimming gaits corresponding to different cyclic variations in camber over time. The foil begins at rest and oscillates at the same frequency in each case, and the three snapshots depict the same instant in time. Shed vortices appear as dots in the foil's wake, scaled according to strength and shaded according to sign. In all three cases shown, the emergence of steady swimming corresponds to the development of the wake structure depicted in [16] – an *inverse Kármán vortex street* – comprising rolled-up collections of vortices of alternating sign.

The sole input to the model is the rate of change $\dot{r}$ of the point in the shape manifold corresponding to the foil's camber. The three gaits shown have been normalized so that the integral of $|\dot{r}|^2$ over one period of oscillation, roughly measuring economy of deformation, is the same in each case. Persistent oscillations in camber always drive the foil toward steady rectilinear swimming, but the asymptotic swimming speed and the detailed structure of the steady-state wake vary from gait to gait.

4. Locomotion Under Heading Control. Among the authors' motivations for developing an idealized model for fishlike swimming is the prospect of using the model as a platform for developing control strategies for a fishlike robotic vehicle. In the remaining pages, we outline a simple approach to the control of the swimming Joukowski foil.

The swimming foil may be viewed as a control system with $\dot{r}$ as the input and the position and orientation of the foil relative to a spatially fixed frame of reference – collectively, an element of $SE(2)$ – as outputs. The *underactuated* nature of this system presents a significant challenge; the desire for the foil to navigate using bounded, cyclic changes in shape complicates the problem further.

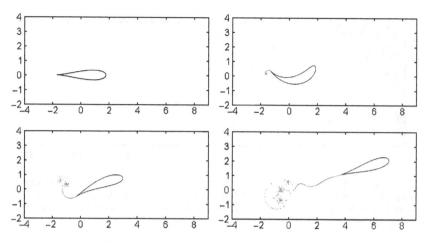

FIG. 6. *A 0.3-radian turn executed from rest using PID control to vary the camber of the foil as a function of the error between its current heading and the desired heading*

Fig. 6 demonstrates that the orientation of the foil can be controlled using feedback. The foil is at rest initially, oriented with no relative rotation between the body frame and the spatial frame. We refer to the angle between the two frames as the foil's *heading*, and specify a desired heading as a function of time. We denote the discrepancy between the desired heading and the actual heading as $\varepsilon(t)$ and implement a standard *PID controller* such that

$$\dot{r}(t) = k_p\varepsilon(t) + k_i \int_0^t \varepsilon(\tau)d\tau + k_d\dot{\varepsilon}(t). \qquad (8)$$

Constant values can be assigned to k_p, k_i, and k_d – respectively, the *proportional, integral,* and *derivative gains* – to realize closed-loop systems with different dynamic properties.

Fig. 6 depicts snapshots of the foil's response to an abrupt change in the desired heading from 0 to 0.3 radians. The controller gains have been chosen so that oscillations in the heading about the desired value decay over time. As a result of the maneuver, $\varepsilon(t)$ approaches zero asymptotically and the foil settles into a steady coast in the desired direction.

4.1. Single-Input Planar Navigation. Experiments indicate that large step changes in the desired heading of the foil can be tracked with no steady-state error using a variety of different controller gains. In particular, a purely proportional controller obtained by setting $k_i = k_d = 0$ in (8) is generally sufficient to guarantee that the foil's time-averaged heading will tend to the desired value, though oscillations in the foil's camber, and thus in the instantaneous heading, may persist in this case.

Under purely proportional control, the average translational speed of the foil as it completes its response to a step change in desired heading varies with the value of the proportional gain k_p. This is illustrated in

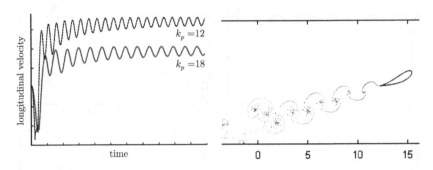

FIG. 7. *Right: A directional change under proportional heading control, leading to steady average translation at a speed determined by the proportional gain k_p. Left: The longitudinal velocity of the foil as it executes such a maneuver, as a function of time, for two different values of k_p*

Fig. 7, which shows the foil responding to such a step change alongside a plot of the foil's forward speed – in other words, the component of velocity in the body-fixed positive x direction – versus time for different choices of k_p. The foil begins not at rest but coasting steadily from left to right, parallel with the bottom of the figure.

Proportional heading control thus forms the basis for a simple method whereby the foil can navigate throughout the plane. Desired changes in swimming direction are achieved through feedback between heading and camber; changes in the steady-state swimming speed are achieved through changes in the feedback gain.

4.2. Energy Harvesting. A nonzero background flow can complicate the variations in camber required for the foil to track a desired heading, but need not increase the control effort required to attain a desired heading and asymptotic speed. Fig. 8 illustrates the use of PID control for *energy harvesting* from an array of vortices corresponding to the fully rolled-up wake of a foil executing the sinusoidal gait from Fig. 5.

The controller gains are chosen to damp oscillations in the foil's camber and the desired heading remains equal to the foil's initial heading. As the ambient vortices begin to rotate the foil, it responds in a manner that stabilizes its motion through the middle of the street. Without deformation, the foil would be ejected laterally [6]. With control, it attains a translational kinetic energy greater than that achievable with the same economy of deformation in a quiescent fluid, while the total *interaction energy* [12] among vortices decreases.

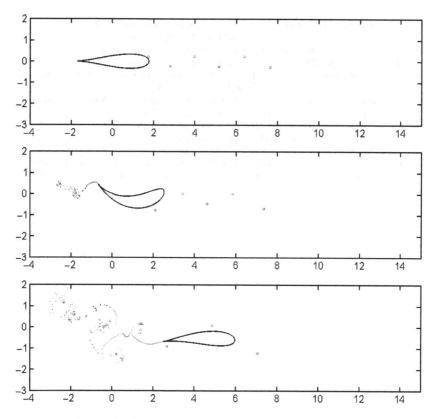

FIG. 8. *Drafting the wake of a preceding foil via PID control*

REFERENCES

[1] Birkhoff G (1978) Hydrodynamics: a study in logic, fact, and similitude. Green-wood, Westport

[2] Bloch AM, Krishnaprasad PS, Marsden JE, Murray RM (1996) Nonholonomic mechanical systems with symmetry. Arch Ration Mech Anal 136:21–99

[3] Childress S (1981) Mechanics of swimming and flying. Cambridge University Press, Cambridge

[4] Kelly SD (1998) The mechanics and control of robotic locomotion with applications to aquatic vehicles, Ph.D. thesis, California Institute of Technology

[5] Kelly SD, Murray RM (1995) Geometric phases and robotic locomotion. J Robot Syst 12:417–431

[6] Kelly SD, Pujari P (2010) Propulsive energy harvesting by a fishlike vehicle in a vortex flow: computational modeling and control. In: Proceedings of the 49th IEEE conference on decision and control, Atlanta

[7] Kobayashi S, Nomizu K (1963) Foundations of differential geometry, vol 1. Inter-science Publishers, New York

[8] Sir Lamb H (1945) Hydrodynamics. Dover, New York

[9] Marsden JE, Ratiu TS (1999) Introduction to mechanics and symmetry, 2nd edn. Springer, New York

[10] Marsden JE, Weinstein A (1983) Coadjoint orbits, vortices, and Clebsch variables for incompressible fluids. Physica 7D:305–323

[11] Milne-Thomson LM (1996) Theoretical hydrodynamics. Dover, New York

[12] Newton PK (2001) The N-vortex problem. Springer, New York

[13] Purcell E (1977) Life at low Reynolds number. Am J Phys 45:3–11

[14] Shashikanth BN (2005) Poisson brackets for the dynamically interacting system of A 2D rigid cylinder and N point vortices: the case of arbitrary smooth cylinder Shapes. Regul Chaotic Dyn 10(1):1–14

[15] Streitlien K, Triantafyllou MS (1995) Force and moment on a Joukowski profile in the presence of point vortices. AIAA J 33(4):603–610

[16] von Kármán T, Burgers JM (1934) General aerodynamic theory: perfect fluids. In: Aerodyn theory, vol II. Berlin, Springer

[17] von Mises R (1959) Theory of flight. Dover, New York

[18] Xiong H (2007) Geometric mechanics, ideal hydrodynamics, and the locomotion of planar shape-changing aquatic vehicles, Ph.D. thesis, University of Illinois at Urbana-Champaign

SLITHERING LOCOMOTION

DAVID L. HU($\boxtimes$)[||] AND MICHAEL SHELLEY[**]

Abstract. Limbless terrestrial animals propel themselves by sliding their bellies along the ground. Although the study of dry solid-solid friction is a classical subject, the mechanisms underlying friction-based limbless propulsion have received little attention. We review and expand upon our previous work on the locomotion of snakes, who are expert sliders. We show that snakes use two principal mechanisms to slither on flat surfaces. First, their bellies are covered with scales that catch upon ground asperities, providing frictional anisotropy. Second, they are able to lift parts of their body slightly off the ground when moving. This reduces undesired frictional drag and applies greater pressure to the parts of the belly that are pushing the snake forwards. We review a theoretical framework that may be adapted by future investigators to understand other kinds of limbless locomotion.

Key words. Snakes, friction, locomotion

AMS(MOS) subject classifications. Primary 76Zxx

1. Introduction. Animal locomotion is as diverse as animal form. Swimming, flying and walking have received much attention [1, 9] with the latter being the most commonly studied means for moving on land (Fig. 1). Comparatively little attention has been paid to limbless locomotion on land, which necessarily relies upon sliding. Sliding is physically distinct from pushing against a fluid and understanding it as a form of locomotion presents new challenges, as we present in this review.

Terrestrial limbless animals are rare. Those that are multicellular include worms, snails and snakes, and account for less than 2% of the 1.8 million named species (Fig. 1). Many are long as well as flexible, enabling them to enter crevices of dimension much smaller than their body length [43]. Such creatures can slither over or burrow through mechanically complex environments such as sand [29], soil [40], grass, or the insides of other organisms such as their intestines or muscle tissue. Investigators are studying locomotion through other complex media, such as viscoelastic or wet granular materials (see [12, 25, 26, 41] and references therein). Snails and many worms propel themselves by virtue of using wet surfaces [7]. Conversely, terrestrial snakes rely upon dry solid-solid friction for propulsion.

Avoiding a harmful tumble or fall is a requirement for moving on land. Because their heights are at most a few centimeters, limbless locomotors have a short gravitational time scale of falling $\tau = \sqrt{L/g} \sim 0.3\,\mathrm{s}$. An outstretched and unconscious snake can thus easily be rolled onto its back.

[||]School of Mechanical Engineering, Georgia Institute of Technology, Atlanta, GA 30318, USA

[**]Courant Institute of Mathematical Sciences, New York University, New York, NY, USA

S. Childress et al. (eds.), *Natural Locomotion in Fluids and on Surfaces*, IMA 155, DOI 10.1007/978-1-4614-3997-4_8,
© Springer Science+Business Media New York 2012

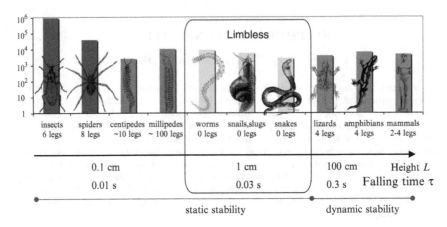

FIG. 1. *Classes of terrestrial animals, arranged according to their size and number of identified species*

To avoid flipping over, they tend to keep their bodies sprawled, such as in the familiar S-shape during slithering. Smaller organisms like the insects maintain stability using their many legs. Larger organisms have sufficient falling time that they may rely on the use of limbs and on gaits such as the trot or gallop, in which an airborne phase occurs. Such is rare for limbless organisms.

One reason for the rarity of limbless organisms may be the cost of abrasion due to sliding against the ground. This is less of a problem for legged organisms, which are in static contact via hard materials such as hooves and nails. A material's resistance to wear is characterized by a wear coefficient $k = V/(ND)$ where V is the volume of the material worn after sliding a distance D under an applied normal force N [42]. Our measurements of the sloughed skin of a 30-cm corn snake and a 2-m red-tailed boa indicate that their skin thicknesses are comparable (0.05 mm). Snakes do not heal their skin, but instead shed and replace their skin periodically. If snake skin thickness is a constant across snakes, then the volume V of the belly skin scales as the surface area of the belly. Therefore, the maximum distance a snake of length L can travel before wearing away its ventral skin scales as $D \sim V/Nk \sim L^2/L^3 \sim L^{-1}$. This scaling indicates that the maximum distance that snakes can travel is inversely proportional to their length, making wear avoidance an important constraint for large snakes.

1.1. Snakes: Movement Using Dry Solid–Solid Friction. Snakes, suborder *Serpentes*, are the most successful class of non-microscopic terrestrial limbless organisms. Numbering over 2,900 species, they have evolved to occupy two orders of magnitude in length-scale, from 10-cm threadsnakes to 10-m long anacondas (Fig. 2). All possess the same basic

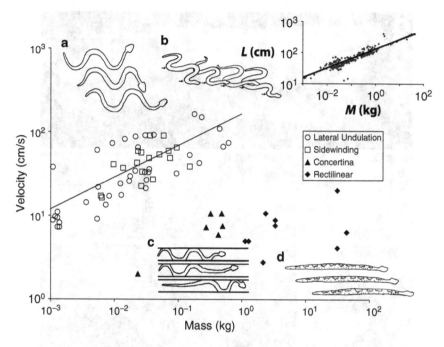

FIG. 2. *The relation between maximum speed U and body mass M for 140 speci-mens of snakes. The fit line suggests that $U \sim M^{1/3}$. The inset, showing the relation between mass M and length L, indicates that snakes are isometric: $M \sim L^3$. Taken together, this suggests that $U \sim L$. Snake modes of locomotion are shown in the insets: (a) lateral undulation or slithering involving two-dimensional undulation; (b) Sidewind-ing, resulting from helical motion; (c) Concertina motion. (d) Rectilinear progression resulting from contraction and extension along a single axis. Anatomical and kinematic measurements were compiled from existing data [13, 16, 21, 24, 27, 32, 35, 38, 39]*

body design: A flexible tube of tissue and skeleton covered in hardened scales (Fig. 3b). Moreover, our measurements indicate that their bodies are isometric, meaning that their proportions are generally independent of size (inset of Fig. 2). Their body form lends them tremendous versatility: They can slither up tree trunks, transition from slithering to swimming without changing gait, or slither across land with surprising rapidity (the red racer [34] of length 60–135 cm can slither at speeds of 130 cm/sec). Such abilities make snakes the champions of terrestrial limbless locomotion.

Limbed animals such as horses have several gaits, such as the walk, trot and gallop [1]. They will readily change gaits in turn as they increase speed, analogous to a car changing gears. We can also ascribe to snakes four principal "gaits", according to the pattern of placements of their limbless body on the ground. They are the undulatory gaits (involving traveling waves) and the ratcheting gaits (involving extensile-contractile motions) (Fig. 2a–d; [3, 11, 15, 17, 24, 33]). The most common is lateral undula-tion, or sinuous slithering, in which the body propagates a 2-D traveling wave from head to tail, in the manner of a swimming eel. The addition

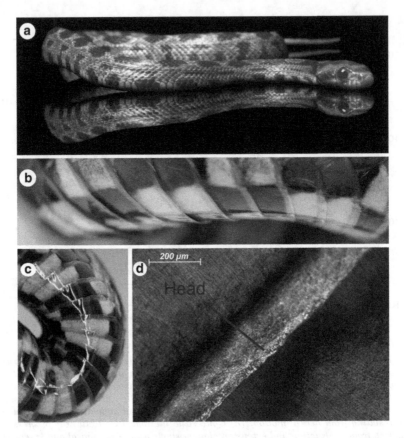

FIG. 3. *(a) A corn snake lifting its body while slithering on a mirrored surface. (b) The ventral scales of the corn snake. Snakeskin adheres to, and folds in and out of, the overlapping lamellae, forming a directionally anisotropic frictional surface. Friction is least when sliding from head to tail, and greatest when sliding towards the flanks. (c), Bending by the snake causes the scales to fan radially outward like a fan, also locally increasing lateral friction. (d) Close-up of the edge of a scale shows micro-ridges that may increase lateral friction*

of a vertical traveling wave to this gait yields a helical body motion called sidewinding, whereby snakes roll along like wheels without axles, and rely on static contact similar to walking. A snake may also progress rectilinearly in the manner of worms by one-dimensional contraction and extension of its belly muscles. Finally, by folding laterally like a sheet of paper, a snake may progress in an accordion-like fashion referred to as concertina. Unlike horses, snake gaits are not so directly related to body speed. Snakes will transition between these gaits in turn as the friction coefficient with the underlying surface is increased [38]. Choice of gait also appears to depend on other factors, such as their body type and the surrounding terrain conditions (flat ground or narrow passageways). In modeling of snake locomotion, we hope to ultimately understand the underlying mechanical reasons (stability, speed, efficiency) leading to the snake's choice of gait.

An attractive feature of limbless locomotion is that its cost of transport is no greater than that of limbed animals. Oxygen consumption is used to measure the energetic net cost of transport (NCT), which has units of energy consumed per mass of animal per distance travelled. By measuring the oxygen consumption of snakes on treadmills, Walton et al. [44] found that the NCT of a slithering snake is 23 J/kg m, which is near that of a similarly-sized running mammal or bird. This result drew attention in its time because biologists had hypothesized that snakes should have a lower NCT than legged organisms because of their energetic savings due to a lower height and lower inertial losses from swinging limbs. Evidently, the frictional costs of sliding trump these gains.

Further treadmill studies by Secor [39] showed that sidewinding has an NCT of 8 J/kg m. Combining the results of Secor and Walton et al., Alexander [1] reports a hierarchy of snake efficiencies: Sidewinding is most efficient, with nearly a third the NCT of lateral undulation; concertina is the least efficient (170 J/kg m) with nearly seven times the NCT of lateral undulation. The NCT of rectilinear motion has yet to be measured. These measurements suggest that future snake robots may have the same efficiencies as legged ones.

The reported efficiencies are reflected in the maximum speeds associated with each gait. The regime diagram Fig. 2 shows the relation between snake speed U and body mass M for 140 species of snakes. Each limbless gait occupies a distinct region in the speed-weight parameter space. Among the undulatory gaits, we find that snake speed scales with body length: $U \sim M^{1/3} \sim L$. This is distinct from Froude's Law ($U \sim L^{1/2}$), known for birds and fish [6]. Presumably, this difference results from the use of frictional rather than fluid dynamic forces. As yet no supporting theoretical models explain these trends.

Size is the clearest indicator of what gait the snake will use. As shown in Fig. 2, snakes lighter than $M \approx 1$ kg prefer undulatory gaits, while those heavier generally prefer ratcheting. One reason for this is the diminishing force-to-weight ratio of animals with increasing size. For a snake to propel itself from rest, it must overcome the static friction force $F_f = \mu M g \sim L^3$ where μ is the coefficient of static friction and $M \sim L^3$ is due to isometry (Fig. 2 inset). The maximum force a snake can generate scales as $F_{max} \sim \sigma L^2$, the product of the peak muscular stress σ and the cross-sectional area of its muscles, which for an isometric snakes, scales as L^2. Small snakes with $F_{max} > F_f$ have no problems slithering, and are quick to escape if startled. However, a sufficiently large snake, for which $L^2 < (\mu g/\sigma)L^3$, only musters enough strength to move individual parts of its body at-a-time, rather than simultaneously. Thus, the largest snakes would tend to use concertina or rectilinear motion to move, which they do.

1.2. Previous Snake Motion Modeling. Models of snake locomotion are generally idealized, without taking directional differences in sliding friction fully into account [19, 28, 37]. Many were developed as part of motion-planning schemes for wheeled snake-robots [5, 8, 10, 22, 31, 36]. These models rely on the high frictional anisotropy provided by passive wheels beneath the body. Despite their reliance on wheels, these models work well to describe the motion of snakes slithering through arrays of rocks, which act as lateral push-points. However snakes may also encounter natural planar surfaces such as bare rock or sand without adequate push-points. Over such surfaces, sliding friction and the frictional properties of snake scales need to be considered [18, 20].

In 2009, we tested unconscious snakes and found that on sufficiently rough surfaces, like stretched cloth, the snakes' overlapping ventral scutes gave them a preferred direction of sliding [23]. Sliding is resisted most in the lateral direction, where snake scales catch in asperities of the underlying surface (Fig. 3b). Using a theoretical model (Fig. 3c), we showed that the level of frictional anisotropy presented by the snake's scales, when coupled with the snake's motion kinematics (undulation and lifting), was sufficient to predict some of the observed snake speeds. We note that our friction measurements were done with unconscious snakes that were laid out straight. As is discussed in the caption of Fig. 3 and in the Discussion, lateral friction is likely increased by bending of the snake body and by active control of individual scales [30].

In the following, we provide a more extensive presentation of our theoretical work. In Sect. 2, we introduce our experimental measurements of sliding resistance in snakes. In Sect. 3, we present our theoretical model based on our friction measurements. We follow in Sect. 4 with descriptions of the snake body-lifting and how lifting augments the body speed. In Sect. 5, we present the implications of our work and suggestions for future work.

2. Snake Experiments. We performed experiments with juvenile milk and corn snakes (methods described in our previous work [23]). We crudely characterize our system by snake length $L = 30$ cm, period of undulation $\tau = 2$ s, mass per unit length ρ and measured friction coefficients $\mu_f = 0.11, \mu_b = 0.14, \mu_t = 0.19$ associated with the snake sliding in the forward, backwards and transverse directions (see [23] and Fig. 4). The forces available to the snake include inertia, scaling as $\rho L^2/\tau^2$, gravitational force ρg, and friction in three directions in the plane, which we scale as $\mu_f \rho g L$ where μ_f is the coefficient of friction for forward sliding. The relative magnitudes of these forces are calculated using corn and milk snake experiments on cloth:

$$\mu_f = \frac{\text{forward friction}}{\text{gravity}} \sim 0.11$$

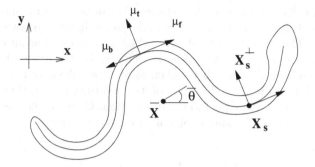

FIG. 4. *Schematic diagram for our theoretical model, where $\bar{X}$ denotes the snake's center of mass, $\bar{\theta}$ its mean orientation, and $\hat{s}$ and $\hat{n}$ the tangent (pointing towards the head) and normal vectors to the body (Taken from Hu et al. [23])*

$$Fr = \frac{L}{\mu_f g \tau^2} = \frac{\text{inertia}}{\text{friction}} \sim 10^{-3}$$

$$An_{\parallel} = \frac{\mu_b}{\mu_f} = \frac{\text{backward friction}}{\text{forward friction}} \sim 1.3 \tag{1}$$

$$An_{\perp} = \frac{\mu_t}{\mu_f} = \frac{\text{transverse friction}}{\text{forward friction}} \sim 1.7.$$

These values will vary according to snake species and surface of choice. We note that for corn and milk snakes on cloth, the frictional anisotropies ($An_{\perp}$ and $An_{\parallel}$) are comparable. For most surfaces, the Froude number Fr is small, indicating that frictional (and gravitational) forces are greatly in excess of inertial forces. Physically this means that snakes do not need "brakes," to decelerate on horizontal surfaces: Cessation of slithering will cause them to quickly come to a halt. When in motion, the most important ratio governing speed is the transverse-to-forward frictional anisotropy $An_{\perp} = \mu_t/\mu_f$. This is the ratio of a snake's resistance to sliding sideways versus sliding towards its head.

We observed that snakes move best on surfaces that provide low abrasion to the snake, but sufficient roughness so that the scales can "catch" and provide frictional anisotropy. We ultimately settled upon 2 test materials for our experiments. The first is a cloth whose stitches are such that the characteristic length scale of roughness (0.2 mm) is comparable with the thickness of the snakes belly scales (0.1 mm). The second is a smooth fiberboard (table top), whose scale of roughness 20 μm is one-fifth that of the snakes scales. For milk snakes on cloth, the transverse-to-forward anisotropy is approximately 2; on smoother fiberboard, it is nearly unity. Higher anisotropy values ($An_{\perp} = 10$) can be achieved by employing wheels, as is done in snake robots [22].

3. The Kinematic Snake Model. We present here a simple kinematic model of serpentine lateral undulation, originally reported in our study [23]. The virtual snake, shown in Fig. 4, is modeled as an inextensible one-dimensional curve $X(s,t) = (x(s,t), y(s,t))$ of length L ($0 \leq s \leq L$) and mass per unit length ρ, here taken as uniform. We describe the shape and dynamics of the snake in terms of its curvature, $\kappa(s,t)$, that is, without reference to absolute position or orientation. Given κ, the position and orientation are given by simple planar geometric relations,

$$X(s,t) = \bar{X}(t) + I_0[X_s](s,t)$$
$$\theta(s,t) = \bar{\theta}(t) + I_0[\kappa](s,t) \tag{2}$$

where $X_s = (\cos\theta, \sin\theta)$ is the unit tangent vector, θ is the tangent angle to the x-axis, and I_0 the mean-zero antiderivative of its argument.[1] Hence $\bar{X}$ is the center of mass and $\bar{\theta}$ is the average orientation. The normal vector is given by $X_s^{\perp} = (-\sin\theta, \cos\theta)$ where $(x,y)^{\perp} = (-y,x)$.

Taking a derivative with respect to time yields

$$X_t = \dot{\bar{X}} + I_0[X_s^{\perp}\theta_t]$$
$$\theta_t = \dot{\bar{\theta}} + I_0[\kappa_t], \tag{3}$$

or

$$X_t = \dot{\bar{X}} + I_0\left[X_s^{\perp}(\dot{\bar{\theta}} + I_0[\kappa_t])\right]. \tag{4}$$

Taking another derivative yields

$$X_{tt} = \ddot{\bar{X}} + I_0\left[-X_s(\dot{\bar{\theta}} + I_0[\kappa_t])^2\right] + I_0\left[X_s^{\perp}(\ddot{\bar{\theta}} + I_0[\kappa_{tt}])\right]. \tag{5}$$

By Newton's second law, the dynamics of the snake is prescribed by the point-wise force balance,

$$\rho X_{tt}(s,t) = F(s,t) + f(s,t) \tag{6}$$

where F and f are the external and internal forces per unit length, respectively. We assume that the total internal forces and torques are zero:

$$\int_0^L f \, ds = 0 \quad \text{and} \quad \int_0^L (X - \bar{X})^{\perp} \cdot f \, ds = 0. \tag{7}$$

External forces on the snake are given entirely by frictional forces acting on its ventral surface. We neglect static friction and address the validity

---

[1] $I_0[f](s,t) = \int_0^s f(s',t)ds' - \frac{1}{L}\int_0^L ds \int_0^s ds' f(s,t).$

of this assumption in our results section. We use a sliding friction law that builds in the measured directionally anisotropic friction in the forward $(\boldsymbol{X}_s)$, backwards $(-\boldsymbol{X}_s)$, and lateral directions $(\boldsymbol{X}_s^{\perp})$ relative to the local direction of motion $\hat{u} = \boldsymbol{X}_t(s,t)/|\boldsymbol{X}_t(s,t)|$:

$$
\begin{aligned}
\boldsymbol{F} = -\rho g \Big(& \mu_t(\hat{u} \cdot \boldsymbol{X}_s^{\perp})\boldsymbol{X}_s^{\perp} \\
& + \Big[\mu_f H(\hat{u} \cdot \boldsymbol{X}_s) + \mu_b(1 - H(\hat{u} \cdot \boldsymbol{X}_s))\Big](\hat{u} \cdot \boldsymbol{X}_s)\boldsymbol{X}_s \Big)
\end{aligned}
\tag{8}
$$

where the Heaviside step function $H = \frac{1}{2}[1 + \text{sgn}(x)]$ is used to distinguish the components in the $\boldsymbol{X}_s$ and $-\boldsymbol{X}_s$ direction.

Scaling $\boldsymbol{X}$ on L, t on the undulation period τ, and the internal force f on ρg, we have

$$
\begin{aligned}
Fr\boldsymbol{X}_{tt} = \boldsymbol{f} - \Big(& \mu_t(\hat{u} \cdot \boldsymbol{X}_s^{\perp})\boldsymbol{X}_s^{\perp} + \Big[\mu_f H(\hat{u} \cdot \boldsymbol{X}_s) \\
& + \mu_b(1 - H(\hat{u} \cdot \boldsymbol{X}_s))\Big](\hat{u} \cdot \boldsymbol{X}_s)\boldsymbol{X}_s \Big).
\end{aligned}
\tag{9}
$$

We close our system, Eqs. 2 and 9, by applying constraints (7) to eliminate the internal forces. Following algebraic manipulation, we derive the governing equations for $\ddot{\boldsymbol{X}}$ and $\ddot{\theta}$:

$$
\begin{aligned}
Fr\ddot{\boldsymbol{X}}(t) = {}& \int_0^1 \boldsymbol{F}\ ds \\
Fr\ddot{\theta}(t) = {}& -\frac{1}{J} \int_0^1 (\boldsymbol{X} - \bar{\boldsymbol{X}})^{\perp} \cdot \boldsymbol{F}\ ds \\
& + Fr\frac{1}{J}\int_0^1 I_0[\boldsymbol{X}_s^{\perp}] \cdot I_0\Big[\boldsymbol{X}_s(\dot{\theta} + I_0[\kappa_t])^2\Big] \\
& - I_0[\boldsymbol{X}_s] \cdot I_0\Big[\boldsymbol{X}_s I_0[\kappa_{tt}]\Big]\ ds
\end{aligned}
\tag{10}
$$

where $J = \int_0^1 (\boldsymbol{X} - \bar{\boldsymbol{X}})^2 ds$ is the moment of inertia. The right hand side is a function of $\bar{\theta}, \dot{\bar{\theta}}, \dot{\bar{\boldsymbol{X}}}$ as well as the prescribed curvature κ and its derivatives. We turn to numerical solutions of Eq. 10.

4. Numerical Results. We observed that the body shape of a slithering snake can be well fit with the traveling wave of curvature

$$
\kappa(s,t) = \epsilon \cos(k\pi(s+t)),
\tag{11}
$$

where $\epsilon = 7.0$ is the maximum radius of curvature of the snake and $k = 2.0$ its wavenumber. These numerical values for (ϵ, k) are used throughout our simulations unless otherwise specified.

We characterized how well a snake performs using two quantities of interest, the average speed in the x-direction and a mechanical efficiency:

$$
\bar{U}_{avg} = \frac{1}{T}\int_0^T \bar{\boldsymbol{U}}(t) \cdot \hat{x}\ dt
\tag{12}
$$

$$\eta = \frac{\mu_f \bar{U}_{avg}}{\frac{1}{T} \int\limits_0^T \int\limits_0^1 \boldsymbol{F} \cdot \dot{\boldsymbol{X}}(s,t) \, ds \, dt} \, . \tag{13}$$

Here, $T = 2/k$ is the period for the sinusoidal curvature given in Eq. 11 and $\bar{U} = \dot{\bar{X}}$ is the speed of the center of mass. There are several ways to define efficiency. For a limbless organism in sliding, the minimum cost of transport, per mass of snake, is $\mu_f U_{avg}$, the power consumed while dragging a straight snake along the ground at a speed U. We define the efficiency η as the ratio of this minimum cost of transport to the snake's mechanical power dissipated during sliding. This ratio is inevitably less than unity because of the snake's serpentine path along the ground.

4.1. Numerical Techniques. Numerical integration of our system of Eqs. 10 is accomplished using a standard second-order Adams-Bashforth scheme. Integrals were evaluated to second order using the trapezoidal rule. A temporal and spatial resolution of $\Delta t = 10^{-3}$ and $\Delta s = 1/300$ were sufficient to obtain accurate results for $(\bar{X}, \bar{Y}, \bar{\theta})$. We assumed that steady-state was established when $\bar{U}_{avg}$ changed by less than 1% between periods. Generally, we found that the virtual snake was found to relax to this steady-state within five periods when released with a speed of 1.

Among certain snake gaits, such as sidewinding, rectilinear, and concertina locomotion, the snake's belly clearly experiences points of instantaneous rest. We could not determine in our experiments of slithering whether points of the belly pass through rest. In fact, an assumption of our model is that no points on the snake's belly experience instantaneous rest, and so sliding friction, rather than static friction, is acting on the snake throughout its motion. This assumption was checked for consistency by determining whether any points on the snake were at rest during the simulation. We decided that rest occurred at a time t_0 if instantaneous velocities $U(t)$ and $V(t)$ along the snake both changed sign at t_0. In our simulations, we found that static contact was experienced only for the lowest snake body speeds. This posed no problem for our modeling because we are interested in peak speed and efficiency, in which rest does not occur. Nevertheless, our model does not include the effects of static friction as its inclusion would be very complicating.

4.2. Results. We performed simulations of our kinematic model to characterize its predictions of peak speed and efficiency of slithering. We tested various waveforms and weight distributions in an *ad hoc* search of an "optimal" slithering gait. We also characterized snake speed in terms of the frictional properties of the underlying surface. We found in our simulations and experiments that snakes rarely slid backwards. Correspondingly, the backwards friction coefficient μ_b had little effect on our results and its effect on snake speed is not presented. Presumably, μ_b would become more important when the snake climbs uphill.

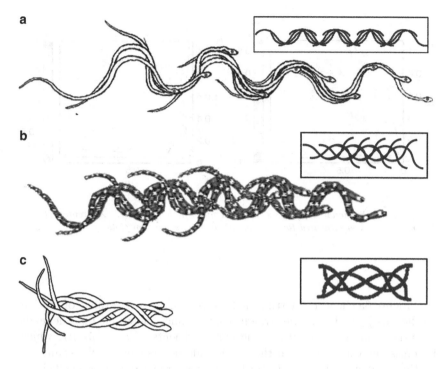

FIG. 5. *Time-lapse trajectories of corn and milk snakes with insets of simulations.* *(a) During "sprinting" on cloth surfaces, both corn and milk snakes seem to be pushing off microscopic push-points, generating a trajectory with near perfect wave efficiency (with wave efficiency defined in Maladen et al. [29]). A nearly matching simulation is generated using* $\mu_t = 10\mu_f$. *(b) On the same cloth surface, and at more leisurely snake speeds, slipping is evident. Simulation is generated using* $\mu_t = 2\mu_f$ *and* $A = 0.2$ *in our weight-redistribution model. (c) On smooth fiberboard, the same milk snake slithers in place, advancing very slowly. Simulation is generated using* $\mu_t = \mu_f$

Figure 5 shows time-lapse trajectories for three snakes from our experiments, which represent useful test cases for our model. Figure 5a shows a corn snake performing lateral undulation at high speed ($\bar{U} = 0.4$) on cloth. Milk snakes are also able to attain such high speeds, but we are unable to account for it with our kinematic model or its elaboration to include lifting, discussed below. However, a trajectory of a virtual snake with a high degree of anisotropy ($\mu_t/\mu_f = 10$; shown in inset at right) shows a qualitatively similar motion. Figure 5b shows a milk snake, again on cloth, moving more slowly. Slipping is clearly evident; we discuss the accompanying simulation for this snake in the next section. In Fig. 5c, the same snake performs poorly on a smooth fiberboard surface; it struggles in vain to move forward. On this surface the friction is nearly isotropic ($\mu_f = \mu_t$), and our model (inset at right) accounts well for the lack of forward motion. Clearly, snake motion is highly dependent on frictional anisotropy.

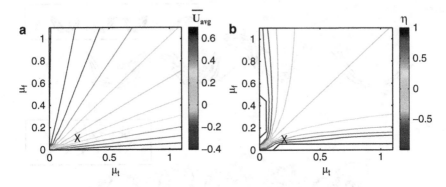

FIG. 6. *Frictional dependence of speed and efficiency. The "X" represents the friction anisotropy measured for experiments of milk snakes on cloth*

Figure 6a shows the virtual snake's speed $\bar{U}_{avg}$ over a range of friction coefficients μ_t and μ_f. The straight contour lines show that $\bar{U}_{avg}$ depends essentially on the ratio of the friction coefficients, $An_\perp = \mu_t/\mu_f$, which is consistent with Eqs. 10 in the low Froude number limit. We examined friction coefficients between 0 and 1, the typical range of friction coefficients for dry solids [2, 45]. The snake geometry was fixed at $(k, \epsilon) = (2.0, 7.0)$, the values observed in our experiments. There are clearly two regimes in this contour plot, separated by the line $\mu_t = \mu_f$. Snakes below this line, for which $\mu_t > \mu_f$, move forward; snakes above this line, $\mu_t < \mu_f$, move in the opposite direction (backwards).

Figure 6b shows the mechanical efficiency η over a range of friction coefficients. For forward motion, we see that regions of high efficiency closely match those of high speed (Fig. 6a), as shown by the similarity in the two plots. Using the geometries observed for the milk snake on cloth, the efficiency of the virtual snake is $\eta = 0.25$. The highest possible speed is 0.6 for the highest frictional anisotropies, which corresponds to the snake moving near its wave speed.

Figure 7a shows the range of possible snake waveforms for $k < 10$ and $\epsilon < 20$. Figure 7b, c shows the speed and efficiency as a function of geometry for anisotropies of $An_\perp = 2$ and 10. For these numerical experiments, the forward friction coefficient μ_f is maintained constant at 0.1 and the transverse coefficient is increased (from 0.2 in Fig. 7a to 1.0 in Fig. 7b). The important features of these contour plots are the position and height of the peaks associated with maximum speed and efficiency. It is interesting that for an anisotropy of $An_\perp = 2$, there are two geometries

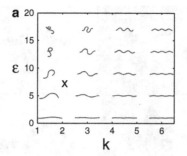

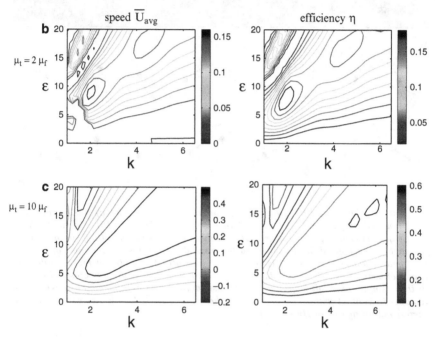

FIG. 7. *Geometric dependence of speed and efficiency. (a) Range of snake wave-forms as a function of k and amplitudes ε. Note that some at high ε are physically unrealizable due to body crossing. The "X" marks the observed snake waveform from our experiments. (b and c) Speed and efficiency for two levels of anisotropy. Dark blue regions occurring at the borders of the speed plots indicate where the snake body passes through a static position (zero speed in x and y directions). Here the model breaks down as static friction forces should then be generated, which we signify by making the body speed zero*

associated with peak speed and efficiency: $(k, \epsilon) = (2, 8)$, as used here as characteristic of snakes, and $(\kappa, \epsilon) = (5, 18)$. As the anisotropy parameter is increased, these peaks coalesce.

These contour plots also reveal the regime of validity of our assumption of sliding contact. We find that certain slowly moving snakes exhibit

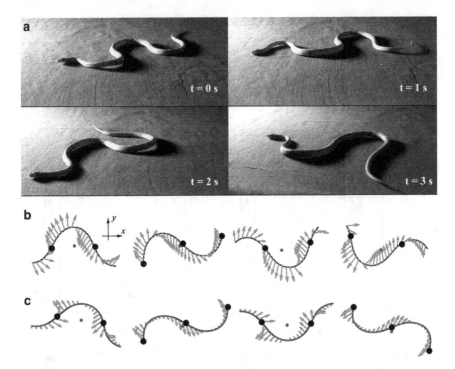

FIG. 8. *(a) Body-lifting during lateral undulation. (b and c) Visualization of the simulated propulsive forces on a virtual snake with uniform (b) and nonuniform (c) weight distribution. Arrows indicate the direction and magnitude of the propulsive frictional force applied by the snake to the ground. Red lines indicate sections of the body with a normal force $N < 1$; the red dot indicates the center of mass. Inflection points of body shape, shown in black, show in (c) where the load is greatest. Note that in these simulations, although the weight distribution is nonuniform, the snake's body remains in contact with the ground everywhere along its body. (b and c) corrects the corresponding figures in Hu et al. [23]*

points of rest, for which our model may not be accurate. These regions are indicated by having zero speed and efficiency (the dark blue areas). They occur at the borders of the contour plots, for snakes with either very low amplitude ϵ or very high wavenumber.

4.3. Weight Redistribution. Thus far, we have assumed that the snake presses its belly uniformly along the ground. This assumption appears to be false in several of our experiments on slithering and clearly in the sidewinding gait studied by other investigators. Figure 8 shows a snake lifting the peaks and troughs of its undulatory wave, while maintaining the majority of its body in sliding contact, as it slithers forward. It is possible that points of the snake pass through instantaneous rest. To consider the effects of a non-uniform weight distribution, we modify our friction force law, Eq. 8, by replacing the weight per unit length ρg by a normal force function $\rho g N(s/L, t/\tau)$. We investigate the effects of the snake unloading

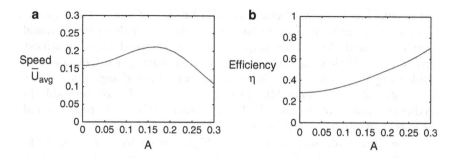

FIG. 9. *The calculated speed and efficiency of the virtual snake as a function of the lifting amplitude A*

its weight in these key areas by assigning its normal force to the approximate square wave centered at the snake's peaks and troughs,

$$N(s,t) = \frac{e^{-A^2\kappa^2}}{\int_0^1 e^{-A^2\kappa^2} ds} \tag{14}$$

where the "unloading" parameter A gives qualitatively the "width" of the lifted region. This function increases normal force where curvature $\kappa = \epsilon \cos(k\pi(s+t))$ is zero (the inflection points) and lifts the snake at regions of high curvature (the peaks and troughs of the body wave). To explore the effect of this unloading parameter A, we first fix the snake's geometry and anisotropy and change the degree the snake lifts. Figure 9 shows the dependence of the snake's speed and efficiency on the unloading parameter A. We use the usual values for anisotropy and waveform ($An_\perp = 2$, $k = 2.0$ and $\epsilon = 7.0$). Moderate weight redistribution by the snake ($A = 0.2$) results in speeds of $\bar{U}_{avg} = 0.21$, which are 35% higher than the speeds at zero unloading. Moreover, redistributing weight causes efficiency to increase nearly up to 50% from $\eta = 0.3$ to 0.55 (see Fig. 9b, inset at right). Note that our definition of efficiency does not account for the cost of raising portions of the body.

5. Discussion. In summary, we have recorded and quantified the motion of snakes on various types of flat surfaces (smooth fiberboard and rough cloth) and developed mathematical models to account for surface texture inducing motion through anisotropic friction. We highlighted the use of dynamic body lifting in increasing locomotion speed and efficiency. By performing a brief optimization – scanning through A to maximize speed — we found that snakes can increase their speed up to 30% and efficiency by 50% by lifting their bodies as they slither.

In legged locomotion on flat rough surfaces, there is often little slipping, and small losses due to friction or air resistance [1]. As a result, metabolic energy is consumed by kinetic energy (swinging of the limbs) and gravitational energy (vertical motion of center of mass). Gaits (walking or running) are chosen according to the ratio of kinetic to gravitational energies [1]. Snake locomotion is quite different because sliding is of utmost importance. Froude numbers are always low, indicating that inertia is negligible compared to friction. Center of masses do not change appreciably in height, unless the snake lifts its body. In our study, we were able to predict the motion of the body by keeping track of the forces resisting and the energy dissipated during sliding.

The efficacy of snake locomotion is highly dependent on the medium the snake moves upon. For example, snakes on smooth surfaces such as hard fiberboard cannot slither forward because their scales can generate insufficient frictional anisotropy. We note that most snakes will quickly learn to rely on their other gaits (or lift their bodies) if slithering does not avail them. We also observed that snakes can slither as quickly on cloth as on peg boards. This is because the asperities in the ground act as microscopic push points to the snake's scales. The idea that snakes can use microscopic push points is a new one and bears consideration in modeling of snakes on all surfaces, not just flat ones.

The anatomical structure and physiological responses of snakeskin is very crudely captured in our model and friction experiments. We include no abilities of the skin to actively modulate its friction. For example, when the trunk of the snake bends, the scales reorient themselves with respect to each other, thus changing their contact orientation with the ground. Moreover, anatomical observations [4, 11] suggest that snakes also have control over individual scales. Modeling features such as these may be necessary to fully account for the range of observed snake speeds.

Our numerical method complemented our experiments, particularly for understanding the effect of body-lifting. Weight redistribution is difficult to experimentally quantify, and we were only able to roughly do so when the degree of lifting was extreme. Use of a photoelastic gelatin, as has been done to study cockroach locomotion [14], is too adhesive for snakes to move naturally (nonetheless, see Fig. 10). The deployment of arrays of small pressure sensors might be useful in this regard.

Can a human ever move like a snake? Infants, who must learn to crawl and eventually walk, begin their motile lives with an inch-worming motion. A full body-suit allowing an adult to slither must reduce abrasive wear and provide frictional anisotropy. There are a number of man-made devices that rely on frictional anisotropy, such as roller-blades and ice skates. Certain toys resembling two skateboards linked together allow one to shuffle one's legs, creating a traveling wave. Coordinated motion of several people standing on a series of devices may generate a sufficiently long traveling wave so as to look snake-like.

Body-lifting shows that limbless locomotion can have similarities to walking. Shifting weight from the left to the right side of the body is a common strategy for legged locomotion, as animals clearly prefer to lift

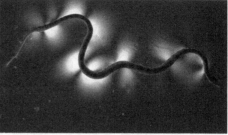

FIG. 10. *A snake attempting to slither on photoelastic gelatin. The luminescent areas indicate regions of highest applied force*

rather than drag their legs. It also appears to be important for snakes, as shown by our predictions of large gains in speed and efficiency. However, we also did not estimate the internal energetic costs of lifting the body, for example, as is expended by a weight-lifter to hold a static load. For large snakes, this may account for a large portion of the energy budget. Such considerations will be especially important for snake-like robots, whose batteries will likely make them heavier than their natural counterparts, and for whom efficiency will be of the utmost importance.

Supplementary Movies.

Movie 1. A corn snake slithering on cloth. Previous models on snake locomotion could not account for the forward motion of the body because there are no apparent push points for the snake's flanks. Body length, 30 cm. `http://youtu.be/urhXl_prdkE`.

Movie 2. A corn snake slithering in place on smooth fiberboard. The snakes is unable to slither forward because its scales cannot gain purchase. `http://youtu.be/YYAmNllYtzQ`.

Movie 3. This sequence of videos shows a milk snake slithering up a cloth-covered incline, increased from 0° to 12°. At 7° of inclination the snake slithers in place, and at higher inclinations, slides backwards. `http://youtu.be/U3qH8hcHZos`.

Movie 4. Viewing a slithering corn snakes from the side, we see that they may lift parts of the body from the ground. The snake's weight is concentrated on the remaining areas of contact. When we incorporated this behavior into our theoretical model, we found increases in both body speed and efficiency. `http://youtu.be/rfbaOJY3lHI`.

Acknowledgements. We acknowledge Grace Pryor, Jasmine Nirody and Terri Scott for assistance with experiments. This work was partially supported by the Lilian and George Lyttle Chair in Applied Mathematics. DLH acknowledges the support of NSF grant PD08-7246.

REFERENCES

[1] Alexander RM (2003) Principles of animal locomotion. Princeton University Press, Princeton.

[2] Avallone EA, Baumeister T III (eds) (1996) Marks' standard handbook for mechanical engineers. McGraw-Hill, New York, pp 3–23

[3] Bellairs A (1970) Life of reptiles, vol 2. Universe books, New York, pp 283–331.

[4] Buffa P (1905) Ricerche sulla muscolatura cutanea dei serpenti e considerazioni sulla locomozione di questi animali. Atti Acad Ven Trent 1:145–237

[5] Burdick JW, Radford J, Chirikjian GS (1993) A 'sidewinding' locomotion gait for hyper-redundant robots. In IEEE international conference on robotics and automation, Los Alamitos, CA, pp 101–106

[6] Bush JWM, Hu DL (2006) Walking on water: biolocomotion at the interface. Ann Rev Fluid Mech 38:339–369

[7] Chan B, Balmforth N, Hosoi A (2005) Building a better snail: lubrication and adhesive locomotion. Phys Fluids 17:113101

[8] Chernousko FL (2003) Snake-like locomotions of multilink mechanisms. J Vib Cont 9:235–256

[9] Childress S (1981) Mechanics of swimming and flying. Cambridge University Press, Cambridge

[10] Choset HM (2005) Principles of robot motion: theory, algorithms and implementation. MIT Press, Cambridge

[11] Cundall D (1987) Functional morphology. In: Siegel RA, Collins JT, Novak SS (eds) Snakes: ecology and evolutionary biology. Blackburn press, Caldwell NJ, pp 106–140

[12] Dorgan KM, Jumars PA, Johnson B, Boudreau BP, Landis E (2003) Burrow elongation by crack propagation. Nature 433:475

[13] Ernst CHZ, Zug GR (1996) Snakes in question. Smithsonian, Washington, DC

[14] Full R, Yamauchi A, Jindrich D (1995) Maximum single leg force production: Cockroaches righting on photoelastic gelatin. J Exp Biol 198:2441–2452

[15] Gans C (1962) Terrestrial locomotion without limbs. Amer Zool 2:167–182

[16] Gasc JP, Gans C (1990) Tests on locomotion of the elongate and limbless lizard anguis fragilis (Squamata: Anguidae), Copeia, pp 1055–1067

[17] Gray J (1946) The mechanism of locomotion in snakes. J Exp Biol 23:101–120

[18] Gray J, Lissman HW (1950) The kinetics of locomotion of the grass-snake. J Exp Biol 26:354–367

[19] Guo ZV, Mahadevan L (2008) Limbless undulatory locomotion on land. Proc Natl Acad Sci U S A 105:3179–3184

[20] Hazel J, Stone M, Grace MS, Tsukruk VV (1999) Nanoscale design of snake skin for reptation locomotions via friction anisotropy. J Biomech 32:477–84

[21] Heckrote C (1967) Relations of body temperature, size and crawling speed of the common garter snake, Thamnophis s. sirtalis. Copeia 4:759–763

[22] Hirose S (1993) Biologically inspired robots: snake-like locomotors and manipulators. Oxford University Press, Oxford.

[23] Hu DL, Nirody J, Scott T, Shelley MJ (2009) The mechanics of slithering locomotion. Proceedings of the national academy of sciences, USA, 106:10081–10085

[24] Jayne BC (1986) Kinematics of terrestrial snake locomotion. Copeia 22:915–927.

[25] Juarez G, Lu K, Sznitman J, Arratia P (2010) Motility of small nematodes in wet granular media. Europhys Lett 92:44002

[26] Jung S (2010) Caenorhabditis elegans swimming in a saturated particulate system. Phys Fluids, 22:031903

[27] Lissman HW (1950) Rectilinear locomotion in a snake (Boa occidentalis). J Exp Biol 26:368–379

[28] Mahadevan L, Daniel S, Chaudhury MK (2004) Biomimetic ratcheting motion of a soft, slender, sessile gel. Proc Natl Acad Sci U S A 101:23–26

[29] Maladen R, Ding Y, Li C, Goldman D (2009) Undulatory swimming in sand: subsurface locomotion of the sandfish lizard. Science 325:314

[30] Marvi H, Hu D (2012) Friction Enhancement in Concertina Locomotion of Snakes. Journal of the Royal Society Interface (In Press)

[31] Miller G (2002) Snake robots for search and rescue. In: Ayers JDJ, Rudolph A (eds) Neurotechnology for biomimetic robots. Bradford/MIT Press, Cambridge, pp 269–284

[32] Moon BR, Gans C (1998) Kinematics, muscular activity and propulsion in gopher snakes. J Exp Biol 201:2669–2684

[33] Mosauer W (1932) On the locomotion of snakes. Science 76:583–585

[34] Mosauer W (1935) How fast can snakes travel? Copeia 1935:6–9

[35] Netting MG (1940) Size and weight of a boa constrictor. Copeia 4:266

[36] Ostrowksi J, Burdick J (1996) Gait kinematics for a serpentine robot. In: IEEE international conference on robotics and automation, Minneapolis, minnesota, pp 1294–1299

[37] Rachevsky N (1938) Mathematical biophysics: physico-mathematical foundations of biology, vol 2. Dover, New York, pp 256–261

[38] Renous S, Hofling E, Gasc JP (1995) Analysis of the locomotion pattern of two microteiid lizards with reduced limbs, Calyptommatus leiolepis and Nothobachia ablephara (Gymnophthalmidae). Zoology 99:21–38

[39] Secor SM, Jayne BC, Bennett AC (1992) Locomotor performance and energetic cost of sidewinding by the snake crotalus cerastes. J Exp Biol 163:1–14

[40] Summers AP, O'Reilly JC (1997) A comparative study of locomotion in the caecilians Dermophis mexicanus and Typhlonectes natans (Amphibia: Gymnophiona). Zool J Linn Soc 121:65–76

[41] Teran J, Fauci L, Shelley M (2010) Viscoelastic fluid response can increase the speed and efficiency of a free swimmer. Phys Rev Lett 104:038101

[42] Tong J, Ma Y-H, Ren L-Q, Li J-Q (2000) Tribological characteristics of pangolin scales in dry sliding. J Mater Sci Lett 19:569–572

[43] Trueman ER (1975) The locomotion of soft-bodied animals. Edward Arnold, London

[44] Walton M, Jayne BC, Bennett AF (1990) The energetic cost of limbless locomotion. Science 249:524–527

[45] Zmitrowicz A (2006) Models of kinematics dependent anisotropic and heterogenous friction. Int J Solids Struct 43:4407–4451

CONTRIBUTED PAPERS

SHARK SKIN BOUNDARY LAYER CONTROL

AMY LANG[*], PHILIP MOTTA[†], MARIA LAURA HABEGGER[‡], AND
ROBERT HUETER[§]

Abstract. An investigation into the separation control mechanisms found on the
skin of fast-swimming sharks, with a particular focus on the shortfin mako (*Isurus
oxyrinchus*) which is considered to be one of the fastest pelagic shark species, was
carried out. Previous researchers have reported a bristling capability of the scales, or
denticles, in certain species of sharks. This study identified that bristling angle is highly
dependent on body location, with some scales easily erectable to angles in excess of 50°.
The flexibility of the scale appears to be due to a reduction in the size of the base of the
scale where anchored into the skin. It is hypothesized that the scales act as a passive,
flow-actuated mechanism as a means of controlling flow separation.

Key words. Shark skin, flow separation, drag reduction

1. Introduction.

1.1. The Shark Skin.
Shark skin is composed of both a collagenous
sublayer and the epidermis that is covered with minuscule scales known
as placoid scales or dermal denticles because of their tooth-like character.
Each scale has a hard, enameloid covering and is secured at the base of the
scale to the underlying stratum spongiosum of the dermis. The shark's surface exposed to water, therefore, consists of an interlocking array of crowns
from each scale. Beginning in the late 1970s and into the 1980s, researchers
began investigating shark skin for its drag reducing capabilities [1]. Initial
focus, from an engineering standpoint, was on the small streamwise riblets,
or keels, observed on the top of each scale crown [2–4]. The streamwise
grooves formed between the keels on shark scales were later found experimentally to reduce turbulent skin friction drag when mimicked and applied
to man-made surfaces [5].

To date one previous study investigated the possible hydrodynamic
advantage of scale bristling, but only from the standpoint that skin friction could be further reduced by bristling. Bechert et al. [6] built a shark
skin replica consisting of an array of individual shark scales (replicas of the
skin of the smooth hammerhead, *Sphyrna zygaena*) and tested it in their oil
tunnel facility for skin friction drag measurements. Cases of both where the
scales were laid flat and were wholly bristled at a collective angle of attack
of approximately 12° were evaluated. With the scales laid flat, results were
consistent with that found on other riblet surfaces the only difference being
that drag reduction had a lower maximum value of about 3% as opposed

---
[*]University of Alabama, Tuscaloosa, AL, USA alang@eng.ua.edu

[†]Department of Integrative Biology, University of South Florida, 4202 East Fowler
Ave Tampa, FL 33620 USA motta@usf.edu

[‡]University of South Florida, Tampa, FL, USA mhabegger@mail.usf.edu

[§]Mote Marine Laboratory, Sarasota, FL, USA rhueter@mote.org

S. Childress et al. (eds.), *Natural Locomotion in Fluids
and on Surfaces*, IMA 155, DOI 10.1007/978-1-4614-3997-4_9,
© Springer Science+Business Media New York 2012

to 9.9% for man-made surfaces consisting of long, streamwise grooves [5]
The small gaps and other imperfections in the shark skin model, preventing a smoother surface, were stated as the likely cause for the decrease in
performance. Cases where the scales were bristled wholly, with variation in
spring stiffness and damping for each scale, resulted only in increased drag.
If scale bristling is to be advantageous to the shark from a hydrodynamic
standpoint, therefore, it must be a mechanism that is activated only as
needed which would be consistent with a flow-actuated mechanism.

Based on these observations, we hypothesized that scale bristling is
used as a means to control flow separation. Past researchers have theorized
that shark skin bristling functions in a manner similar to vortex generators,
a well-known technique to control flow separation used in applications since
the 1940s [2, 6]. Vortex generators must be placed at a precise location
within the boundary layer to be most effective, and in general upstream of
the point of separation [7]. A single vortex generator will produce a wake
consisting of long, streamwise vortices which energize the flow near the
surface. On a more global scale, a method consisting of movable flaps has
also been shown to control flow separation on the surface of an airfoil; in one
study these were placed close to the trailing edge ($\sim$10% of chord length)
and delayed the onset of stall resulting in overall greater lift. Furthermore
when the trailing edge of the flap was given a 3-D jagged nature, it was
found to act even more effectively [6]. We have theorized that flexible shark
scales work in a manner similar to movable flaps but occurring at microscale levels in that the preferred flow direction over the skin is critical (i.e.
flow reversal as occurs during flow separation will initiate scale bristling)
[8]. More recent studies on the shark skin herein described corroborate this
hypothesis.

1.2. Flow Separation. Arguably the most important type of drag to
be minimized by a swimming animal is that due to differences in pressure
around the body, and is often referred to as form drag. Flow separation
from the body leads to regions of low pressure resulting in the generation of
high pressure drag. The first means to reduce separation is to streamline
the body, hence the streamlined shape of most fishes. However, a certain girth and length is a requisite characteristic of large predatory sharks
and, due to the side-to-side sweep of the body and tail during swimming,
high lateral body curvature also occurs. Video evidence of a swimming
shortfin mako shark (*Isurus oxyrinchus*) pursuing bait shows the shark's
ability to turn in one direction and then change direction before the body
completes the initial turn. This type of turning behavior, labeled herein
as contragility, is a maneuver that requires not only large muscular effort
but also low form drag (Frank Fish, personal communication, February 18,
2008).

Separation of the boundary layer to from a body typically occurs in
vicinities where the flow is decelerating due to passage of the flow over

a body, resulting in an adverse pressure gradient. As a result separation typically occurs in areas posterior to the maximum body thickness or point of minimum pressure via the Bernoulli principle. Incipient separation is characterized by regions of decreasing skin friction approaching zero, and consequent reversal of the flow at the surface. When three-dimensionality and unsteadiness are added to the flow kinematics, boundary layer separation does not always coincide with a point of zero shear stress at the wall. In fact, the shear stress may vanish only at a limited number of points along the separation line, and a convergence of skin-friction lines onto a particular separation line is required for separation to occur. As a result, 3D boundary layers can be more capable of overcoming an adverse pressure gradient without separating [9]. If the denticles on shark skin prevent this required convergence of skin friction lines, they might passively act to keep the flow attached, thereby reducing pressure drag and increasing the performance of control surfaces (e.g. fins). When considering the effects of flow separation, whether the flow is laminar, transitioning, or turbulent is of obvious importance. However, this flow condition is difficult to predict given the wide range of swimming speeds (1 m/s up to in excess of 10 m/s), and a separation control mechanism that can work over a range of Re and unspecified separation location would be of obvious benefit to the shark. Imparting momentum or maintaining higher velocity flow near the surface is critical in any mechanism working to control or delay flow separation, as evidenced by the fact that a turbulent boundary layer is less prone to separate than its laminar counterpart [9].

For this investigation of shark skin boundary layer control, we chose to study the shortfin mako based on two factors. First, the shortfin mako is one of the fastest, if not the fastest, species of pelagic sharks [10]. This mackerel shark (family Lamnidae, which also includes the white shark *Carcharodon carcharias*) is one of the more derived species (appearing approximately 55 million years ago) in the long history of shark evolution dating back more than 400 million years [11]. Second, it is one of two shark species, the other being the smooth hammerhead studied by Bechert et al. [6], previously reported in literature as having flexible scales over large portions of its body [12]. Of these two species the shortfin mako was more obtainable off the Atlantic coast of the U.S., where it is not currently considered to be overfished or otherwise in a vulnerable state for conservation purposes [13].

2. Materials and Methods.

2.1. Collection, General Measurements and Sampling Areas.
Two subadult shortfin mako shark specimens (female total length [TL] 192 cm, fork length [FL] 171.5 cm; male TL 158 cm, FL 150 cm) from U.S. Atlantic waters were used. Sharks were frozen within hours of capture and thawed prior to measurement. Following Reif [14] denticles at 16 regions along the body were marked in order to sample scales over various body regions (Fig. 1). A total of three 1 cm^2 samples from each location were used for testing, two for histological study and one for scanning electron

microscopy (SEM) views of the placoid scales. Because swimming sharks
have superambient subcutaneous pressure ranging as high as 100–200 kPa
increasing skin stiffness [15, 16], scale erection angles were recorded with
and without subcutaneous pressure.

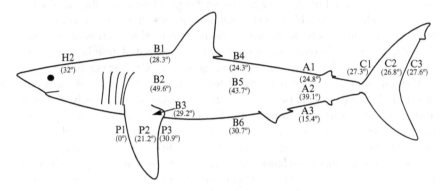

FIG. 1. *Outline of a representative shortfin mako shark Isurus oxyrinchus showing
the 16 regions sampled for scale photography, angle measurement, and histology. The
mean erection angle for five scales in each region is given beside the nomenclature for
each region. Flank scales (B2, B5, A2) are capable of greater erection angles than
dorsal (B1, B4, A1) and ventral (B3, B6, A3) regions. All regions on the pectoral fin
(P1, P2, P3) differed in the degree of erection possible with the leading edge scales
(P1) incapable of erection. Erection angle did not differ among the three regions on the
caudal fin (C1, C2, C3).*

Scale bristling angle was measured using two methods. The shark specimen was placed under a Meiji EMZ-TR stereo microscope at 135X magnification. An incision, of approximately 8 cm width and approximately
0.5 cm under the muscle, was made through the skin in 7 of the 16 regions
(B1, B2, B4, B5, A1, A2, A3, Fig. 1); this procedure was not possible on
the fins, dorsal head, or over the body cavity. An infant aneroid sphygmo-
manometer, which consists of a 7 × 8 cm inflatable cuff, was covered with
a hard plastic sheath and placed under the skin into the pocket formed by
the above incision, and attached to a Wika[©] 113.13, 0–30 psi calibrated
pressure gauge. The cuff was then inflated to 15 psi (103 kPa); this was the
maximum allowable pressure without resulting in damage of the underlying muscle tissue. A microscope with an ocular calibrated reticle was used
to measure the crown length of 1–3 individually marked scales (as viewed
orthogonally to the scale). Measurements were obtained both before and
after inflation of the cuff; if scale bristling occurred due to increased skin
tension, the apparent crown length would decrease. The angle of scale erection was then calculated using trigonometry. Bristling did not occur for
either mako specimen in any of the regions tested due to applied pressure;
testing then proceeded further. Under the same subcutaneous pressure
of 103 kPa, five randomly selected scales were gently manipulated, using
a fine acupuncture needle, to the maximum possible bristling angle whilst

continuing to remain in place; the bristling angle was measured again. This procedure allowed for a measurement of the minimum erection angle. For all cases, once manually bristled the scales were left to remain in place with no additional manipulation before they were measured. Also, no scales were manually erected to the degree such that the base was removed from the underlying skin. Finally, the cuff was deflated and the angle correspondingly measured on the flaccid skin. An a priori test was carried out and it was found that stretching the skin in this procedure did not change the non-pressurized erection angle.

At all additional regions (H2, B3, B6, P1, P2, P3, C1, C2, C3, Fig. 1), where subcutaneous pressure testing was not possible, the erection angles were measured by gently manipulating the scales again with the acupuncture needle and measuring their pre- and post-erection crown length. As a final more global investigation, after all measurements were taken, 35 equidistant sampling locations encompassing the entire dorsal, left lateral, and ventral surfaces of each shark were selected and 1–3 scales in each region manually bristled as before without subcutaneous pressure. Considering the results from the previous test, an extreme angle approaching 50° was noted. The scale angle in each new location was thus subjectively assessed as being either greater or less than 50° by two researchers independently. If the angle was close to 50° but ambiguous, the angle was calculated by measuring crown length before and after manipulation. In this manner, a more comprehensive map of the most erectable scales on the body was determined without undue damage to the skin. From histological and fresh tissue samples of the shark flank (B2, B5, A2), scale length (168 scales) and riblet spacing (68 scales) were measured using an ocular micrometer and SEM photographs with Sigma Scan Pro 4 software (Systat Software Inc.).

2.2. Tissue Processing SEM. Skin samples with denticles were first rinsed in an ultrasonic cleaner with water for approximately 30 s and further cleaned several times with 0.1 PBS (Phosphate buffered saline) at 37°C to remove any biological residues. Tissue samples were fixed for 1–7 days with a 2.5% glutaraldehyde in 0.1 M phosphate buffer (pH 7.2 at 4°C) solution. To remove buffer salts after fixation, samples were rinsed multiple times with water. To dehydrate the tissue samples, a consecutive series of increased concentrations of ethanol (35%, 70%, 95% and 100%) was applied for approximately 15 min for each treatment. After dehydratation, the samples were immersed in 100% Hexamethyldisilazane (HMDS) for 10 min; this procedure was repeated two more times and the tissue was air-dried under vacuum. SEM pictures were acquired with a JEOL JSM-35 scanning electron microscope at 10 kv (using Fuji FP-100B45 film at 100 and 200X magnification) after the samples were sputter-coated.

2.3. Tissue Processing Histology. Samples were stored in 10% buffered formalin upon removal from the frozen shark. Upon fixation, samples were decalcified for 7–10 days in a formic acid (50% HCOOH, 50% H_2O) and sodium citrate (500 g $NaH_2C_6H_5O_7$, 2,500 ml H_2O) solution. Thin $4\mu m$ sections were cut using a Biocut microtome (Leica/Reichert Jung, model 2030) and stained by means of the Verhoeff-Van Gieson (VVG) Staining Protocol (Sigma-Aldrich$^©$, Accustain elastic stain Procedure No. HT25). Using an Olympus IM compound microscope with a MD 900 AmScope digital camera, slides were observed and photographed. The experiments and observations conducted during this research comply with the Animal Care and Use Committee of the University of South Florida protocol numbers T 3253 and T 3839.

2.4. Statistics. The measurement data from three main areas of the body (dorsal B1, B4, A1; lateral B2, B5, A2; and ventral region B3, B6, A3) from two mako sharks were collected and erection angles statistically compared. Additional statistical comparisons were performed among all fin regions as well. Due to the deficiency of normality even after transformation, non parametric tests were carried out for the body and pectoral fins. In these cases a Kruskal-Wallis one-way ANOVA on ranks was followed by a Tukey's pairwise multiple comparison. A one-way ANOVA was used for all caudal fin data. All statistical analyses were conducted using SigmaPlot software (version 11.0, Systat Software Inc.).

3. Results. From the 16 body regions the average erection angles ranged from $0°$ to $49.6° \pm 1.77$ SE (Table 3 and Fig. 1). The lateral (flank) scales (B2, B5, A2) had significantly greater angles (mean angle = $44° \pm 1.44$) than both the ventral (mean angle = $25.1° \pm 1.9$ SE) and dorsal regions (mean angle = $25.8° \pm 0.78$ SE). The dorsal region erection angles did not vary significantly from that of the ventral region (H = 53.173, df = 2, P = < 0.001). Comparing pectoral fin scales, those on the trailing edge (P3) had the largest erection angles (mean angle = $30.9° \pm 1.67$ SE) as compared to the leading edge (P1) (mean angle = $0°$) and the middle region of the fin (P2) (mean angle = $21.2° \pm 0.85$ SE) and all regions were significantly different from each other (H = 26.289, df = 2, P =< 0.001). In contrast, for the caudal fin the mean angles were not significantly dissimilar among the three regions (C1 mean angle = 27.3 ± 1.84 SE, C2 mean angle = 26.8 ± 0.88 SE, C3 mean angle = 27.3 ± 1.84 SE) (F = 0.0614, df = 2, p = 0.941).

Scales capable of erection to at least $50°$ were found in a region along the flank, where scale flexibility was estimated for 35 regions from the dorsal to ventral surface of the shark (Fig. 2). For the flexible flank region (B2, B5, A2), scales had a mean crown length of 0.18 ± 0.0016 mm (n = 168) and a keel spacing of 0.041 ± 0.0005 mm (n = 69). Preliminary examination of these scales found those in the extremely flexible areas (B2, B5, A2) to possess comparatively larger crown length/base length ratios than those located in the dorsal regions (B1, B4, A1). The ventral body scales (B3, B6, A3) have a similar crown length/base length ratio as the flexible flank

TABLE 3

Mean angles (degrees) for the 16 regions samples from two mako sharks combined

Region	Mean angle ± SE
H2	32 ± 1.02
B1	28.3 ± 1.12
B2	49.6 ± 1.77
B3	29.2 ± 1.67
B4	24.3 ± 1.42
B5	43.7 ± 2.38
B6	30.7 ± 1.05
A1	24.8 ± 1.22
A2	39.1 ± 2.24
A3	15.4 ± 3.72
P1	0 ± 0.00
P2	21.2 ± 0.85
P3	30.9 ± 1.67
C1	27.3 ± 1.83
C2	26.8 ± 0.88
C3	27.6 ± 1.96

scales, but are considerably shorter overall (Fig. 3). In addition to being more flexible, the scales located at the trailing edge of the pectoral fin (P3) are considerably smaller and less firmly anchored in the dermis than those of the inflexible leading edge scales (P1) (Fig. 4).

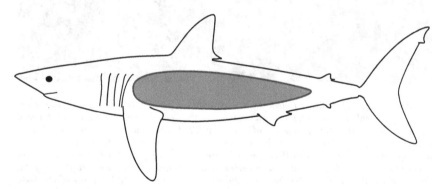

FIG. 2. *Outline of a representative shortfin mako shark Isurus oxyrinchus showing the approximate region on the flank with the most flexible scales capable of erection to at least 50°.*

4. Discussion. Our previous work to discern the flow occurring over a bristled shark skin model confirmed the presence of cavity vortices induced to form between the scales [8]. An example flow visualization image showing vortex formation is shown in Fig. 5. In this previous study an extreme

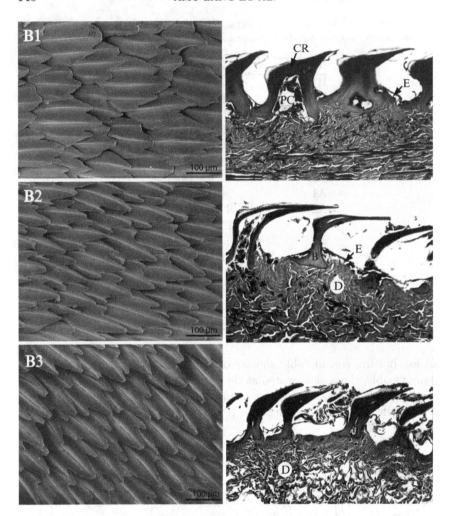

FIG. 3. (Left) Scanning electron micrographs of placoid scales (dermal denticles) from the dorsal, lateral, and ventral regions of the body between the pectoral fin and dorsal fin (B1, B2, B3 in Fig. 1), of a 192 cm TL female shortfin mako shark Isurus oxyrinchus. Three riblets or keels are visible on each scale. Anterior is to the left although the placoid scales in these pictures are not exactly oriented to the longitudinal axis of the shark. Placoid scale shape varies among regions of the body. (Right) Sagittal section through placoid scales from the dorsal, lateral, and ventral regions of the body between the pectoral fin and dorsal fin (B1, B2, B3 in Fig. 1) of a 158 cm TL male shortfin mako shark. The bases of the scales (B) are anchored in the dermis (D), and the thin epidermis (E) is visible between the scales. The pulp cavity (PC) is visible below the crown (CR) in some sections. Anterior is to the left.

angle of bristling, corresponding to upright or vertical scale placement, was used as the current results herein presented had not yet provided the correct angle for scale bristling to be tested. Current work underway involves

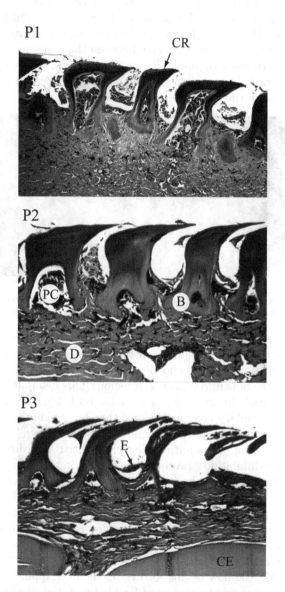

FIG. 4. *Sagittal section through placoid scales from the leading edge, mid-fin, and trailing edge regions of the pectoral fin (P1, P2, P3 in Fig. 1) of a 158 cm TL male shortfin mako shark. The bases of the scales (B) are anchored in the dermis (D), and the thin epidermis (E) is visible between the scales. The pulp cavity (PC) is visible in some of the scales, and not all scales are sectioned through their center, resulting in some crown (CR) lengths appearing shorter than others. Ceratotrichia (CE) fin rays are visible in region P3 because this part of the fin is very thin. Anterior is to the left.*

building models with the herein reported extreme angles of bristling to measure the induced cavity flow fields for both reversed and forward flow over the shark skin model. It can be assumed, however, given the angles observed in the range of 50°, that flow between the scales nevertheless will result in the formation of embedded cavity vortices. We thus hypothesize that the bristled shark skin geometry has the potential to delay flow separation through three possible mechanisms.

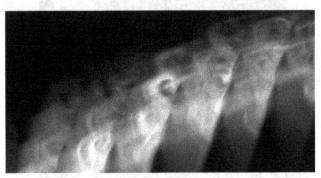

FIG. 5. *Flow visualization of the vortices forming between the erected scales of a bristled shark skin model. Flow proceeds over the model from left to right.*

The first separation flow control mechanism is with respect to the formation of embedded vortices. Their presence within the cavities formed between the scales obviates the no-slip condition which results in increased momentum of the flow in the near wall region. This deters flow separation and maintains globally attached flow under both laminar, transitioning and turbulent boundary layer conditions, thereby working similarly to the mechanism by which dimples on a golf ball control flow separation and reduce pressure drag. In addition, the preferential flow direction of the geometry aids to inhibit flow reversal and, unlike dimples on a golf ball, this property of the surface results in a second separation control mechanism. This unique surface micro-structure may also lead to minimal drag penalty, a byproduct typically associated with passive techniques [9]. Laid flat the scales do not present any significant protrusion into high-momentum flow, and upon flow reversal the denticles would act as flow-activated, movable micro-flaps. Working in this capacity, if at any point adjacent to the surface of the shark skin the flow begins to reverse, the overlapping nature of the microgeometry channels this reversed flow between the riblets (which also appear to keep the upstream denticles from coming into complete contact with the downstream denticles and perhaps assist in this flow-actuated scenario). This initial flow reversal, which forms due to the presence of an adverse pressure gradient resulting in an upstream suction pressure, will cause a denticle(s) in that region to erect.

It should be noted here that observations involving scale manipulations during this study found that moving one denticle would also cause local neighboring denticles upstream to rotate upward because of the overlap between scales. The formation of embedded vortices in the skin, between the now bristled region of scales, is thereby initiated and self-sustained under adverse boundary layer conditions. Furthermore, these embedded vortices will be more energized in the turbulent boundary layer case, due to the exchange of fluid into and out of the cavities induced by the turbulent nature of the flow naturally occurring above the cavities. This results in increased partial slip velocities and higher Reynolds stresses in the near wall region. Thus, a third separation control mechanism, through passive turbulence amplification [9], is also theorized to result under transitioning/turbulent boundary layer conditions. This is the most likely scenario given the thin boundary layers and correspondingly high Re over most portions of the shark's body.

Further experimental hydrodynamic testing, using both real shark skin specimens as well as scaled up, rapid-prototyped shark skin models, is currently investigating these three hypothesized mechanisms whereby flow separation is controlled. Our ultimate goal is to gain a fundamental understanding of the mechanisms such that bio-inspired surfaces can be engineered for technological application to control flow separation in both air and water environments.

5. Summary. Results found that high scale flexibility, including bristling angles up to a range of 50°, corresponds with regions on the body where flow separation control is likely to be most advantageous to the shortfin mako shark. These regions are on the flank and trailing edges of the pectoral fin. In the former case, this is where high curvature of the body will result from the shark's lateral swimming motions and is also in the vicinity of the point of maximum girth. In the latter case it is hypothesized that the flexible scales can lead to the control of flow separation over the pectoral fin. Controlling flow separation in both regions will lead to increased swimming speeds with high contragility for the shark. The increased scale flexibility of the flank scales appears to be a result of reduction in the relative size of the base where the scales in these regions are anchored into the skin. Future work will lead to the bio-inspired application of the shark skin microgeometry whereby a passive, flow-actuated surface patterning can be manufactured for applications where flow separation control is desired.

Acknowledgements. Funding for this work received through collaborative NSF grants (0932352, 0744670 and 0931787) to A. Lang, P. Motta, and R. Hueter to support both the engineering and biological work is gratefully acknowledged. We also thank Jessica Davis for assisting in the shark measurements and Candy Miranda for preparing the histological samples. We also thank Edward Haller for assistance with acquiring the SEM shark skin images. Finally, we wish to express our gratitude to Paul and Jane Majeski and crew, Captain Mark Sampson, Captain Al VanWormer, Philip Pegley, Jack Morris, and Mote Marine Laboratory for providing shark specimens; and to Lisa Natanson for her aid as well in obtaining specimens.

REFERENCES

[1] Reif W (1978) Protective and hydrodynamic function of the dermal skeleton of elasmobranchs. Neues Jahrbuch fur Geologie und Palaontologie Abhandlungen 157:33–141

[2] Bushnell D, Moore K (1991) Drag reduction in nature. Ann Rev Fluid Mech 23:65–79

[3] Walsh M (1980) Drag characteristics of V-groove and transverse curvature riblets. In: Hough GR (ed) Viscous flow drag reduction. Progress in astronautics and aeronautics, (AIAA) New York, vol 72, pp 168–184

[4] Bechert D, Hoppe G, Reif W (1985) On the drag reduction of the shark skin. Am Inst Aeronaut Astronaut (AIAA) Paper No. 85-0546, pp 1–18

[5] Bechert D, Bruse M, Hage W, Van der Hoeven J, Hoppe G (1997) Experiments on drag-reducing surfaces and their optimization with an adjustable geometry. J Fluid Mech 338:59–87

[6] Bechert D, Bruse M, Hage W, Meyer R (2000) Fluid mechanics of biological surfaces and their technological application. Naturwissenschaften 80:157–171

[7] Lin J (2002) Review of research on low-profile vortex generators to control boundary-layer separation. Prog Aero Sci 38:389–420

[8] Lang A, Motta P, Hidalgo P, Westcott M (2008) Bristled shark skin: a microgeometry for boundary layer control? Bioinspiration Biomim 3:046005

[9] Gad-el Hak M (2000) Flow control: passive, active and reactive flow management. Cambridge University Press, Cambridge, UK

[10] Stevens J (2009) The biology and ecology of the shortfin mako shark, *isurus oxyrinchus*. In: Camhi MD, Pikitch EK, Babcock EA (eds) Sharks of the open ocean: biology, fisheries and conservation. Blackwell, Oxford

[11] Naylor G, Martin A, Mattison E, Brown W (1997) The inter-relationships of lamniform sharks: testing phylogenetic hypotheses with sequence data. In: Kocher TD, Stepien C (eds) Molecular systematics of fishes. Academic, New York, pp 199–217

[12] Bruse M, Bechert D, van der Hoeven J, Hage W, Hoppe G (1993) Experiments with conventional and with novel adjustable drag-reducing surfaces. Proceeding of the international conference on near-wall turbulent flows, Tempe, AZ, pp 719–738

[13] NMFS (2010) Final amendment 3 to the consolidated Atlantic highly migratory species fishery management plan. National Oceanic and Atmospheric Administration, National Marine Fisheries Service, Office of Sustainable Fisheries, Highly Migratory Species Management Division, Silver Spring, MD. Public Document. pp. 632

[14] Reif W (1985) Squamation and ecology of sharks. Courier forschungs-Institut senckenberg, vol 78. Cour. Forsch. -Inst. Senchenberg. Frankfurt a.M

[15] Wainwright S, Vosburgh F, Hebrank J (1978) Shark skin: Function in locomotion. Science 202:747–749

[16] Martinez G, Drucker E, Summers A (2002) Under pressure to swim fast. Integr Comp Biol 42(6):1273–1274

NUMERICAL MODELING OF THE PERFORMANCE OF RAY FINS IN FISH LOCOMOTION

QIANG ZHU(✉)* AND KOUROSH SHOELE†

Abstract. This is a review of our recent investigation on the structure versus performance of ray fins via a potential-flow based fluid-structure interaction model. The kinematics and dynamic performance of two structurally idealized fins, a caudal fin and a pectoral fin, are considered. The numerical method includes a boundary-element model of the fluid motion and a fully-nonlinear Euler-Bernoulli beam model of the embedded rays. Using this model we studied thrust generation and propulsion efficiency of the fins at different combinations of parameters. Effects of kinematic as well as structural properties are examined. It has been illustrated that the fish's capacity to control the motion of each individual ray, as well as the anisotropic deformability of the fins determined by the architecture of the rays (especially the detailed distribution of ray stiffness), are essential to high propulsion performance.

Key words. Ray fin, fluid-structure interaction, numerical simulation

1. Introduction. Fins of bony fishes are characterized by a skeleton-reinforced membrane structure consisting of a soft collagen membrane strengthened by embedded flexible rays [1–4]. Morphologically, each ray is connected to a group of muscles so that the fish can control the rotational motion of each ray individually. This design enables multi-degree of freedom control over the fin motion and deformation.

Compared with the flapping foil design used in current biomimetic propulsion devices, the skeleton-reinforced membrane design of a fish fin possesses the following advantages: (1) enhanced efficiency: a flexible fin with 3D deformability yields better performance than a rigid one, manifested in the increased efficiency, reduced sensitivity to kinematic parameters, and diminished force in the transverse direction; (2) controllability and versatility: the structural properties (e.g. the anisotropic flexibility) of the composite membrane can be adjusted and its motion can be controlled in detail. In a ray-strengthened fin, the motion of each ray as well as its curvature can be controlled so that it can easily achieve complicated movements to generate different forces (thrust, lift, etc.) required in cruising, bursting, and maneuvering; (3) structural strength and lightness: The biomimetic composite membrane provides a light structure with high strength; (4) deployability: The skeleton-reinforced membrane resembles deployable structures such as cable roof buildings. These structures can be easily folded and unfolded, making them an ideal design as a portable device.

*Department of Structural Engineering, University of California San Diego, La Jolla, CA 92093, USA, qizhu@ucsd.edu. This work is funded by the National Science Foundation under CBET-0844857.

†Department of Structural Engineering, University of California San Diego, La Jolla, CA 92093, USA, kshoele@ucsd.edu.

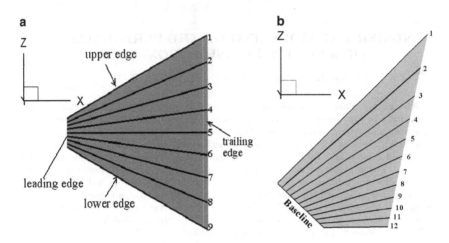

FIG. 1. *Numerical model of geometrically and structurally simplified ray fins:* (a) *caudal fin [10],* (b) *pectoral fin [12]*

In recent years, there have been extensive efforts to experimentally characterize the performance of ray fins by observations of live fish [5], or by direct measurements using mechanical duplicates of caudal or pectoral fins [6–8]. Corresponding to these experimental investigations, a numerical model based upon the immersed boundary algorithm has been developed and the results have been shown to be consistent with experiments [9]. In this numerical study, however, rather than allowing fully coupled fluid-structure interactions, the fin motion and deformation are prescribed based upon observations of live fish swimming.

In order to numerically predict detailed structure versus locomotion performance of the ray fins, and to extract key structural characteristics that contribute to performance enhancement, we have developed a fluid-structure interaction model to simulate dynamics of geometrically and structurally simplified caudal and pectoral fins [10–12]. For the labriform swimming using pectoral fins, both lift-based and drag-based locomotion modes have been considered.

The numerical model combined a boundary-element algorithm of fluid dynamics and a fully nonlinear Euler-Bernoulli beam depiction of rays. These computational investigations indicate that the anisotropic deformability of the fin attributed to the flexibility and distribution of the rays is essential to high efficiency propulsion.

2. Materials and Methods. As shown in Fig. 1, we created numerical models of both a caudal fin and a pectoral fin. Both fins are depicted as soft membranes supported by embedded rays. Despite the fact that the real fin rays are non-uniform, anisotropic, and have curvatures actively controlled by offsets of tendons linked to their bases, for simplicity in the

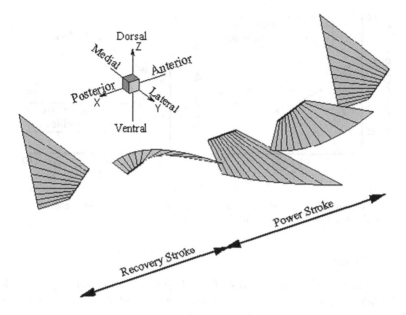

FIG. 2. *Kinematics of a pectoral fin during labriform swimming [12]*

present investigation we model the rays as uniform Euler-Bernoulli beams with circular cross-sections and passive deformability. These rays are able to sustain stretching, bending, and twisting loads. We further idealize the rest of the fin as a membrane that can sustain stretching/compression but not bending, and replace it by distributions of spring-dampers between neighboring rays.

Kinematically, the motion of the fins is achieved by prescribing the motion and rotation of the rays at their basal ends. In the caudal fin, the basal ends of the rays undergo sway and yaw motions. As illustrated in Fig. 2, during labriform swimming the pectoral fin undergoes a combination of back-and-forth rowing motion (drag based) and up-and-down flapping motion (lift based). These motions are activated by rotations of the rays as well as the reorientation of the baseline (the line formed by the basal ends of the rays). Each locomotion period T is separated into two sub-periods, a recovery stroke (abduction) and a power stroke (adduction).

A fluid-structure interaction model is developed to study the dynamics of the aforementioned ray fin. This model contains three inter-coupled models: the rays are modeled as Euler-Bernoulli beams that can sustain bending and stretching loads; the membrane between neighboring rays are represented by distributions of linear springs; and the flow field around the fin is assumed to be potential and mathematically described by using the boundary-integral equations. The fluid and structural models are numerically coupled through an iteration algorithm.

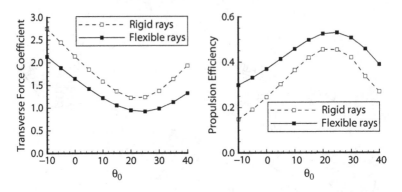

FIG. 3. *Lateral force (left) and propulsion efficiency (right) of the caudal fin in homocercal swimming with rigid and flexible rays [10]*

3. Results.

3.1. Dynamics of Caudal Fin. Figure 3 displays the amplitude of the lateral force coefficient (the force perpendicular to the direction of swimming) as well as the propulsion efficiency (defined as the forward speed times the mean thrust divided by the mean power expenditure) of the caudal fin in homocercal mode i.e. the motion is symmetric in the dorsal-ventral direction). Two cases, one with rigid rays and the other with flexible rays (the rays have the same stiffness), are considered. The values are plotted as functions of the yaw amplitude. The Strouhal number based on the forward speed, the sway amplitude, and the frequency of oscillation is 0.3.

A pronounced advantage of the fin with flexible rays is that it displays significantly reduced transverse force. According to our simulations, an estimated 10–40% reduction in the transverse force coefficient is recorded. With regard to the propulsion efficiency, the flexible fin outperforms the rigid fin by a large margin throughout the whole range. In addition, the flexible fin also achieves good performance within a much broader parameter range than the rigid fin.

Thus, we conclude that by using a flexible fin, we can greatly reduce the lateral force generation (so that the fish can easily keep on a straight line), increase the propulsion efficiency, and reduce the sensitivity of its propulsion performance on kinematic parameters.

3.2. Pectoral Fin During Labriform Swimming. In order to study the effects of structural flexibility upon the performance of the pectoral fin, we examine the capacity of force generation by three different fins: in the first one the rays are all rigid; in the second one (referred to as fin A) the rays are flexible and the mechanical properties of all the rays are identical; in the third one (fin B) the ray at the leading edge (Ray 1) is strengthened so that its effective diameter is twice as large as other

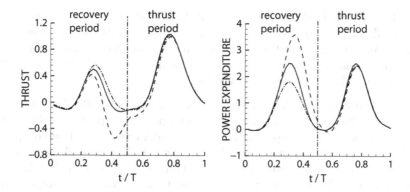

FIG. 4. *Thrust force (left) and power expenditure (right) of the pectoral fin in one period. Three different fins are considered: a fin with rigid rays (solid lines), a fin with flexible rays and the rays have the same stiffness (dashed line), and a fin with flexible rays and strengthened leading edge (dashdot lines) [12]*

rays i.e. the bending stiffness of the first ray is 15 times larger than that of other rays). The configuration of the third fin is based upon morphological observations that the first two rays at the leading edge are often connected, creating a much stiffer element than other rays [13]. In Fig. 4 we plot time histories of thrust generation and power expenditure of these three fins during one motion period. The flexible fins with and without reinforcement at the leading edge display completely different behaviors.

As shown in Fig. 4, in comparison with the fin with rigid rays, fin B achieves a slight increase in thrust coefficient. Fin A, on the other hand, generates a much smaller thrust, especially near the end of the recovery stroke when it produces negative thrust. In terms of power expenditure, large increase and decrease are observed in fin A and fin B, respectively. Most of the change occurs during the recovery stroke. This suggests significant efficiency enhancement by fin B and decrease by fin A. Indeed, for the particular cases shown in Fig. 4, the propulsion efficiencies of the fin with rigid rays, fin A, and fin B are 0.25, 0.11, and 0.30, respectively.

The different performances of fin A and fin B correlate with different fin deformations near the leading edge. In fin A, the leading edge bends slightly upwards against the direction of flapping motion. This is owing to the fact that the rays near the leading edge are the longest and thus least rigid. It increases the effective angle of attack at the leading edge. In fin B, By strengthening Ray 1 it is found that the leading edge bends downwards towards the direction of flapping so that the effective angle of attack at the leading edge is significantly reduced (Fig. 5). To accurately predict the effects of leading edge strengthening, future studies with accurate rendition of leading edge vortices are required.

3.3. Flow Field. Typical flow fields behind the caudal fin and the pectoral fin, visualized through the iso-surfaces of vorticity, are shown in Fig. 6. In both cases vortex ring structures are observed. These rings are connected and they induce jet flows, which contribute to the generation of thrust, lateral, and lift forces by the fin.

4. Conclusions. We have numerically investigated the force generation by a flexible fin made of a thin membrane reinforced by embedded rays by using a fully coupled fluid-structure interaction model based upon the boundary element method and a nonlinear Euler-Bernoulli beam formulation. By studying two cases, a caudal fin and a pectoral fin, it is found that fins with flexible rays outperform rigid ones in increased propulsion efficiency, reduced lateral force generation, and reduced dependency on kinematic parameters. Through this skeleton-reinforced membrane design, a fish achieves multi-degree-of-freedom control over its fins since each fin ray can be controlled individually. In addition, as a biological structure ray fins may have other advantages, e.g. structural lightness and deployability.

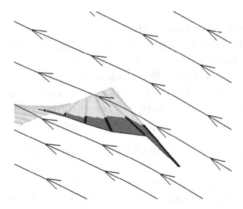

FIG. 5. *Deformation of fin B at the leading edge during the recovery stroke and its relation with the incoming flow*

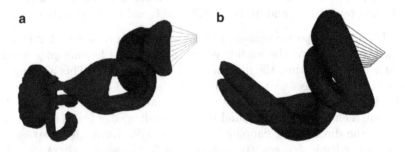

FIG. 6. *Iso surfaces of vorticity behind (a) the caudal fin undergoing homocercal motions [10], and (b) the pectoral fin during labriform swimming*

REFERENCES

[1] Lauder GV (1989) Caudal fin locomotion in ray-finned fishes: historical and functional analyses. Am Zool 29:85–102

[2] Lauder GV, Drucker EG (2004) Morphology and experimental hydrodynamics of fish fin control surfaces. IEEE J Oceanic Eng 29(3): 556–571

[3] Fish FE, Lauder GV (2006) Passive and active flow control by swimming fishes and mammals. Ann Rev Fluid Mech 38:193–224

[4] Alben S, Madden PG, Lauder GV (2007) The mechanics of active fin-shape control in ray-finned fishes. J R Soc Interface 4:243–256

[5] Lauder GV, Madden PGA, Mittal R, Dong H, Bozkurttas M (2006) Locomotion with flexible propulsors I: experimental analysis of pectoral fin swimming in sunfish. Bioinspir Biomim 1:S25–S34

[6] Tangorra JL, Davidson SN, Hunter IW, Madden PGA, Lauder GV, Dong H, Bozkurttas M, Mittal R (2007) The development of a biologically inspired propulsor for unmanned underwater vehicles. IEEE J Oceanic Eng 32: 533–550

[7] Tangorra JL, Lauder GV, Madden PG, Mittal R, Bozkurttas M, Hunter IW (2008) A biorobotic flapping fin for propulsion and maneuvering. In: IEEE International Conference on Robotics and Automation, Pasadena, pp 700–705

[8] Tangorra JL, Lauder GV, Hunter IW, Mittal R, Madden PGA, Bozkurttas M (2010) The effect of fin ray flexural rigidity on the propulsive forces generated by a biorobotic fish pectoral fin. J Exp Biol 213: 4043–4054

[9] Dong H, Bozkurttas M, Mittal R, Madden P, Lauder GV (2010) Computational modelling and analysis of the hydrodynamics of a highly deformable fish pectoral fin. J Fluid Mech 645:345–373

[10] Zhu Q, Shoele K (2008) Propulsion performance of a skeleton-strengthened fin. J Exp Biol 211(13):2087–2100

[11] Shoele K, Zhu Q (2009) Fluid-structure interactions of skeleton-reinforced fins: performance analysis of a paired Fin in lift-based propulsion. J Exp Biol 212(16):2679–2690

[12] Shoele K, Zhu Q (2010) Numerical simulation of a pectoral fin during labriform swimming. J Exp Biol 213:2038–2047

[13] Westneat M, Thorsen DH, Walker JA, Hale M (2004) Structure, function, and neural control of pectoral fins in fishes. IEEE J Oceanic Eng 29:674–683

FORMATION OF OCEAN SURFACE PATTERNS BY CETACEAN FLUKE OSCILLATIONS

RACHEL LEVY(✉)* AND DAVID UMINSKY†

Abstract. This paper presents a theory describing the fluid mechanics of whale flukeprints. It contains excerpts of a longer paper recently published in the International Journal of Non-Linear Mechanics special issue on biological structures. Whale flukeprints are smooth oval-shaped water patches that form on the surface of the ocean behind a swimming or diving whale. The prints persist up to several minutes and can be used to track whales cruising near the ocean surface. The motion of the fluke provides a mechanism for shedding vortex rings. The subsequent interaction of the vortex ring with the ocean surface damps the short wavelength capillary waves in the print. The theory suggests that the role of natural surfactants are of secondary importance in the early formation of flukeprints. We describe potential directions for future research, including collection of quantitative data from real flukeprints.

Key words. Whale, flukeprint, footprint, vortex ring, vortex shedding, wave damping, breakwater, surfactant

1. Introduction. Flukeprints are a visible pattern that appears on the surface of the ocean when a whale is swimming at a shallow depth or beginning a terminal dive. Whale-watchers can easily observe these striking oval-shaped prints on the surface of the ocean. The outer edge of the print is accentuated by small ridges, where wave-breaking may occur. The interior of the print is smooth compared to the surface outside the print, since very few capillary (wind-driven) waves are visible as in Fig. 1. The print grows radially and may remain visible for as long as several minutes, depending on ocean and wind conditions. A popular name for this phenomenon is "whale footprint" since the prints can be used to track whales over long distances as they migrate. While smaller swimming animals such as dolphins and manatees create surface prints, whales create the largest prints with the longest duration.

There is no extensive research on the phenomenon of whale prints. However, existing theories from fluid mechanics can provide an accurate perspective on this interesting phenomenon. The characteristics of the flukeprints can be explained as a result of the hydrodynamic shedding of powerful vortex rings off the edge of the whale's fluke during swim-

---

*Department of Mathematics, Harvey Mudd College, 301 Platt Blvd., Claremont, CA 91711, USA. The work of the first author was supported in part by ONR grant N000141010641, HMC internal funding and the HMC Center for Environmental Studies. Any findings, conclusions, opinions, or recommendations are those of the authors, and do not necessarily reflect the views of HMC or ONR.

†Department of Mathematics, UCLA, 520 Portola Plaza Box 951555, Los Angeles, CA 90095-1555 USA. The work of the second author was supported in part by DMS-0902792 and UC Lab Fees Research Grant 09-LR-04-116741-BERA. Any findings, conclusions, opinions, or recommendations are those of the authors, and do not necessarily reflect the views of the NSF.

S. Childress et al. (eds.), *Natural Locomotion in Fluids and on Surfaces*, IMA 155, DOI 10.1007/978-1-4614-3997-4_11,
© Springer Science+Business Media New York 2012

FIG. 1. *Typical flukeprint from humpback whale (Megaptera novaeangliae). Inside the print, the surface is smooth due to a lack of short capillary waves. Long wavelengths persist. At the boundary, wavebreaking has occurred (Photo courtesy of Kuanyin Moi)*

ming/diving and the ring's subsequent interaction with the ocean surface. This paper contains excerpts of a longer paper recently published in the International Journal of Non-Linear Mechanics special issue on biological structures [1], which contains flow visualization experiments using an oscillating robotic fluke and more detailed information on the role of surfactants in print formation.

2. Oceanographic Evidence. Our theory is based on observations by marine biologists at Cascadia Research of real flukeprints made by whales and hydrodynamic theory from the literature.

The first observation is that prints are created when whales are swimming horizontally below the surface or diving. The second observation is that a buoyant object (such as an orange) thrown into a flukeprint consistently moves to the outer edge of the print. This indicates that there is steady movement of water from the center of the print to the edge. The third observation is one made by divers filming whales underwater. They observed vortices being shed from a blue whale fluke as the whale began a terminal dive. The motion indicated strong hydrodynamic forces originating below the surface when the whale was fluking. The fourth observation, captured in Fig. 1, is wave-breaking at the boundary of the smooth area which remains throughout the visible duration of the flukeprint. A fifth observation, also visible in Fig. 1, is that while small wavelength capillary waves (small surface ripples) are absent in the print, long wavelength waves are still evident. This phenomenon is typical of whale flukeprints observed in the ocean and has been experimentally observed by Evans [2] in the context of pneumatic breakwaters, discussed in Sect. 5. A sixth observation is from another study conducted concurrently with this one. Aerial photographs show that the water temperature in flukeprints is lower than that of the surrounding water, indicating that the water in the print has been brought from below the surface [3].

The six observations regarding whales in natural habitats confirm that fluke motion brings water to the surface and induces an outward surface current. The same observations were made in a laboratory setting using a robotic fluke to visualize vortex shedding and surface flow in [4]. The remainder of this paper will describe the non-linear hydrodynamics of print formation based on our theory that prints are formed by vortex ring shedding and the subsequent interaction of the ring with the free surface. In other words, we will explain the phenomenon of whale flukeprints as a process, starting at the fluke and ending with a smooth surface region. This process occurs in roughly three stages: swimming and vortex shedding, ring-surface interactions, and finally, ring-induced surface currents and dampening of short waves.

3. Swimming and Vortex Shedding. The fluid dynamics of swimming has been heavily studied in mathematics, engineering and biology literature [5–11]. In the context of dolphins and humans, Mittal et al. [8, 9] use detailed three-dimensional simulations which demonstrate vortex ring shedding during swimming. In these simulations, vortex rings are shed on the up and down strokes of a cetacean fluke or human legs performing a dolphin kick (swimming on one's back underwater with legs together and kicking). Borazjani and Sotiropoulos [5] also numerically demonstrate the shedding of vortex ring-like structures by simulating a carangiform shape during swimming. Precise measurement and analysis of vortex ring shedding in the context of fish swimming was done by Lauder and collaborators [12–14]. Lauder et al. were able to carefully record and measure the full three-dimensional velocity fields using digital particle image velocimetry (DPIV) in the wake of sunfish and sun perch and clearly demonstrate the three-dimensional vortex ring shedding off the back of the tail fin. Moreover, a relationship of body angle to vortex ring ejection angle was demonstrated showing a three-dimensional analog of the reverse von Karman street associated with improved thrust and more efficient swimming locomotion. In the next section we discuss how vortex rings interact with the surface to produce whaleprints.

4. Vortex Ring Collisions with a Free Surface. Theoretical, numerical, and experimental methods have explored vortex ring collisions with a free surface including head-on collisions, see [15–18], and oblique collisions [19–21]. The angle of incidence affects the shape of the resulting

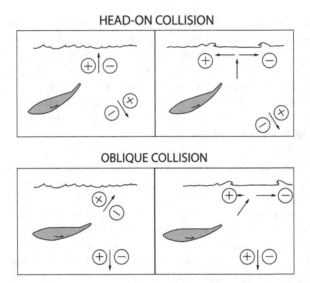

FIG. 2. *A schematic of head-on and oblique collisions. From above, prints from head-on collisions would look round or oval, whereas the uneven timing of the oblique collision leads to a heart-shaped print (See [20], Song et al. [21] and [4] for details)*

surface pattern. Both types of collisions are accompanied by an upwelling jet of fluid through the center of the upward-moving rings. When the fluid jet hits the surface it immediately creates a strong radially symmetric outward current, directly above the ring. Outside of this area, the induced radial velocity drops off rapidly as the radial surface current is circulated underneath the surface. This strong radial surface velocity persists while the core of the ring approaches the surface, deforms and moves radially outward itself. As the ring slows and breaks up, secondary and surface normal vorticity develops and attaches to the surface. A schematic illustrating the early dynamics of head-on and oblique collisions can be found in Fig. 2.

We note here that the associated Reynolds number (Re) for swimming whales, computed using the swim speed and the fluke length, falls into the range of Re 10^4–10^6. Most of the experimental and numerical literature cited above fall in the range of Re 10^1–10^4. Practically speaking, as the Reynolds number increases so does the complexity and turbulent behavior of the fluid which can create computational difficulties. Fortunately, even though the Reynolds numbers found in the literature are smaller than those realistic for whales, the observed collision dynamics in flukeprints share common features characteristic of head-on and oblique collisions.

5. Damped and Lengthened Gravity Waves. As the vortex ring collides with the surface, the jet of fluid from the center of the ring creates the surface pattern known as whale flukeprints. As mentioned above, the most striking feature of a flukeprint is the smooth patch on the surface of the ocean, where the short capillary waves have been damped out. The damping is caused by a breakwater, described in companion papers by Evans [2] and Taylor [22] who developed theory and experiments. They considered a rising jet of fluid, which in turn created a strong surface current that disrupted oncoming waves. The key feature of the theory is a dispersion relation that explains what velocity is required for a current of a certain depth h to damp waves of a given wavelength λ_0 or shorter.

First let us define a wave of length $\lambda = 2\pi/k$ to have frequency $\sigma/2\pi$ and let U be the uniform velocity of the current over a finite depth h. Enforcing continuity of the fluid and balancing the pressure Taylor derives the following dispersion equation:

$$\frac{(1-Y)^2(1-\alpha Y)}{(1-Y)^4 - \alpha Y} = -\tanh\left(\frac{ZY}{\alpha}\right), \tag{1}$$

where Y, Z, and α are the non-dimensional variables defined as:

$$Y = \frac{kU}{\sigma}, \quad \alpha = \frac{g}{U\sigma}, \quad Z = \frac{hg}{U^2}, \tag{2}$$

and g is the gravitational constant.

In the wave-stopping regime, in which current is directed against oncoming waves, $U < 0$ and hence the sign of α and Y are both negative. Remarkably, from (1), for a fixed Z, Y may take on any value in $(-\infty, 0)$ but $-\alpha$ always has a minimum value $-\alpha_m$ which falls in the range of $(0, 4)$. In the context of flukeprints, we assume the water is deep. Thus a wave with frequency $\sigma/2\pi$ has wavelength $\lambda = 2\pi g/\sigma^2$ and by the definition of α, Taylor concludes that all waves of length less than

$$\lambda_0 = \frac{2\pi(U\alpha_m)^2}{g}$$

cannot propagate against a current of velocity U with depth h. For example, to stop a wave with a wavelength of 100 ft (a very long wave) Taylor calculates that one would need an opposing velocity of 45 ft/s with a depth of 0.34 ft or 15 ft/s at a depth of 1.6 ft, both of which represent a tremendous amount of power. On the other hand, if one wants to stop all waves of 2 ft or less (still much longer than a wind-driven capillary wave), one only needs to generate a uniform current of $U = -1$ ft/s at depth of 0.1 ft, which is well within the ability of a humpback whale at cruising speeds [3]. It is also clear from the dispersion relation (1) that while waves can be disrupted by a horizontal current initiated by a vertical jet of fluid, for reasonable current velocities only the shorter wavelengths are easily eradicated.

Brevik [23] combines the work of Taylor and Evans, as well as the theory of Longuet-Higgins and Stewart [24], to classify the phenomena of wave interaction with opposing currents into three distinct regions. Region I is characterized by wave amplification in which "oncoming waves meet the current and become shortened as the current becomes stronger." In

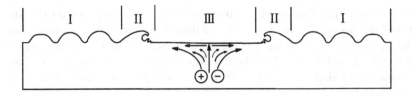

FIG. 3. *A schematic drawing of the three regions corresponding to those described by Brevik in [23] and applied to a wave packet interacting with an opposing, radially outward current. Region I represents the region just outside of the flukeprint. Region II is the edge of the flukeprint where wavebreaking occurs. Region III is the smooth interior of a flukeprint*

addition, "...short waves increase in amplitude more rapidly than long waves." Region II is marked by wave-breaking which is wavelength dependent (shorter waves break more easily). Region III is marked by the maximum velocity of the current where rapid wave dampening occurs. A schematic of the three regions can be seen in Fig. 3. He then compares this deep water theory to large scale pneumatic breakwater experiments conducted at a finite depth, D. Brevik notes that the theory shows reasonable agreement when λ is approximately the same order as D. It is also observed that "...for the shortest wavelengths it is easier to damp the waves than the foregoing theory predicts..." which he attributes to turbulent dissipation absorbing a larger fraction of the wave energy.

To summarize, pneumatic breakwater experiments and theory are consistent with oceanographic observations of flukeprints. The theory provides the connection between vortex rings and the surface signature. Vorticity is created during the downstroke on the underside of the fluke and then shed on the upstroke toward the surface. As the vortex ring drifts toward the surface, it creates a powerful jet of fluid that is circulated from below the ring, through the center and then out of the top of the ring. In addition to damping short waves, this jet creates the long-lasting and dramatic temperature gradients on the ocean surface found in flukeprints (as shown in [3]) by bringing colder water from below the ring to the surface efficiently and in large quantities.

6. Conclusion. In this paper we have described the origin of whale flukeprints, integrating results from several areas of fluid dynamics: vortex ring generation from swimming, interaction of vortex rings with surfaces, and disruption of surface waves by oncoming currents. The motion of the whale's fluke sheds vortex ring structures that create a powerful jet of fluid, circulated up through the center of the ring. When fluid from the jet contacts the surface, an outward radial surface current induces a breakwater. This in turn damps the wind-driven capillary waves in the center of the print and causes the wavebreaking observed at its edge.

We hope that this investigation sparks further research, to answer open questions about the formation, shape and duration of flukeprints. In the laboratory setting, there is much need for further investigation of the role of surfactants, forward velocity (using a flow tank or moving fluke), and wind generated capillary waves (perhaps created in a tank by blowing air across the surface). In addition, observations of cetaceans could provide quantitative data from real flukeprints. We hope the results of such experiments will be compared to the theoretical calculations described above. Further collaboration between mathematicians, experimentalists and biologists can help answer these fundamental questions.

Acknowledgments. We would like to thank John Calambokidis of Cascadia Research for helpful discussions about his observations of whales.

REFERENCES

[1] Levy R, Uminsky D, Park A, Calambokidis J (2011) A theory for the hydrodynamic origin of whale flukeprints. Int J Non Linear Mech 46(4): 616–626

[2] Evans JT (1955) Pneumatic and similar breakwaters. Proc R Soc London Ser A 231(1187):457–466

[3] Churnside J, Ostrovsky L, Veenstra T (2009) Thermal footprints of whales. Oceanography 22: 206–109

[4] Levy R, Uminsky D (2011) A theory for the hydrodynamic origin of whale flukeprints. Int J Non Linear Mech 46(4): 616–626

[5] Borazjani I, Sotiropoulos F (2008) Numerical investigation of the hydrodynamics of carangiform swimming in the transitional and inertial flow regimes. J Exp Biol 211:1541–1558

[6] Dong H, Mittal R, Najjar FM (2006) Wake topology and hydrodynamic performance of low-aspect-ratio flapping foils. J Fluid Mech 566:309–343

[7] Eldredge JD (2007) Numerical simulation of the fluid dynamics of 2D rigid body motion with the vortex particle method. J Comput Phys 221(2):626–648

[8] von Loebbecke A, Mittal R, Fish F, Mark R (2009) A comparison of the kinematics of the dolphin kick in humans and cetaceans. Human Movement Science 28:99–112

[9] Loebbecke AV, Mittal R, Mark R, Hahn J (2009) A computational method for analysis of underwater dolphin kick hydrodynamics in human swimming. Sports Biomech 8:60–77

[10] Wang ZJ (2000) Vortex shedding and frequency selection in flapping flight. J Fluid Mech 410:323–341

[11] Lighthill M (1969) Hydromechanics of aquatic animal propulsion. Ann Rev Fluid Mech 1(1):413–446

[12] Lauder G, Drucker EG (2002) Forces, fishes, and fluids: Hydrodynamic mechanisms of aquatic locomotion, News Physiol Sci 17:235–240

[13] Drucker E, Lauder GV (2002) Experimental hydrodynamics of fish locomotion: functional insights from wake visualization. Integr Comp Biol 42(2):243–257

[14] Fish F, Lauder G (2006) Passive and active flow control by swimming fishes and mammals. Ann Rev Fluid Mech 38(1):193–224

[15] Song M, Bernal LP, Tryggvason G (1992) Head-on collision of a large vortex ring with a free surface. Phys Fluids A Fluid Dyn 4(7):1457–1466

[16] Chuijie W, Qiang F, Huiyang M (1995) Interactions of three-dimensional viscous axisymmetric vortex rings with a free surface. Acta Mech Sin 11(3):229–238

[17] Chu C, Wang C, Hsieh C (1993) An experimental investigation of vortex motions near surfaces. Phys Fluids A Fluid Dyn 5(3):662–676. URL http://link.aip.org/link/?PFA/5/662/1

[18] Hirsat A, Willmarth W (1994) Measurements of vortex pair interaction with a clean or contaminated free surface. J Fluid Mech 259:25–45

[19] Ohring S, Lugt HJ (1991) Interaction of a viscous vortex pair with a free surface. J Fluid Mech 227:47–70

[20] Gharib M, Weigand A (1996) Experimental studies of vortex disconnection and connection at a free surface. J Fluid Mech 321:59–86

[21] Song M, Kachman N, Kwon J, Bernal L, Tryggvason G (1991) Vortex ring interaction with a free surface. In: 18th Symposium on Naval Hydrodynamics. The National Academies Press, Washington, DC, pp 479–489

[22] Taylor GI (1955) The action of a surface current used as a breakwater. Proc R Soc Lond A 231(1187):466–478

[23] Brevik I (1976) Partial wave damping in pneumatic breakwaters. J Hydraulics Div 102(9):1167–1176

[24] Longuet-Higgins MS, Stewart RW (1961) The changes in amplitude of short gravity waves on steady non-uniform currents. J Fluid Mech 10(4):529–549

UNSOLVED PROBLEMS IN THE LOCOMOTION
OF MAMMALIAN SPERM

SUSAN S. SUAREZ(✉)*

Abstract. Infertility is a significant health problem. On the other hand, unintended pregnancies also remain a major concern for women worldwide. Improved methods could be developed for diagnosing and treating infertility, as well as for contraception, if more were known about how sperm move through the female reproductive tract. Such information would also benefit dairy production, because fertility of cattle has been declining. Four major questions remain about sperm movement: (1) How do sperm pass through the cervix? (2) How do sperm pass through the uterotubal junction? (3) How are sperm stored and released in the oviduct? (4) Are sperm guided by chemotaxis to the egg? One important aspect of these unknowns is how physical features of the female tract affect the movement of sperm. Expertise in microfluidics and the modeling of movement at low Reynolds numbers would help biologists immensely in addressing these questions.

Key words. Sperm, spermatozoa, cervix, uterus, oviduct, fallopian tube, fertilization

1. Motivation. Infertility is a significant human health problem. According to a 2002 survey by the US CDC, 7% of 2.1 million married couples in which the woman was of reproductive age reported that they had not used contraception for 12 months and yet the woman had not become pregnant (http://www.cdc.gov/ART/). It is estimated that roughly half of the cases of infertility are due to a male factor, such as abnormal sperm motility [28]. The developments of in vitro fertilization (IVF) and intracytoplasmic sperm injection (ICSI) have done much to alleviate infertility problems; however, these procedures are associated with health risks. When a large cohort of siblings in which one was conceived naturally and the other by IVF/ICSI were compared, the IVF/ICSI babies were more likely to have been born prematurely and with low birth weight [15].

In the US dairy industry, 58% of farms use artificial insemination to impregnate cows [22]; however, it produces the undesired outcome of producing more male dairy calves than does natural mating [1]. Technology has been developed to produce "sexed semen", which produces about 90% female calves [24]. The high cost of sexed semen has pushed the industry to use fewer sperm per insemination, despite the fact that the lower dose reduces fertility [31, 32]. Fertility has also been declining significantly over the past few decades, even without the use of sexed semen. While some of this is blamed on difficulties with controlling hormonal cycles of cows,

*Department of Biomedical Sciences, Cornell University, Ithaca, NY 14853, USA, sss7@cornell.edu. This research was supported in part by NIH grant 1R03HD062471 and National Research Initiative Competitive Grant no. 2008-352-19031 from the USDA National Institute of Food and Agriculture.

S. Childress et al. (eds.), *Natural Locomotion in Fluids and on Surfaces*, IMA 155, DOI 10.1007/978-1-4614-3997-4_12, © Springer Science+Business Media New York 2012

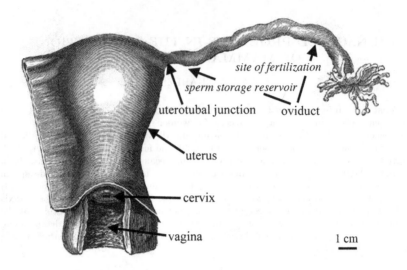

FIG. 1. *The human female reproductive tract, modified from Gray's anatomy of the human body, 1918 (Wikimedia Commons)*

some of the problems could be overcome if more information were known about sperm movement through the female tract. This could be used to optimize the process of artificial insemination.

Thus, there is much to be done to improve the outcome of reproductive technologies for human couples and for domestic cattle. These technologies are also being adapted by zoos and conservationists in order to save endangered species.

Contraception also remains a major human health problem. In the US, about half of pregnancies are unintended [16]. According to the Guttmacher Institute, the contraceptive method most used in the US is the birth control pill, which is currently used by almost 11 million women (http://www.guttmacher.org). The use of the combined ethinylestradiol/progestagen pill significantly increases the risk of venous thromboembolism and myocardial infarction [16], and impacts the development of osteoporosis by reducing bone mass acquisition in young women [29]. There is still an urgent need to develop contraceptives that are highly effective, do not produce harmful side effects, are reversible and easy to use, and also reduce risk of contracting sexually transmitted diseases. Greater understanding of the mechanisms of sperm movement through the female tract can contribute to the improvement of contraceptive technology.

In mammals, fertilization takes place in the oviduct (fallopian tube), deep within the body of the female (Fig. 1). The route from the site where the male deposits his semen to the site of fertilization is complex and requires sperm to pass through the cervix, uterus, uterotubal junction, and oviduct. Success in passing through each of these compartments is based on various interactions betweenthe sperm and the female reproductive tract that are partly physical and partly chemical in nature. None of the interactions are well understood, particularly those of a physical nature. Experts in fluid flow and movement in circumstances dictated by small Reynolds numbers are needed to work with biologists to elucidate these mechanisms.

Information about how sperm interact with the female reproductive tract could inspire development of new methods to diagnose and treat infertility, as well as development of new safe and effective contraceptives. In this review, unsolved problems of the four main events of sperm interaction with the mammalian female reproductive tract will be considered.

2. Problem 1: How Do Sperm Pass Through the Cervix? One of the major functions of the cervix is to guard the female reproductive tract against invasion by infectious microbes. At the same time, it must allow the passage of sperm and also serve as the birth canal. Some species solve the dilemma of getting sperm through the cervix by bypassing it altogether. In the pig, the penis is shaped like a corkscrew and the cervix contains complementary furrows. During copulation, the penis screws into the cervix to inseminate semen directly into the uterus [10, 11].

In humans and other primates, as well as domestic cattle, however, the male deposits semen in the vagina at the entrance to the cervix and the sperm must swim through the cervix reach the uterus.

While it is not yet technically possible to observe sperm directly as they pass through the cervix, the evidence strongly indicates that the sperm do it by swimming rather than being pushed or pulled by the female tract. However, there is also some evidence that the female provides preferential pathways specifically designed for sperm that guide them through the cervix. Nevertheless, although the existence of preferential pathways for sperm was proposed more than 20 years ago, the hypothesis is yet to be tested, due to lack of the technology to do it.

The preliminary evidence for a preferential pathway arose from a careful study of the gross and microscopic anatomy of the bovine cervix, which was made by Mullins and Saacke in 1989 [21]. They produced graphic reconstructions of tracings of serial cross sections of bovine cervixes and used these to follow the course of folds in the interior surface of the cervical canal. They found large primary folds that branched into smaller secondary folds and ran the length of the cervix. Small (micro) grooves, on the order of 10 μm in width, were seen in the surfaces of the folds. Many of the micro grooves that cut into the tissue surfaces at the base of the folds could be traced long distances through the cervix, whereas grooves near the apices of the folds were much shorter.

Histochemical stains revealed that the cervical canal was filled with mucus; however, the mucus in the basal micro grooves was chemically different from that which filled the rest of the cervical lumen [21]. Many cilia could be seen in the micro grooves and their orientation indicated

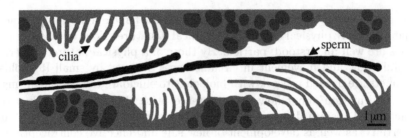

FIG. 2. *Diagram of a transmission electron micrograph of bull sperm in a micro groove of the bovine cervix, based on [21]. The sperm were sectioned through the broad surfaces of their paddle-shaped heads. Dark gray shapes in the walls of the groove represent granules of mucus in the oviductal epithelium, prior to secretion into the lumen*

that they would direct fluid flow in the grooves toward the vagina. After insemination, large numbers of sperm were found in the basal micro grooves. Because sperm orient themselves into a flow and because there was a different kind of mucus in the micro grooves, Mullins and Saacke proposed that these grooves form preferential pathways for sperm. Figure 2 illustrates the orientation of sperm in the micro grooves. Bull sperm have paddle-shaped heads that are 10 μm long, 5 μm wide, and 2 μm thick. Their tails (flagella) bring the total length to about 50 μm [4].

Sperm may also be guided through the cervix by the microarchitecture of the cervical mucus itself, which is comprised of long, flexible, linear molecules. It has been proposed that these long molecules are aligned as they are secreted into the micro grooves and their alignment guides sperm. Human and bovine sperm orient themselves along the long axis of threads of bovine cervical mucus [3, 35] and human sperm swim through cervical mucus in a straighter path than they do in seminal plasma or medium [18].

The proposal that preferential pathways guide sperm through the cervix is intriguing, because it could account for the ability of large numbers of sperm to travel through the cervix while infectious microbes are discouraged from doing so. The size, shape, and mechanics of sperm movement are quite different from those of the common infectious organisms, such as bacteria, viruses, and flagellate protozoa.

Understanding how sperm swim through the cervix could inspire development of new contraceptives for humans and improve the success rate of artificial insemination in dairy cattle.

3. Problem 2: How Do Sperm Pass Through the Uterotubal Junction? The diameter of junction between the uterus and oviduct is much smaller than that of the cervix. A scanning electron microscope study of the linings of bovine uterotubal junctions that had been cut open revealed that the inner surface is comprised of folds in the pattern of cul de sacs. Such an arrangement would seem to steer sperm into dead ends.

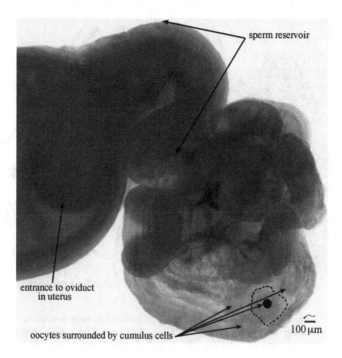

FIG. 3. *The transilluminated oviduct of the mouse. Mouse sperm are 124 μm long (Image by S. Suarez, modified from [5], with permission)*

However, if one considers how these folds would be positioned in the intact junction, one could imagine that they form funnels directing sperm through the center of the junction into the oviduct.

Even more curious than the enigma of the shape of the folds is the observation that mouse sperm require certain proteins on their surface in order to pass into the oviduct. Mice are used extensively as model species for the study of mammalian reproduction, because they can be genetically manipulated in order to examine the functions of various genes. In this case, a strain was developed in which the gene that produces the protein ADAM3 was inactivated. ADAM3 is a protein that is expressed on the front surface of the mouse sperm head [19]. Male mice lacking an active ADAM3 gene are infertile, because their sperm cannot pass into the oviduct nor penetrate the proteinaceous shell around the egg [23, 38]. The question is: how does ADAM3 enable sperm to pass through the uterotubal junction?

The mouse uterotubal junction and oviduct are so small and the walls are so thin, that one can observe sperm inside by transillumination (Fig. 3). To do this, females are euthanized after mating and the oviducts are removed to a chamber on a warm microscope stage [14, 30]. When this is done, large numbers of sperm can be seen moving in narrow channels.

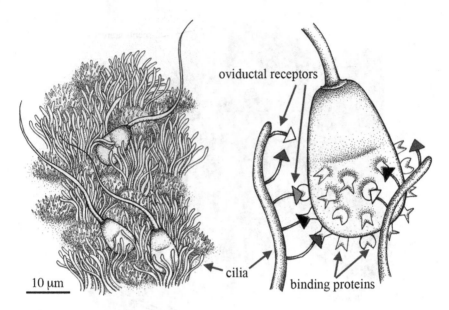

FIG. 4. *Left: bull sperm binding to ciliated cells of bovine oviductal epithelium. Right: enlarged diagram illustrating three types of binding proteins on sperm and four types of receptor proteins on cilia (Drawing by R. Gottlieb, modified from [33])*

Some appear to stick lightly and sporadically to the walls of the channels. The dissected uterotubal junction contracts intermittently, sweeping unattached sperm along the channels.

Unfortunately, this approach is limited because one can only watch sperm for a few minutes, because the oviduct is disconnected from its blood supply. Consequently, the sperm seen in the uterotubal junction cannot be followed through to the oviduct to see exactly how they move through it. We propose that ADAM3-induced sporadic sticking of sperm to the walls enables sperm to advance, but we do not know how this works. Development of mathematical and microfluidics models may help to elucidate the process.

4. Problem 3: How Are Sperm Stored and Released in the Oviduct? Most sperm that manage to pass through the uterotubal junction do not simply continue to ascend the oviduct. Instead, they are trapped and held in a storage reservoir. They are held because proteins on the sperm heads recognize and bind to receptor proteins on the epithelial cells lining the oviductal lumen (Fig. 4) [7, 8, 17]. The adhesive interaction not only holds sperm, but also maintains their fertility during storage [26, 27].

As the time approaches that the egg will be released from the ovary into the distal end of the oviduct, sperm are gradually released from the storage reservoir. The gradual release assists in preventing more than one sperm from reaching an egg simultaneously and fertilizing it, which would completely disrupt subsequent embryonic development [12, 25].

There is some evidence that sperm are released from the reservoir by shedding the proteins that bind them to the oviductal epithelium [7] and are further assisted by hyperactivation of flagellar movement. Hyperactivated sperm produce high-amplitude bends on one side of the flagellum. These deep bends appear to rip the sperm off of the epithelium [14]. In transilluminated oviducts, sperm are seen to detach and reattach several times and are thus thought to gradually work their way out of the reservoir [6].

Three different binding proteins have been identified on bull sperm and four different receptor proteins have been identified in the bovine oviduct (Fig. 4) [8, 17]. Each type of binding protein on sperm can act alone to enable sperm to adhere to oviductal epithelium. Thus, the release and subsequent advances of sperm may depend on various combinations of interactions between the three sperm proteins and the four oviductal receptors, plus appropriate triggering of hyperactivation. As with the other problems of sperm movement, understanding the release of sperm requires elucidation of the physical aspects of sperm interactions with the oviduct and begs for the attention of biophysicists.

5. Problem 4: Are Sperm Guided by Chemotaxis to the Egg?

It has been well established that sperm of various species of marine invertebrates respond to chemotactic signals from eggs via Ca^{2+}-mediated changes in the degree of flagellar beat asymmetry [20].

In human sperm, odorant receptors have been detected at the base of the flagellum and the sperm have been reported to respond chemotactically to the floral odorant bourgeonal (the odor of lilies of the valley) [34]. The problem is that bourgeonal is a floral product and the human homologue has yet to be identified, despite several years of searching.

Progesterone has also been implicated in human sperm chemotaxis [13, 36, 37]. Progesterone is secreted by the cumulus cells around the oocyte [9] and could therefore direct sperm toward the site of fertilization. The level of progesterone that acts as chemoattractant for human sperm is 0.01–10 nM [36]. A rise in sperm Ca^{2+} levels has been implicated in a flagellar response of human sperm to progesterone, particularly when sperm are exposed to a progesterone gradient [2] Nevertheless, only low percentages of sperm have been reported to orient into gradients of progesterone and extensive statistical analysis has been required in the attempt to demonstrate chemotaxis [9, 36]. Whereas, the response of sea urchin sperm to specific chemoattractants from the egg can be seen convincingly in videos, the response of human sperm to progesterone is so low that it cannot be directly observed in videos.

Despite the lack of strong data in support of the existence of sperm chemotaxis in humans and other mammals, it is generally thought that chemotaxis must exist, because the egg is such a small target for sperm within the oviduct (Fig. 3) and is only viable for a short time.

Because it has not been possible as yet to follow sperm movement continuously from the storage reservoir in the lower oviduct to the egg in the upper oviduct or to duplicate the physical environment of the oviduct in vitro, biologists have been frustrated in their attempts to determine whether and how sperm are guided to the site of fertilization. The talents of biophysicists and bioengineers are badly needed to attack this problem.

REFERENCES

[1] Berry DP, Cromie AR (2007) Artificial insemination increases the probability of a male calf in dairy and beef cattle. Theriogenology 67:346–352

[2] Bedu-Addo K, Barratt CL, Kirkman-Brown JC, Publicover SJ (2007) Patterns of [Ca2+](i) mobilization and cell response in human spermatozoa exposed to progesterone. Dev Biol 302:324–332

[3] Chretien FC (2003) Involvement of the glycoproteic meshwork of cervical mucus in the mechanism of sperm orientation. Acta Obstet Gynecol Scand 82:449–461

[4] Cummins JM, Woodall PF (1985) On mammalian sperm dimensions. J Reprod Fertil 75:153–175

[5] Chang H, Suarez SS (2010) Rethinking the relationship between hyperactivation and chemotaxis in mammalian sperm. Biol Reprod 83:507–513

[6] Demott RP, Suarez SS (1992) Hyperactivated sperm progress in the mouse oviduct. Biol Reprod 46:779–785

[7] Gwathmey TM, Ignotz GG, Suarez SS (2003) PDC-109 (BSP-A1/A2) promotes bull sperm binding to oviductal epithelium in vitro and may be involved in forming the oviductal sperm reservoir. Biol Reprod 69:809–815

[8] Gwathmey TM, Ignotz GG, Mueller JL, Manjunath P, Suarez SS (2006) Bovine seminal plasma proteins PDC109, BSPA3, and BSP30kDa share functional roles in storing sperm in the oviduct. Biol Reprod 75:501–507

[9] Guidobaldi HA, Teves ME, Unates DR, Anastasia A, Giojalas LC (2008) Progesterone from the cumulus cells is the sperm chemoattractant secreted by the rabbit oocyte cumulus complex. PLoS ONE 3:e3040

[10] Hunter RHF (1981) Sperm transport and reservoirs in the pig oviduct in relation to the time of ovulation. J Reprod Fertil 63:109–117

[11] Hunter RHF (1982) Reproduction of farm animals. Longman, London, p 149

[12] Hunter RHF, Leglise PC (1971) Polyspermic fertilization following tubal surgery in pigs, with particular reference to the role of the isthmus. J Reprod Fertil 24:233–246

[13] Harper CV, Barratt CL, Publicover SJ (2004) Stimulation of human spermatozoa with progesterone gradients to simulate approach to the oocyte. Induction of [Ca(2+)](i) oscillations and cyclical transitions in flagellar beating. J Biol Chem 279:46315–46325

[14] Ho K, Wolff C, Suarez SS (2009) CatSper null mutant sperm are unable to ascend beyond the oviductal sperm reservoir. Reprod Fertil Dev 21:345–350

[15] Henningsen AK, Pinborg A, Lidegaard O, Vestergaard C, Forman JL, Andersen AN (2011) Perinatal outcome of singleton siblings born after assisted reproductive technology and spontaneous conception: Danish national sibling-cohort study. Fertil Steril. 95:959–963

[16] Hugon-Rodin J, Chabbert-Buffet N, Bouchard P (2010) The future of women's contraception: stakes and modalities. Ann N Y Acad Sci 1205:230–239

[17] Ignotz GG, Cho MY, Suarez SS (2007) Annexins are candidate oviductal receptors for bovine sperm surface proteins and thus may serve to hold bovine sperm in the oviductal reservoir. Biol Reprod 7:906–913

[18] Katz DF, Mills RN, Pritchett TR (1978) The movement of human spermatozoa in cervical mucus. J Reprod Fertil 53:259–265

[19] Kim E, Nishimura H, Iwase S, Yamagata K, Kashiwabara S, Baba T (2004) Synthesis, processing, and subcellular localization of mouse ADAM3 during spermatogenesis and epididymal sperm transport. J Reprod Dev 50:571–578

[20] Kaupp UB, Hildebrand E, Weyand I (2006) Sperm chemotaxis in marine invertebrates–molecules and mechanisms. J Cell Physiol 208:487–494

[21] Mullins KJ, Saacke RG (1989) Study of the functional anatomy of bovine cervical mucosa with special reference to mucus secretion and sperm transport. Anat Rec 225:106–117

[22] NAHMS (2009) Dairy 2007, Part IV: reference of Dairy cattle health and management practices in the United States, 2007. #N494.0209. Centers for Epidemiology and Animal Health, USDA:APHIS:VS, Fort Collins

[23] Nishimura H, Kim E, Nakanishi T, Baba T (2004) Possible function of the ADAM1a/ADAM2 fertilin complex in the appearance of ADAM3 on the sperm surface. J Biol Chem 279:34957–34962

[24] Norman HD, Hutchison JL, Miller RH (2011) Use of sexed semen and its effect on conception rate, calf sex, dystocia, and stillbirth of Holsteins in the United States. J Dairy Sci 93:3880–3890

[25] Polge C, Salamon S, Wilmut I (1970) Fertilizing capacity of frozen boar semen following surgical insemination. Vet Rec 87:424–428

[26] Pollard JW, Plante C, King WA, Hansen PJ, Betteridge KJ, Suarez SS (1991) Fertilizing capacity of bovine sperm may be maintained by binding of oviductal epithelial cells. Biol Reprod 44:102–107

[27] Pacey AA, Hill CJ, Scudamore IW, Warren MA, Barratt CL, Cooke ID (1995) The interaction in vitro of human spermatozoa with epithelial cells from the human uterine (fallopian) tube. Hum Reprod 10:360–366

[28] Patrizio P, Sanguineti F, Sakkas D (2008) Modern andrology: from semen analysis to postgenomic studies of the male gametes. Ann N Y Acad Sci 1127:59–63

[29] Pikkarainen E, Lehtonen-Veromaa M, Mottonen T, Kautiainen H, Viikari J (2008) Estrogen-progestin contraceptive use during adolescence prevents bone mass acquisition: a 4-year follow-up study. Contraception 78:226–231

[30] Suarez SS (1987) Sperm transport and motility in the mouse oviduct: observations in situ. Biol Reprod 36:203–210

[31] Seidel GE Jr (2003) Economics of selecting for sex: the most important genetic trait. Theriogenology 59:585–598

[32] Seidel GE Jr (2007) Overview of sexing sperm. Theriogenology 68:443–446

[33] Suarez SS (2007) Interactions of spermatozoa with the female reproductive tract: inspiration for assisted reproduction. Reprod Fertil Dev 19:103–110

[34] Spehr M, Gisselmann G, Poplawski A, Riffell JA, Wetzel CH, Zimmer RK, Hatt H (2003) Identification of a testicular odorant receptor mediating human sperm chemotaxis. Science 299:2054–2058

[35] Tampion D, Gibons RA (1962) Orientation of spermatozoa in mucus of the cervix uteri. Nature 194:381

[36] Teves ME, Barbano F, Guidobaldi HA, Sanchez R, Miska W, Giojalas LC (2006) Progesterone at the picomolar range is a chemoattractant for mammalian spermatozoa. Fertil Steril 86:745–749

[37] Teves ME, Guidobaldi HA, Unates DR, Sanchez R, Miska W, Publicover SJ, Morales Garcia AA, Giojalas LC (2009) Molecular mechanism for human sperm chemotaxis mediated by progesterone. PLoS One 4:e8211

[38] Yamaguchi R, Muro Y, Isotani A, Tokuhiro K, Takumi K, Adham I, Ikawa M, Okabe M (2009) Disruption of ADAM3 impairs the migration of sperm into oviduct in mouse. Biol Reprod 81:142–146

COMPUTING OPTIMAL STROKES FOR LOW REYNOLDS NUMBER SWIMMERS

ANTONIO DESIMONE(✉)*, LUCA HELTAI*,
FRANÇOIS ALOUGES†, AND ALINE LEFEBVRE-LEPOT†

Abstract. We discuss connections between low-Reynolds-number swimming and geometric control theory, and present a general algorithm for the numerical computation of energetically optimal strokes. As an illustration of our approach, we show computed motility maps and optimal strokes for two model swimmers.

Key words. Biopropulsion in water and in air, low-Reynolds-number motions, control theory and feedback, locomotion, motility maps

AMS(MOS) subject classifications. 76Zxx, 76Z10, 74F10

1. Introduction. Self–propulsion at low Reynolds number is a problem of considerable biological and biomedical relevance, with a great appeal from the point of view of fundamental science. Starting from the pioneering work by Taylor [11] and Lighthill [6], it has received considerable attention in recent years (see the review paper [7] for a comprehensive list of references).

Both relevance for applications and theoretical interest stem from the small size of the swimmers. Reynolds number $Re = LV/\nu$ gives an estimate for the relative importance of inertial to viscous forces for an object of size L moving at speed V through a newtonian fluid with kinematic viscosity ν. Since in all applications V rarely exceeds a few body lengths per second, if one considers swimming in a given medium, say, water, then Re is entirely controlled by L. At small L, inertial forces are negligible and, in order to move, micro-swimmers can only exploit the viscous resistance of the surrounding fluid. The subtle consequences of this fact (which are rather paradoxical when compared to the intuition we can gain from our own swimming experience) are discussed in [9]. For example, the motion of microswimmers is geometric: the trajectory of a low Re swimmer is entirely determined by the sequence of shapes that the swimmer assumes. Doubling the rate of shape changes simply doubles the speed at which the same trajectory is traversed. As observed in [10], this suggests that there must be a natural, attractive mathematical framework for this problem (which the authors, indeed, unveil).

The geometric structure of low Re swimming emerges easily from a factorization of the variables describing the state of a swimmer. The state is a located shape, namely, a shape plus the assignment of a position and orientation. This leads to a bundle-like foliation of state space

---

*SISSA, Via Bonomea 265, 34136 Trieste, ITALY desimone@sissa.it

†CMAP UMR 7641, École Polytechnique CNRS, 91128 Palaiseau Cedex, FRANCE

S. Childress et al. (eds.), *Natural Locomotion in Fluids and on Surfaces*, IMA 155, DOI 10.1007/978-1-4614-3997-4_13,
© Springer Science+Business Media New York 2012

(leaves are rigid motions at frozen shape) which, in the simplified setting of an axisymmetric swimmer moving along a line considered in this paper, reduces to a trivial fiber bundle (see Fig. 1).

The basic problem of swimming is easy to state: given a (periodic) time history of shapes of a swimmer (a sequence of strokes), determine the corresponding time history of positions and orientations in space. A natural, related question is the following: starting from a given position and orientation, can the swimmer achieve any prescribed position and orientation by performing a sequence of strokes? This is a question of *controllability*. The peculiarity of low Re swimming is that, since inertia is negligible, reciprocal shape changes lead to no net motion, so the question of controllability may become non trivial for swimmers that have only a few degrees of freedom at their disposal to vary their shape. The well known scallop theorem [9] is precisely a result of non-controllability. Once controllability is known, one can ask the question of how to reach the target at minimal energetic cost. This is a question of *optimal control*.

In spite of the clear connections between low Re self-propulsion and control theory, this viewpoint has started to emerge only recently. In this paper we report on some recent progress in the development of numerical algorithms to find energetically optimal strokes that take advantage of the geometric control viewpoint and exploit the geometric structure underlying low Re swimming, see Refs [1–3].

2. Optimal Swimming with Few Formulas. Consider a swimmer having an axisymmetric shape Ω with axis of symmetry parallel to the unit vector $\vec{\imath}$ and swimming along $\vec{\imath}$. The state **s** of the system is described by one scalar positional parameter c, e.g., the coordinate along the symmetry axis of either the center of mass or of a distinguished point) and by N shape parameters $\boldsymbol{\xi} = (\xi_1, \ldots, \xi_N)$. For simplicity, we will only consider the simplest nontrivial case $N = 2$. The velocity $\bar{\mathbf{v}}$ of each point of the boundary $\partial\Omega$ of the swimmer depends linearly on $\dot{\boldsymbol{\xi}}$ and $\dot{c}$

$$\bar{\mathbf{v}}(x) = \sum_{i=1}^{N} \bar{\mathbf{v}}_i(\boldsymbol{\xi}, x)\dot{\xi}_i + \bar{\mathbf{v}}_{N+1}(\boldsymbol{\xi}, x)\dot{c}, \qquad x \in \partial\Omega, \tag{1}$$

through known functions $\bar{\mathbf{v}}_i$, $i = 1, \ldots N$ specifying the kinematics of the swimmer. In particular,

$$\bar{\mathbf{v}}_{N+1}(\boldsymbol{\xi}, x) \equiv \vec{\imath}, \qquad x \in \partial\Omega. \tag{2}$$

Since the swimmer is self-propelled, its inertia is being neglected, and in view of the axial symmetry, the equations of motion reduce to the vanishing of the total viscous force exerted by the fluid, which we write as

$$0 = \int_{\partial\Omega} \boldsymbol{\sigma}_\xi[\bar{\mathbf{v}}]\mathbf{n}\, dS \cdot \vec{\imath} \tag{3}$$

where $\boldsymbol{\sigma}_\xi[\bar{\mathbf{v}}]$ is the stress tensor in the fluid surrounding Ω, and $\mathbf{n}$ is the outward unit normal to $\partial\Omega$, the boundary of Ω. The viscous stress $\boldsymbol{\sigma}_\xi[\bar{\mathbf{v}}]$ is obtained by solving Stokes equation outside Ω with Dirichlet boundary data $\mathbf{v} = \bar{\mathbf{v}}$ on $\partial\Omega$, and depends linearly on $\bar{\mathbf{v}}$ and non-linearly on the parameters ξ_i describing the shape of Ω.

By linearity of Stokes equations, viscous forces depend linearly on $\bar{\mathbf{v}}$ and hence on $\dot{\boldsymbol{\xi}}$ and $\dot{c}$, see (1). We can thus write (3) as

$$0 = \sum_{i=1}^{N} \varphi_i(\boldsymbol{\xi})\, \dot{\xi}_i + \varphi_{N+1}(\boldsymbol{\xi})\, \dot{c}. \tag{4}$$

The coefficients φ_i relating drag to velocities in (4) are independent of c by translational invariance, and are given by

$$\varphi_j(\boldsymbol{\xi}) = \int_{\partial\Omega} \boldsymbol{\sigma}_\xi[\bar{\mathbf{v}}_j]\, \mathbf{n}\, dS \cdot \vec{i} \quad j = 1, \ldots, N+1 \tag{5}$$

where the functions $\bar{\mathbf{v}}_i$ are given in (1). The coefficient φ_{N+1} of $\dot{c}$ represents the drag due to a rigid translation along the symmetry axis at unit speed, see (2), and it never vanishes. Thus, (4) can be solved for $\dot{c}$ yielding

$$\dot{c} = \sum_{i=1}^{N} V_i(\xi_1, \ldots, \xi_N)\dot{\xi}_i = \mathbf{V}(\boldsymbol{\xi}) \cdot \dot{\boldsymbol{\xi}} \tag{6}$$

where

$$V_i(\boldsymbol{\xi}) = -\frac{\varphi_i(\boldsymbol{\xi})}{\varphi_{N+1}(\boldsymbol{\xi})} \quad i = 1, \ldots, N \tag{7}$$

and the φ_i are given in (5).

A stroke is a T-periodic map $t \mapsto \boldsymbol{\xi}(t)$ describing a closed path γ in the space $\mathcal{S}$ of admissible shapes. In view of (6), swimming requires that

$$0 \neq C := c(T) - c(0) = \int_0^T \sum_{i=1}^{N} V_i\dot{\xi}_i dt \tag{8}$$

i.e., that the differential form $\sum_{i=1}^{N} V_i d\xi_i$ be not exact. For $N = 2$ we can think of γ as the boundary of a region ω in shape space and use Stokes theorem to write (8) as

$$0 \neq C = \int_\omega \mathrm{curl}\, \mathbf{V}(\xi_1, \xi_2) d\xi_1 d\xi_2 \tag{9}$$

where $\mathrm{curl}\, \mathbf{V}$ is the scalar $\partial_1 V_2 - \partial_2 V_1$. Clearly, $\mathrm{curl}\, \mathbf{V}(\xi_0) \neq 0$ is sufficient for local controllability near ξ_0. This is nothing but Chow's theorem of geometric control theory [5]. Indeed, setting $q = (q_1 := \xi_1, q_2 := \xi_2, q_3 := c)$ we can write (6) as a control system

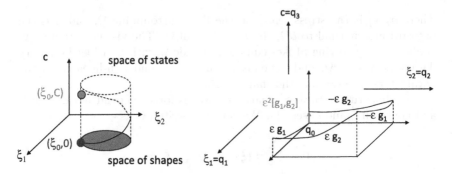

FIG. 1. *Fiber bundle structure and Lie brackets for a* $N = 2$ *swimmer*

$$\dot{q} = \begin{bmatrix} 1 \\ 0 \\ V_1(q_1, q_2) \end{bmatrix} u_1 + \begin{bmatrix} 0 \\ 1 \\ V_2(q_1, q_2) \end{bmatrix} u_2 = \sum_{i=1}^{2} g_i(q) u_i. \qquad (10)$$

The Lie bracket $[g_1, g_2]$ of the coefficients g_1, g_2 is

$$[g_1, g_2] := (\frac{\partial}{\partial q} g_2) g_1 - (\frac{\partial}{\partial q} g_1) g_2 = \begin{bmatrix} 0 \\ 0 \\ \text{curl } \mathbf{V} \end{bmatrix} \qquad (11)$$

and we have that

$$\det [g_1 | g_2 | [g_1, g_2]] \, (q_1, q_2, q_3) = \text{curl } \mathbf{V}(q_1, q_2) \qquad (12)$$

and the condition $\text{curl } \mathbf{V}(q) \neq 0$ implies that the Lie algebra generated by the coefficients g_i of the system (10) and their Lie brackets is of full rank (three, in the example considered). Figure 1 illustrates the geometric structure underlying (10) and the physical meaning of (11).

We are interested in strokes that realize a given net displacement C at minimal expended energy

$$\int_0^T \int_{\partial\Omega} \sigma_\xi[\bar{\mathbf{v}}] \, \mathbf{n} \cdot \bar{\mathbf{v}} \, dS \, dt = \int_0^T \mathbf{G}(\boldsymbol{\xi}(t)) \dot{\boldsymbol{\xi}}(t) \cdot \dot{\boldsymbol{\xi}}(t) \, dt, \qquad (13)$$

where we have used the linearity of Stokes equations and the linear dependence on $\dot{\boldsymbol{\xi}}$ of the velocity $\bar{\mathbf{v}}$ at an arbitrary point of the swimmer, which is a consequence of (1) and (6). The entries of the $N \times N$ symmetric and positive definite matrix $\mathbf{G}(\boldsymbol{\xi})$ are independent of c by translational invariance, and are given by

$$G_{ij}(\boldsymbol{\xi}) = \int_{\partial\Omega} \sigma_\xi[\bar{\mathbf{w}}_j] \, \mathbf{n} \cdot \bar{\mathbf{w}}_i \, dS, \quad i, j = 1, \dots, N \qquad (14)$$

where

$$\bar{\mathbf{w}}_i(\boldsymbol{\xi}, x) = \bar{\mathbf{v}}_i(\boldsymbol{\xi}, x) + V_i(\boldsymbol{\xi})\bar{\mathbf{v}}_{N+1}(\boldsymbol{\xi}, x), \quad x \in \partial\Omega \tag{15}$$

while $\bar{\mathbf{v}}_i$ and V_i are given by Eqs. 1 and 7.

Minimizers of (13) subject to the constraint (9) are solutions of the Euler-Lagrange equations (written for $N = 2$)

$$-\frac{d}{dt}(\mathbf{G}\dot{\boldsymbol{\xi}}) + \frac{1}{2}\left(\begin{array}{c} \partial_1 \mathbf{G}\dot{\boldsymbol{\xi}} \cdot \dot{\boldsymbol{\xi}} \\ \partial_2 \mathbf{G}\dot{\boldsymbol{\xi}} \cdot \dot{\boldsymbol{\xi}} \end{array}\right) + \lambda\,\mathrm{curl}\,\mathbf{V}(\boldsymbol{\xi})\,\dot{\boldsymbol{\xi}}^{\perp} = 0 \tag{16}$$

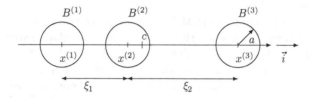

FIG. 2. *Three-sphere (3S) swimmer [8]*

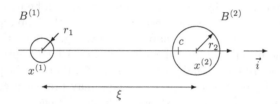

FIG. 3. *Pushmepullyou (PMPY) swimmer [4]*

where $\dot{\boldsymbol{\xi}}^{\perp} = (-\dot{\xi}_2, \dot{\xi}_1)$ and λ is a Lagrange multiplier associated with (9). Optimal strokes that start and end at a given shape $\boldsymbol{\xi}_0$ are solutions of Eq. 16 such that $\boldsymbol{\xi}(0) = \boldsymbol{\xi}(T) = \boldsymbol{\xi}_0$. They can be computed by solving (16) numerically, e.g. with a shooting method, provided that the coefficients $\mathbf{G}$ and $\mathbf{V}$ are known. These coefficients encode all hydrodynamic interactions between the swimmer and the surrounding fluid, and can be obtained by solving numerically the outer Stokes problem at each point $\boldsymbol{\xi}$ of a discrete parametrization of the space of admissible shapes. Further optimization with respect to $\boldsymbol{\xi}_0$ delivers the strokes realizing the net displacement C at maximal energetic efficiency.

Extensions of the geometric approach described here to general swimmers (e.g., multiple rigid bodies in relative motion, such as the 3-link swimmer [9]) are possible. This requires replacing (4) with force and torque balance, leading to six ODEs determining the instantaneous translational and angular velocity of the swimmer. The general full rank controllability condition is required in this setting, rather than its simplified form $\mathrm{curl}\,\mathbf{V}(q) \neq 0$, see (12).

3. Applications. We have computed optimal strokes for the two model swimmers shown in Figs. 2 and 3. We always consider swimmers in water ($\eta = 10^{-3}$ Pa s at room temperature).

For the three-sphere-swimmer (3SS) [8] made of balls of radius $a = 0.05$ mm which can change the distances between their centers ξ_1, ξ_2, three strokes starting and ending at shape $\boldsymbol{\xi}_0 = (0.3\,\text{mm}, 0.3\,\text{mm})$, and inducing a displacement $C = 0.01\,\text{mm}$ in $T = 1\,\text{s}$ are shown in Fig. 4. The energy consumption for the optimal stroke (full curve), the small and the large square loops – are, respectively, 0.229×10^{-12}J, 0.278×10^{-12}J, and 0.914×10^{-12}J. The instantaneous velocity of the center of mass along the optimal stroke is shown in Fig. 5.

For the pushmepullyou (PMPY) [4], made of two spheres which can change the distance ξ between their centers and the relative volume θ (defined as the ratio between the volume of one sphere and the total, constant volume), optimal strokes corresponding to net displacements $C = 0.03\,\text{mm}$ and $C = 0.9\,\text{mm}$ are shown in Fig. 4. For this second one, the instantaneous velocity of the center of mass along the stroke is shown in Fig. 5.

Figure 4 shows the level curves of curl $\mathbf{V}$ in the space of shapes. Since, in view of (9), the net displacement is the flux of curl $\mathbf{V}$ across the region of shape space enclosed by the stroke, these graphs can be used as *motility maps* giving immediately the distance traveled once the loop representing a stroke is traced on the space of shapes. Finally, Fig. 6 shows the impact of prescribing the initial and final shape on the geometry of optimal strokes. If these are prescribed through $\boldsymbol{\xi}(0) = \boldsymbol{\xi}(T) = \boldsymbol{\xi}_0$, then the loop representing the optimal stroke in the space of shape may present a corner at $\boldsymbol{\xi}_0$. Optimizing over $\boldsymbol{\xi}_0$ produces a stroke with smaller energy consumption, and represented by a smooth closed curve.

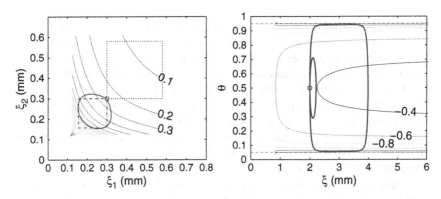

FIG. 4. *Motility maps (i.e., level curves of curl* $\mathbf{V}$ *in the space of shapes) and optimal strokes for 3S (left) and PMPY (right)*

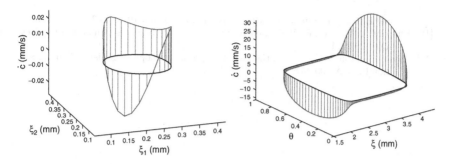

FIG. 5. *Instantaneous velocity of the center of mass along optimal strokes for 3S (left) and PMPY (right)*

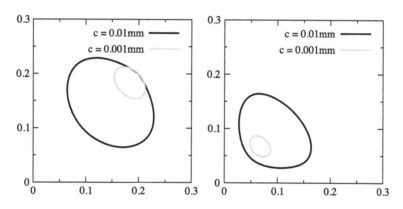

FIG. 6. *Optimal strokes for 3S: initial and final shape are prescribed in the left panel ($\xi(0) = \xi(T) = \xi_0$) while, on the right, a further optimization over ξ_0 has been performed*

REFERENCES

[1] Alouges F, DeSimone A, Lefebvre A (2008) Optimal strokes for low Reynolds number swimmers: an example. J Nonlinear Sci 18:277

[2] Alouges F, DeSimone A, Lefebvre A (2009) Optimal strokes for axisymmetric microswimmers. Eur Phys J E 28:279

[3] Alouges F, DeSimone A, Heltai L (2011) Numerical strategies for stroke optimization of axisymmetric microswimmers. Math Model Method Appl Sci 21:361

[4] Avron JE, Kenneth O, Oakmin DH (2005) Pushmepullyou: an efficient microswimmer. New J Phys 7:234

[5] Jurdjevic V (1997) Geometric control theory. Cambridge University Press, Cambridge/New York

[6] Lighthill MJ (1952) On the squirming motion of nearly spherical deformable bodies through liquids at very small Reynolds numbers. Commun Pure Appl Math 2:109

[7] Lauga E, Powers TR (2009) The hydrodynamics of swimming microorganisms. Rep Prog Phys 72(9): 096601

[8] Najafi A, Golestanian R (2004) Simple swimmer at low Reynolds numbers: three linked spheres. Phys Rev E 69:062901

[9] Purcell EM (1977) Life at low Reynolds numbers. Am J Phys 45:3

[10] Shapere A, Wilczek F (1987) Self-propulsion at low Reynolds numbers. Phys Rev
     Lett 58:2051

[11] Taylor GI (1951) Analysis of the swimming of microscopic organisms. Proc R Soc
     Lond Ser A 209:447

MODELS OF LOW REYNOLDS NUMBER SWIMMERS INSPIRED BY CELL BLEBBING

QIXUAN WANG(⊠)*, JIFENG HU*, AND HANS OTHMER*

Abstract. Eukaryotic cells move through the complex micro-environment of a tissue either by attaching to the extracellular matrix – sometimes degrading it locally – and pulling themselves along, or by squeezing through the matrix by appropriate sequences of shape changes. Some cells can even swim by shape changes, and one mode used is called blebbing, in which a cell creates a small hemispherical protrusion that may grow to incorporate the entire cell volume or may be reabsorbed into the primary volume. Herein we develop and analyze several models for swimming at low Reynolds number inspired by cell blebbing. These models comprise several connected spheres, and each connected pair of spheres can exchange volume with their complement in the pair. We show that the cell can propel itself through the fluid using a suitable sequence of volume exchanges, and we evaluate the efficiency of this mode of swimming.

Key words. Cell protrusion, micro-swimmer, Stokes solution, linked-sphere models

AMS(MOS) subject classifications. Primary 76Z10, 49J20, 92C17, 93B05

1. Introduction. Cell locomotion is essential for embryonic development, angiogenesis, tissue regeneration, the immune response, and wound healing in multicellular organisms, and plays a very deleterious role in cancer metastasis in humans. Locomotion involves the detection and transduction of extracellular chemical and mechanical signals, integration of the signals into an intracellular signal, and the spatio-temporal control of the intracellular biochemical and mechanical responses that lead to force generation, morphological changes and directed movement [10]. While many single-celled organisms use flagella or cilia to swim, there are two basic modes of movement used by eukaryotic cells that lack such structures – mesenchymal and amoeboid [4]. The former, which can be characterized as 'crawling' in fibroblasts or 'gliding' in keratocytes, involves the extension of finger-like filopodia or pseudopodia and/or broad flat lamellipodia, whose protrusion is driven by actin polymerization at the leading edge. This mode relies on strong adhesion to the substrate, and dominates in cells such as fibroblasts crawling on a 2D substrate. In the amoeboid mode, which does not rely on strong adhesion, cells are more rounded and employ shape changes to move – in effect 'jostling through the crowd' or 'swimming'. Recent experiments have shown that numerous cell types display enormous plasticity in locomotion, in that they sense the mechanical properties of their environment and adjust the balance between the mesenchymal and amoeboid modes accordingly by altering the balance between parallel signal transduction pathways [9]. Pure crawling and pure

---

*School of Mathematics, University of Minnesota, Minneapolis, MN 55455, USA
wangx825@umn.edu

S. Childress et al. (eds.), *Natural Locomotion in Fluids and on Surfaces*, IMA 155, DOI 10.1007/978-1-4614-3997-4_14,
© Springer Science+Business Media New York 2012

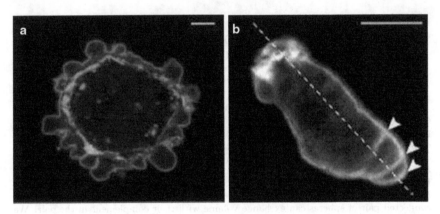

FIG. 1. *(a) Blebbing in a filament-deficient melanoma cell. (b) The actin cortex of a blebbing Dd cell migrating to the lower right. White arrowheads indicate the successive blebs and arcs of the actin cortex (Reproduced with permission from [12] and [13])*

swimming are the extremes on a continuum of locomotion strategies, but cells can sense their environment and use the most efficient strategy in a given context.

Some cells produce membrane 'blisters' called blebs, in which the membrane detaches from the cortex locally and the pressure generated by the cortex forces the membrane outward. Figure 1 shows a *Dictyostelium discoideum* (Dd) cell that uses blebs to extend the leading edge while moving. Blebbing is a specialized form of shape change that may produce movement, but to understand when it does, and to understand movement more generally, one has to integrate the cellular dynamics with the dynamics of the surrounding complex medium, the ECM. Here we begin with swimming, motivated by recent experiments which show that both neutrophils and Dd can swim – in the strict sense of propelling themselves through a fluid without using any attachments – in response to chemotactic gradients [3]. The experimental observations show a very complex sequence of shape changes in Dd, but we first study how shape changes in abstract models of swimmers can lead to movement in a viscous fluid. Interest in this classical problem stems from the description of life at low Reynolds number by Purcell [8]. The essential ideas are as follows – the current state of knowledge is reviewed elsewhere [7].

The governing equations for an incompressible Newtonian fluid of density ρ and viscosity μ are given by

$$\rho\frac{\partial \mathbf{u}}{\partial t} + \rho(\mathbf{u}\cdot\nabla)\mathbf{u} = -\nabla p + \mu\Delta\mathbf{u} + \mathbf{f}, \qquad \nabla\cdot\mathbf{u} = 0 \qquad (1)$$

where **f** is the external force field. The Reynolds number based on a characteristic length scale L and speed scale V is Re = $\rho L V / \mu$, and nondimensionalization of Eq. 1 shows that when Re $\ll$ 1 the acceleration terms can be ignored. This defines a low Reynolds number (LRN) flow. When there are no external force fields, as we will assume hereafter, the equations simplify to the Stokes equations

$$\mu \Delta \mathbf{u} - \nabla \mathbf{p} = \mathbf{0}, \qquad \nabla \cdot \mathbf{u} = \mathbf{0}. \qquad (2)$$

The small size and slow speed of cells considered here leads to LRN flows, and in this regime cells move by exploiting the viscous resistance of the fluid. However, since time is absent from the equations, a time-reversible stroke produces no net motion, which is the content of the famous 'scallop theorem' [8]. Because there is no net force or torque on a swimmer in the Stokes regime, movement is a purely geometric process: the net displacement of a swimmer during a stroke is independent of the rate at which the stroke is executed, as long as the Reynolds number remains small enough. The properties of the exterior fluid come into play only when addressing the efficiency of a stroke.

A *swimming stroke* is defined by a time-dependent sequence of shapes, and a *cyclic swimming stroke* is a swimming stroke for which the initial and final shapes are identical. Let $\mathcal{B}(t)$ and **V** be the boundary and velocity of the swimmer, respectively. **V** can be decomposed into a part **v** that corresponds to the intrinsic shape deformations, and a part **U** that corresponds to a rigid motion. Then most LRN self-propulsion problems can be stated as: *given a cyclic shape deformation by specifying* **v**, *solve the Stokes equations subject to*

$$\sum \mathbf{F}_i = 0, \quad \sum \mathbf{\Gamma}_i = 0, \quad \mathbf{u}|_{\mathcal{B}} = \mathbf{V}, \quad \mathbf{u}|_{\mathbf{x} \to \infty} = \mathbf{0}, \qquad (3)$$

where **u** is the velocity field of the surrounding fluid, and $\mathbf{F}_i$ and $\mathbf{\Gamma}_i$ are forces and torques on the swimmer.

Purcell's model swimmer comprises three connected, rigid links constrained to move in a plane, where adjacent links do a constrained rotation around joints and can thereby change the angle between them. Here the shape is specified by two parameters, the angles between adjacent links, and Purcell showed that one can impose sequences of changes in the angles that produce motion of the swimmer. Despite its geometric simplicity, the relationships between geometric parameters, speed and efficiency of swimming are not simple, and various simpler models such as the three-linked-spheres model and the push-me-pull-you model have appeared since [5].

As an example, consider a swimmer made of three spheres and two rods connecting them, all immersed in a LRN fluid (Fig. 2). Assume that the lengths L_1 and L_2 of the rods connecting them are changed in a prescribed way. Then for a large separation one can use the Oseen tensor, which

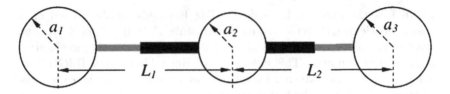

FIG. 2. *The three-linked-spheres swimmer (Reproduced with permission from the authors of [6])*

defines the leading-order term for **u**, to relate the forces f_i on the spheres to the speed u_1 of the first sphere as follows [6]:

$$u_1 = \frac{f_1}{6\pi\mu a_1} + \frac{f_2}{6\pi\mu L_1} + \frac{f_3}{6\pi\mu(L_1 + L_2)}.$$

There are similar equations for u_2 and u_3, and the leading order approximation U_0 to the speed U is the mean speed $U_0 = \sum_i u_i/3$. After specifying the velocities L'_i and using the force-free condition at Eq. 3, one can eliminate the forces, and when all spheres have the same radius a one finds that

$$U_0 = \frac{a}{6}\left[\left(\frac{L'_2 - L'_1}{L_1 + L_2}\right) + 2\left(\frac{L'_1}{L_2} - \frac{L'_2}{L_1}\right)\right] \tag{4}$$

plus terms that average to zero over a cycle [6]. Thus the swimmer can move for suitable choices of the velocities L'_i. The efficiency of swimming for this model has been estimated, and an algorithm for finding optimal strokes exists [1].

2. The Linear *3*-Sphere Swimmer with Volume Exchange. As a generalization of the previous example that is more directly related to cell motility and blebbing, we consider a connected linear chain of spheres, each of which can exchange mass with its neighbors. We assume that the distance between their centers of mass remains unchanged, and that the distance between each pair is much larger than the radii of the spheres. It follows from the scallop theorem that a two-sphere model cannot swim, since it has only one degree of freedom, and therefore a minimal model must comprise at least three spheres, as in the previous example.

Here we restrict attention to the three-sphere swimmer shown in Fig. 3, immersed in a Newtonian fluid with no-slip boundary conditions at the boundary of the spheres. The connecting rods are massless, the distance l between the centers of two adjacent spheres is constant, and for simplicity we assume that during the volume exchange process between adjacent spheres, the volumes remain spherical. Let a_i be the radius of the ith sphere and assume that $\varepsilon \sim a_i/l \ll 1$.

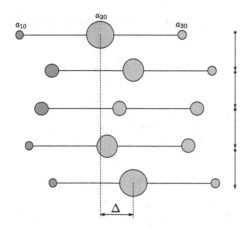

(1) $\delta a_1 = \delta_1 > 0$ while $\delta a_3 \equiv 0$;

(2) $\delta a_3 = \delta_3 > 0$ while $\delta a_1 \equiv 0$;

(3) $\delta a_1 = -\delta_1 > 0$ while $\delta a_3 \equiv 0$;

(4) $\delta a_3 = -\delta_3 > 0$ while $\delta a_1 \equiv 0$.

FIG. 3. *A cycle of the swimmer*

The solution to Eq. 2 for the flow around a single sphere of radius a subject to a force $\mathbf{f}$ and dilated at the rate $\dot{v}$ is

$$\mathbf{u}(\mathbf{r}; a, \mathbf{f}, \dot{v}) = \frac{1}{24\pi\mu r}\left[\left(3 + \xi^2\right)\mathbf{f} + 3(1 - \xi^2)(\mathbf{f} \cdot \hat{\mathbf{r}})\hat{\mathbf{r}}\right] + \frac{\dot{v}}{4\pi r^2}\hat{\mathbf{r}} \qquad (5)$$

where $r = |\mathbf{r}|$, $\hat{\mathbf{r}}$ is the unit direction of $\mathbf{r}$, and $\mathbf{u}(\mathbf{r}; a, \mathbf{f}, \dot{v})$ is the velocity at position $\mathbf{r}$ from the center of the sphere [2] and $\xi \equiv a/r$.

In the linear case, by symmetry, both the net velocities of the spheres and the net forces acting on each sphere should be parallel to the symmetry axis, thus can be taken as scalars. Let f_i be the net force acting on the ith sphere by the surrounding Stokes fluid. It is reasonable to assume that $O(f_1) \sim O(f_2) \sim O(f_3)$. To leading order in $\varepsilon \sim a_i/l$,

$$\begin{cases} U_1 \sim \dfrac{f_1}{6\pi\mu a_1} - \dfrac{\dot{v}_2}{4l^2} - \dfrac{\dot{v}_3}{16l^2} \\[2mm] U_2 \sim \dfrac{f_2}{6\pi\mu a_2} + \dfrac{\dot{v}_1}{4l^2} - \dfrac{\dot{v}_3}{4l^2} \\[2mm] U_3 \sim \dfrac{f_3}{6\pi\mu a_3} + \dfrac{\dot{v}_1}{16l^2} + \dfrac{\dot{v}_2}{4l^2} \end{cases} \qquad (6)$$

where U_i is the velocity at the center of the ith sphere.

The swimming velocity of the whole object is the mean translational velocity. Because we assume that the length of the two connecting arms is always l, we have $\dot{l} = U_2 - U_1 = U_3 - U_2 = 0$, and therefore

$$U_1 = U_2 = U_3 = \bar{U}. \qquad (7)$$

Since the system is force-free $f_1 + f_2 + f_3 = 0$ and because the total volume V is conserved

$$\dot{v}_1 + \dot{v}_2 + \dot{v}_3 = 0 \qquad \text{or} \qquad a_1^2 \dot{a}_1 + a_2^2 \dot{a}_2 + a_3^2 \dot{a}_3 = 0. \tag{8}$$

From the foregoing we find that

$$\bar{U} = \frac{(a_1 + a_2 - \frac{3}{4}a_3)\dot{v}_1 - (a_3 + a_2 - \frac{3}{4}a_1)\dot{v}_3}{4l^2(a_1 + a_2 + a_3)} \tag{9}$$

where $a_2 = \left(3V/(4\pi) - a_1^3 - a_3^3\right)^{1/3}$.

To propel the swimmer with $\dot{v}_i$ given, the power required is

$$P = \frac{\mu}{\pi}\left[\left(\frac{1}{a_1^3} + \frac{1}{a_2^3}\right)\dot{v}_1^2 + \frac{2}{a_2^3}\dot{v}_1\dot{v}_3 + \left(\frac{1}{a_2^3} + \frac{1}{a_3^3}\right)\dot{v}_3^2\right] \tag{10}$$

and the efficiency of a stroke γ is [2]

$$e(\gamma) := \frac{6\pi\mu X^2(\gamma)}{\tau \int_0^\tau P\,dt}.$$

Some analytical results on movement can be obtained – detailed proofs of the folllowing will appear elsewhere [11]. Let $\bar{d}X > 0$ represent an infinitesimal displacement to the right in Fig. 3. Given an infinitesimal shape change (da_1, da_3), it follows from Eq. 9 that

$$\bar{d}X = \frac{\pi}{l^2}\left[a_1^2\left(1 - \frac{7}{4}\frac{a_3}{a_1 + a_2 + a_3}\right)da_1 - a_3^2\left(1 - \frac{7}{4}\frac{a_1}{a_1 + a_2 + a_3}\right)da_3\right]. \tag{11}$$

Using Stokes' theorem, the translation δX associated with a closed loop is

$$\delta X = \frac{7\pi}{4l^2}\left[a_1^2\partial_{a_3}\frac{a_3}{a_1 + a_2 + a_3} + a_3^2\partial_{a_1}\frac{a_1}{a_1 + a_2 + a_3}\right]da_1 \wedge da_3, \tag{12}$$

where $da_1 \wedge da_3$ denotes the signed area enclosed by the loop. Some conclusions that can be drawn from this are as follows:

- When only one da_i is non-zero, the direction of movement is always from the expanding sphere to the contracting one, provided that the center sphere is large enough. An intuitive explanation of this can be found in [2]: *the expanding sphere acts as a source pushing away the shrinking sphere which acts as a sink to pull the expanding sphere.*
- For any stroke γ homotopic to S^1 in the (a_1, a_2) plane, an increase of the stroke amplitude will increase the net translation per stroke, while an increase in the initial radius a_{20} of the central sphere (with a_{10} and a_{30} fixed) will decrease the net translation per stroke. A first-order approximation to the displacement is $|X(\gamma)| \sim \frac{\varepsilon}{7}\text{Area}(\Omega)$, where $\varepsilon \sim a_i/l$, Ω is the region enclosed by γ, and $\text{Area}(\Omega)$ is the signed area of Ω.

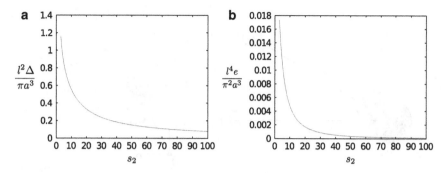

FIG. 4. (a) The relationship between the net translation $l^2\Delta/(\pi a^3)$ and the initial size s_2 of the center sphere, and (b) between the efficiency $l^4e/(\pi^2 a^3)$ and s_2, for $s_3 = r_1 = r_3 = 1$

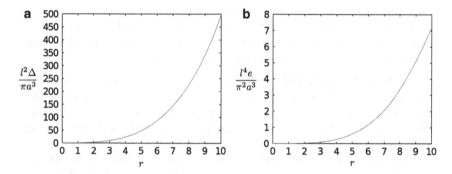

FIG. 5. (a) The relationship between the net translation $l^2\Delta/(\pi a^3)$ and the stroke amplitude r, and (b) between the efficiency $l^4e/(\pi^2 a^3)$ and r, for $s_3 = 1$, $r_2 = r_3 = r$, and $s_2 = 15$

- Increasing the initial radius a_{20} of the central sphere (with a_{10} and a_{30} fixed) decreases the efficiency. Also, for infinitesimal strokes, increasing the stroke amplitude $|da_1 \wedge da_3|$ symmetrically by a factor r (i.e., $\widetilde{da_1} = rda_1$ and $\widetilde{da_3} = rda_3$) will increase the efficiency by r^2. In particular, if we assume that $a_i \sim a$ and $da_i \sim da$, then we have the approximation: $\bar{d}e \sim a|da|^2/l^4$.

For finite-amplitude changes we must compute the displacement and efficiency numerically, and for this we consider the cycle shown in Fig. 3. We suppose that initially $a_{10} = a$, $a_{20} = s_2 a$, and $a_{30} = s_3 a$, where s_i measures the initial relative size. At the end of each step within a stroke, the radius of sphere i ($i = 1, 3$) is either a_{i0} or $a_{i0} + \delta_i$. Let $\delta_1 = r_1 a$, $\delta_3 = r_3 a$, where r_i measures the stroke amplitude, and let Δ be the net translation after one full cycle. Figures 4 and 5 show the effect of varying the initial size of the central sphere and the stroke amplitude on the displacement and efficiency.

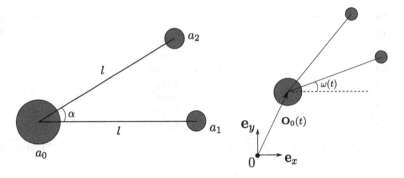

FIG. 6. *The planar 3-sphere swimmer*

Figure 4 shows that increasing the initial volume of the center sphere will decrease both the net translation and the efficiency of the cycle, whereas in Fig. 5 one sees that increasing the amplitude in each step of the cycle will increase both the net translation of the swimmer after a full cycle and its efficiency, as predicted by the analysis.

3. The Planar 3-Sphere Swimmer. Next consider a three-sphere swimmer that can move in the plane as shown in Fig. 6, wherein the first and second spheres can exchange volume with the zeroth sphere while preserving the total volume. We assume that $a_i/l\alpha \ll 1$, and suppose that other constraints and the notation are as before.

The motion of the structure can be uniquely determined by $(\mathbf{O}_0; \omega)$, where $\mathbf{O}_0(t)$ is the position of the center of sphere 0 in a global chart, and $\omega(t)$ measures the rotation of the structure with respect to sphere 0.

If, as before, we solve the Stokes equations plus the volume conservation constraint Eq. 8, we obtain

$$n\dot{\omega} = \frac{a_0\left(a_2\dot{v}_1 - a_1\dot{v}_2\right)\sin\alpha}{4\pi l^3\left[a_0\left(a_1 + a_2\right) + 2a_1a_2\left(1 - \cos\alpha\right)\right]}\left(\frac{1}{8\sin^3\frac{\alpha}{2}} - 1\right). \quad (13)$$

The assumption that $a_i/l\alpha \ll 1$ ensures that $(8\sin^3\frac{\alpha}{2})^{-1} - 1$ will not be too large, and it follows that

$$
\begin{aligned}
\mathbf{U}_0 = &-\left(\frac{\dot{v}_1}{4\pi l^2}\left[a_0 + a_1 + a_2\left(\cos\alpha + \frac{1}{4\sin\frac{\alpha}{2}}\right)\right] + \frac{\dot{v}_2}{4\pi l^2}\left[a_0\cos\alpha\right.\right.\\
&\left.+ a_1\left(1 - \frac{1}{4\sin\frac{\alpha}{2}}\right) + a_2\cos\alpha\right] - a_2l\dot{\omega}\sin\alpha\right)\left(\frac{\cos\omega\,\mathbf{e}_x + \sin\omega\,\mathbf{e}_y}{a_0 + a_1 + a_2}\right)\\
&-\left(\frac{a_2\dot{v}_1}{4\pi l^2}\sin\alpha\left(1 - \frac{1}{8\sin^3\frac{\alpha}{2}}\right) + \frac{\dot{v}_2}{4\pi l^2}\left(a_0\sin\alpha + a_2\sin\alpha\right.\right.\\
&\left.\left.+\frac{a_1}{4\sin\frac{\alpha}{2}\tan\frac{\alpha}{2}}\right) + \left(a_1 + a_2\cos\alpha\right)l\dot{\omega}\right)\left(\frac{-\sin\omega\,\mathbf{e}_x + \cos\omega\,\mathbf{e}_y}{a_0 + a_1 + a_2}\right) \quad (14)
\end{aligned}
$$

where $\omega(t) = \int_0^t \dot{\omega}\, dt$.

As was true for the linear array, some analytical results can be obtained [11]. From Eq. 13, we have

$$\bar{d\omega} = \frac{\sin\alpha}{l^3}\left(\frac{1}{8\sin^3\frac{\alpha}{2}} - 1\right)\frac{a_0 a_1 a_2 (a_1 da_1 - a_2 da_2)}{a_0(a_1 + a_2) + 2a_1 a_2(1 - \cos\alpha)} \quad (15)$$

where $\bar{d\omega} > 0$ represents an infinitesimal rotation counterclockwise. The swimming rotation $\delta\omega$ associated to an infinitesimal closed loop is

$$\delta\omega = \phi\cdot(a_1\partial_{a_2} + a_2\partial_{a_1})\psi\, da_2 \wedge da_1 \quad (16)$$

where

$$\phi(\alpha) = \frac{\sin\alpha}{l^3}\left(\frac{1}{8\sin^3\frac{\alpha}{2}} - 1\right)$$

$$\psi(a_1, a_2; \alpha) = \frac{a_0 a_1 a_2}{a_0(a_1 + a_2) + 2a_1 a_2(1 - \cos\alpha)}.$$

In Eq. 16, $da_2 \wedge da_1$ denotes the signed area enclosed by the loop. From this one can conclude that:

1. Any linear or equilateral triangular swimmer cannot rotate at any time.
2. If the loop is chosen so that $da_2 \wedge da_1 < 0$ and $|2(1-\cos\alpha)\frac{a_1^2 a_2^2}{a_0^4}| < 1$ holds throughout the stroke, then the swimmer rotates clockwise when $\alpha < \pi/3$, and rotates counterclockwise when $\pi/3 < \alpha < \pi$.
3. When α satisfies the standing constraint, we have

$$\delta\omega \sim \frac{\varepsilon}{l^2}\sin\alpha\left(\frac{1}{8\sin^3\frac{\alpha}{2}} - 1\right)|da_2 \wedge da_1| \quad (17)$$

which implies that either increasing a_i or increasing the stroke amplitude $|da_2 \wedge da_1|$ will increase the net rotation $\delta\omega$ of a stroke.

To extend these results to finite changes, consider a full cycle comprising the following four steps:

1. Sphere 1 transfers volume $\delta_1 > 0$ to sphere 0;
2. Sphere 2 transfers volume $\delta_2 > 0$ to sphere 0;
3. Sphere 0 transfers volume δ_1 back to sphere 1;
4. Sphere 0 transfers volume δ_2 back to sphere 2.

We assume that the initial state of a stroke is $(\mathbf{O_0}, \omega)(t = 0) = (\mathbf{0}, 0)$, and let $\Omega = \omega(T)$ and $\mathbf{\Delta} = \mathbf{O_0}(T)$ be the rotation angle and the net translation of the swimmer at the end of the stroke, respectively, where T is the period of the stroke. We express $\mathbf{\Delta}$ in global polar coordinates as $\mathbf{\Delta} = R\cdot\mathbf{e}_r + \Theta\cdot\mathbf{e}_\theta$.

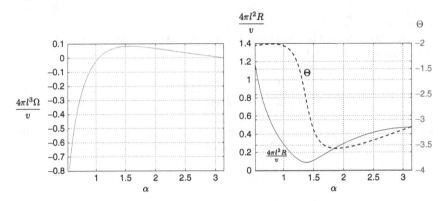

FIG. 7. *The rotation $4\pi l^3 \Omega(\alpha)/v$ (left) and the translation $4\pi l^2 R(\alpha)/v$ and $\Theta(\alpha)$ (right) of the swimmer as a function of the acute angle α*

Figure 7 illustrates some of the computational results for $\frac{4\pi l^3}{v}\Omega(\alpha)$, $\frac{4\pi l^2}{v}R(\alpha)$ and $\Theta(\alpha)$ with $a_{10} = a_{20} = a$, $a_{00} = 2^{1/3}a$, and $\delta_1 = \delta_2 = 0.9v$ for $\alpha \in [\pi/6, \pi]$. From this computational experiment one can conclude the following:

- There are two zeros of $\Omega(\alpha)$, one at $\alpha = \pi/3$ (corresponding to an equilateral triangle), and the other at $\alpha = \pi$ (a straight line, the case discussed earlier). (This confirms the general result (1) stated above.)
- When $\alpha < \pi/3$, $\Omega < 0$ (which corresponds to clockwise rotation), and $|\Omega|$ increases rapidly as α decreases. When $\pi/3 < \alpha < \pi$, $\Omega > 0$ (counterclockwise rotation), and $4\pi l^3 |\Omega|/v$ has a maximum value of about 0.0830 at $\alpha = \pi/2$.
- Under the assumption that $\varepsilon \sim a_i/l \ll 1$, $\Omega \sim o(\varepsilon^3)$, which implies that the effect of rotation is not significant.
- α has complicated effects on both the translation distance and the translation direction. $\frac{4\pi l^2}{v}R$ has a minimum of about 0.0884 at around $\alpha = 0.44\pi$; while Θ has a maximum of about -0.48π at around $\alpha = 0.26\pi$, and a minimum of about -1.13π at around $\alpha = 0.60\pi$.

4. Discussion. In this paper we have briefly sketched some results for a three-sphere model in which movement results from volume exchange between the spheres. Here we only considered square loops in the control space, but an open problem is to determine the optimal loop in the control space. Another generalization under consideration is a model in which volume exchange is coupled with active contraction of the connecting rods to more closely approximate movement in cells. In reality, microorganisms execute much more complicated shape deformations in order to swim, but important insights can be gained from abstract models such as considered here.

Acknowledgments. This research was supported in part by NSF grants DMS-0517884 and DMS-0817529.

REFERENCES

[1] Alouges F, DeSimone A, Lefebvre A (2009) Optimal strokes for axisymmetric microswimmers. Eur Phys J E Soft Matter Biol Phys 28:279–284

[2] Avron JE, Kenneth O, Oaknin DH (2005) Pushmepullyou: an efficient microswimmer. New J Phys 7:234

[3] Barry NP, Bretscher MS (2010) Dictyostelium amoebae and neutrophils can swim. PNAS 107:11376

[4] Binamé F, Pawlak G, Roux P, Hibner U (2010) What makes cells move: requirements and obstacles for spontaneous cell motility. Mol Biosyst 6:648–661

[5] Cohen N, Boyle JH (2010) Swimming at low Reynolds number: a beginners guide to undulatory locomotion. Contemp Phys 51:103–123

[6] Golestanian R, Ajdari A (2008) Analytic results for the three-sphere swimmer at low Reynolds number. Phys Rev E 77:36308

[7] Lauga E, Powers TR (2009) The hydrodynamics of swimming microorganisms. Rep Prog Phys 72:096601

[8] Purcell E (1977) Life at low reynolds number. Am J Phys 45:3–11

[9] Renkawitz J, Schumann K, Weber M, Lämmermann T, Pflicke H, Piel M, Polleux J, Spatz JP, Sixt M (2009) Adaptive force transmission in amoeboid cell migration. Nat Cell Biol 11:1438–1443

[10] Sheetz MP, Felsenfeld D, Galbraith CG, Choquet D (1999) Cell migration as a five-step cycle. In: Lackie JM, Dunn GA, Jones GE (eds) Cell motility: control and mechanism of motility. Princeton University Press, Princeton, pp 233–243

[11] Wang Q (2012) Mathematical and computational studies of cell motility. Ph.D. thesis, University of Minnesota

[12] Yoshida K, Soldati T (2006) Dissection of amoeboid movement into two mechanically distinct modes. J Cell Sci 119:3833–3844

[13] Charras G, Paluch E (2008) Blebs lead the way: how to migrate without lamellipodia. Nature Reviews Molecular Cell Biology. Nature Publishing Group, 9:730–736

A LOW-REYNOLDS-NUMBER TREADMILLING SWIMMER NEAR A SEMI-INFINITE WALL

KIORI OBUSE* AND JEAN-LUC THIFFEAULT(✉)†

Abstract. We investigate the behavior of a treadmilling microswimmer in a two-dimensional unbounded domain with a semi-infinite no-slip wall. The wall can also be regarded as a probe or pipette inserted into the flow. We solve the governing evolution equations in an analytical form and numerically calculate trajectories of the swimmer for several different initial positions and orientations. We then compute the probability that the treadmilling organism can escape the vicinity of the wall. We find that many trajectories in a 'wedge' around the wall are likely to escape. This suggests that inserting a probe or pipette in a suspension of organism may push away treadmilling swimmers.

1. Introduction. The locomotion of microorganisms is an active research area of fluid dynamics and biology (see for instance the reviews [9, 14]). As their motion occurs on very small length scales and speeds, their dynamics is governed by low-Reynolds-number hydrodynamics, where inertial forces are negligible in comparison to the viscous effects of the fluid (Stokes flow).

Many studies deal with such dynamics in unbounded or very large domains [1, 11, 17]. In reality, however, most organisms are in the vicinity of other bodies or boundaries, where hydrodynamic interactions can have a significant effect on their motion. The importance of boundaries has also been suggested by experimental observations. For example, some research suggests that microorganisms are attracted to solid walls [3, 16, 18]. Berke et al. [2] measured the steady-state distribution of *E. Coli* bacteria swimming between two glass plates and found a strong increase of the cell concentration at the boundaries. They also theoretically demonstrated that hydrodynamic interactions of swimming cells with solid surfaces lead to their reorientation in the direction parallel to the surfaces. Lauga et al. [8] showed that circular trajectories are natural consequences of force-free and torque-free swimming and hydrodynamic interactions with the boundary. This leads to a hydrodynamic trapping of the cells close to the surface. Drescher et al. [6] found that when two nearby Volvox colonies swim close to a solid surface, they attract one another and can form stable bound states in which they 'waltz' or 'minuet' around each other. These observations suggest that, in order to obtain a comprehensive understanding of low-Reynolds-number locomotion, it is necessary to study hydrodynamic interactions between microorganisms and boundaries.

*Research Institute for Mathematical Sciences, Kyoto University, Kyoto, 606-8502, Japan.

†Department of Mathematics, University of Wisconsin, Madison, WI 53706, USA, jeanluc@math.wisc.edu. The work of the second author was supported in part by NSF grant DMS-0806821.

S. Childress et al. (eds.), *Natural Locomotion in Fluids and on Surfaces*, IMA 155, DOI 10.1007/978-1-4614-3997-4_15,
© Springer Science+Business Media New York 2012

Some of the phenomena stated above have already been confirmed by numerical simulations [7, 15]. However, not many physical explanations have been given to the locomotion of microorganism near boundaries. Berke et al. [2] have captured the swimming microorganisms' attraction to boundaries by modeling the swimmer as a force dipole singularity. However, contrary to the experimental findings, the microorganism in this model crashes into the boundary in finite time. Or and Murray [13] studied the dynamics of low-Reynolds-number swimming organism near a plane wall. They analyzed the motion of a swimmer consisting of two rotating spheres connected by a thin rod as a simple theoretical model of a 'treadmilling' swimming organism. They found that when the spheres rotate with different velocities their model has a solution with steady translation parallel to the wall, and that under small perturbation the swimmer exhibits a 'bouncing' motions parallel to the wall. These results have recently been verified experimentally on a macroscale robotic prototype swimming in a highly viscous fluid [19]. Furthermore, Crowdy and Or [4] have proposed a singularity model for swimming microorganisms placed near an infinite no-slip boundary. Their model was based on a circular treadmilling organism which has no means of self-propulsion, that is, the organism doesn't move unless it interacts with a boundary. (For example, the organism may be creating a feeding current.) They proposed an appropriate Stokes singularities that represent the flow field created by this treadmilling organism. By studying the interaction between these singularities and the no-slip wall, they formulated explicit evolution equations for the motion of the organism, and fully characterized its motion near the wall. They found trajectories with a periodic bouncing motion along the wall which had remarkable similarity to the trajectories shown in Or and Murray [13]. Crowdy and Samson [5], using the point-singularity model, investigated the dynamics of treadmilling organism near an infinite no-slip boundary with a gap of a fixed size. They employed a conformal mapping technique to avoid the difficulty in treating the image of the treadmilling organism on the wall. They found that the treadmilling organism can exhibit several qualitatively different types of trajectories: jumping over the gap, rebounding from the gap, being trapped near the gap, and escaping the gap region even when the organism has initial position in the gap. They also performed a bifurcation analysis in terms of the model parameters, and demonstrated the presence of stable equilibrium points in the gap region as well as Hopf bifurcations to periodic bound states. This reduced model also exhibited a global gluing bifurcation in which two symmetric periodic orbits merge at a saddle point into symmetric bound states having more complex spatio-temporal structure.

In the present paper we examine the dynamics of a treadmilling organism near a semi-infinite no-slip wall, modeled as a flat plate of zero thickness. Though this is a special case of the model of Crowdy and Samson [5], it deserves separate investigation because of the simpler equations involved, andbecause the semi-infinite wall can be regarded as a probe or pipette inserted in the system, a common situation in microbiology. We also analyze the trajectories in a very different manner to [5], as we attempt to quantify the probability of escape from the wall's vicinity, assuming the treadmillers are randomly oriented.

2. Model of a Treadmilling Microorganism. Following previous authors [4, 5], we consider a two-dimensional model for a microorganism in the (x, y)-plane, which we treat as the complex plane with $z \equiv x + iy$. Our derivation is a special case of [5], who considered a microswimmer near a slit or gap in an infinite wall. Nevertheless, as mentioned in the introduction, the semi-infinite wall is important enough to be treated separately, as it arises in the neighborhood of a probe or a pipette.

The Stokes equations which describe the motion of an incompressible viscous fluid are

$$\nabla p = \eta \Delta u, \qquad \nabla \cdot u = 0, \tag{1}$$

where Δ is the Laplace operator, $u = (u_x, u_y)$ is the fluid velocity, and p and η are the pressure and dynamic viscosity, respectively. As we are considering a two-dimensional flow, we can introduce a stream function ψ, such that the velocity is given by $u_x = \partial \psi / \partial y$, $u_y = -\partial \psi / \partial x$. Then the Stokes equations (1) reduce to the biharmonic equation $\Delta^2 \psi = 0$. The complex velocity is $W = u_x + iu_y = -2i\partial \psi / \partial \bar{z}$, with

$$W = u_x + iu_y = -2i \frac{\partial \psi}{\partial \bar{z}} = f(z) + z\overline{f'(z)} + \overline{g'(z)}, \tag{2}$$

$$\psi = \text{Im}[\bar{z}f(z) + g(z)]. \tag{3}$$

where $f(z)$ and $g(z)$ are called Goursat functions; they are analytic everywhere in the flow domain, except where isolated singularities are introduced to model swimmers. For a treadmilling swimmer, we take $f(z)$ to have a simple pole at $z = z_d$, so that

$$f(z) = \frac{\mu}{z - z_d} + \dots, \qquad g'(z) = \frac{\mu \bar{z}_d}{(z - z_d)^2} + \dots. \tag{4}$$

where $\mu \in \mathbb{C}$ and the form of g' is forced by the requirement that the complex velocity (2) have no higher than a $|z - z_d|^{-1}$ singularity. This solution corresponds to a stresslet of strength μ at z_d. The ellipses in (4) indicate analytic terms. The expansion (4) is the basic solution for a treadmilling swimmer, which does not have any self-propulsion in itself, but can move due to its interaction with boundaries [4, 5, 10, 13, 19].

In a simple model, we assume that the treadmilling organism has a circular body of radius ϵ, with a center at $z_d(t) = x_d(t) + iy_d(t)$. We also assume that, with respect to the angle $\theta(t)$ of the head of the treadmilling organism from the real axis, surface actuators of the treadmilling organism induce a tangential velocity profile given by Crowdy and Or [4]

$$U(\phi, \theta) = 2V \sin(2(\phi - \theta)), \tag{5}$$

where V is a constant and ϕ is the angle measured from the positive x direction.

Next consider the treadmiller near an infinite wall along the x axis. To satisfy the no-slip boundary condition at the wall, we make the expansion

$$f(z,t) = \frac{\mu}{z - z_d(t)} + f_0 + (z - z_d(t))f_1 + \cdots ,$$

$$g'(z,t) = \frac{b}{(z - z_d(t))^3} + \frac{a}{(z - z_d(t))^2} + g_0 + \cdots , \tag{6}$$

$f(z,t)$ having no Stokeslet term and $g(z,t)$ having no rotlet term implying that the treadmilling organism is force-free and torque-free. We use the boundary condition to find

$$\mu = -i\epsilon\bar{c}, \qquad a = \mu\bar{z}_d, \qquad b = \mu\epsilon^2 - i\bar{c}\epsilon^3 = 2\mu\epsilon^2, \tag{7}$$

where $c(t) \equiv -iV \exp(-2i\theta(t))$. We set the time scale of the motion by letting $V = \epsilon^{-1}$ so that $\mu(t) = \exp(2i\theta(t))$. The coefficients f_0, f_1, and g_0 are given in [4].

The time derivative of the position and orientation of the treadmilling organism is obtained by equating the time rate of change of position to the local fluid velocity, and the time rate of change of orientation to half the local vorticity:

$$\frac{dz_d}{dt} = -f_0 + z_d\bar{f}_1 + \bar{g}_0, \qquad \frac{d\theta}{dt} = -2\,\mathrm{Im}\,f_1. \tag{8}$$

Equation 8 can then be solved as a set of three ODEs determining the motion of the treadmiller.

Now we turn to a semi-infinite wall, extending along the negative x axis. The conformal mapping $\zeta = iz^{1/2}$ maps the z plane to upper-half ζ plane, with the negative x-axis of the z plane mapped to the real axis in the ζ plane. In the ζ plane we can use a similar singular expansion as (6), but we must take care to map the boundary conditions to the ζ plane. We omit the lengthy details, which are similar to Crowdy and Samson [5]. See [12] for a more complete derivation. All that is required for simulating the swimmer trajectories are the coefficients that appear in (8), which are

$$f_0 = \frac{\mu}{4z_d} - \frac{\epsilon^2\bar{\mu}}{4\mathcal{Z}^3\bar{z}_d^{3/2}} + \frac{\left(2|z_d|^2 - 2\bar{z}_d^2 - 3\epsilon^2\right)\bar{\mu}}{8\mathcal{Z}^2\bar{z}_d^2}$$

$$+ \frac{\left(2|z_d|^2 + 2\bar{z}_d^2 - 3\epsilon^2\right)\bar{\mu}}{8\mathcal{Z}\bar{z}_d^{5/2}}, \tag{9a}$$

$$f_1 = \frac{1}{12z_d^2}\left(-\frac{3\mu}{4} + \frac{9z_d^{3/2}\epsilon^2\bar{\mu}}{2\mathcal{Z}^4\bar{z}_d^{3/2}} - \frac{3z_d^{3/2}\left(2|z_d|^2 - 2\bar{z}_d^2 - 3\epsilon^2\right)\bar{\mu}}{2\mathcal{Z}^3\bar{z}_d^2}\right.$$

$$\left.-\frac{3z_d^{3/2}\left(2|z_d|^2 + 2\bar{z}_d^2 - 3\epsilon^2\right)\bar{\mu}}{4\mathcal{Z}^2\bar{z}_d^{5/2}}\right), \tag{9b}$$

and

$$g_0 = -\frac{3\mu\bar{z}_d}{16z_d^2} + \frac{10\epsilon^2\mu}{32z_d^3} + \frac{3\epsilon^2\bar{\mu}}{8\mathcal{Z}^4\bar{z}_d} - \frac{(z_d - \bar{z}_d)\bar{\mu}}{4\mathcal{Z}^3\bar{z}_d^{1/2}}$$

$$+\frac{\left(2|z_d|^2 - 6\bar{z}_d^2 - 3\epsilon^2\right)\bar{\mu}}{16\mathcal{Z}^2\bar{z}_d^{5/2}}\left(\bar{z}_d^{1/2} + \mathcal{Z}\right), \tag{9c}$$

where $\mathcal{Z} \equiv z_d^{1/2} + \bar{z}_d^{1/2}$.

3. Results of Numerical Simulations. We now present the results of numerical simulations of the governing evolution equations (8), together with the coefficients (9). The radius of the circular treadmillers is set to $\epsilon = 1$, giving the reference length scale. The time scale is set by $V = \epsilon^{-1}$ in (5).

Figure 1 shows examples of swimmer trajectories for different initial conditions for position z_{d0} and angle θ_0. Some trajectories, such as (a) and (d), end up far away from the wall; we refer to those as escaping trajectories. Others, such as (b) and (c), remain close to the wall for all time, exhibiting the 'bouncing' behavior noted in [4]. A fourth type of trajectory (not shown) crashes into the wall, but this is an unphysical consequence of the boundary condition at the swimmer's surface being only approximately satisfied. It has been verified numerically that the qualitative features of the trajectories are not affected by this approximation (MD Finn, 2011 Private communication).

Both experimental observations and previous theoretical studies suggest that, when there is a no-slip wall near a treadmilling organism, the organism tends to be attracted to its own image and move towards the wall [2, 4, 5, 7, 8, 15, 16, 18]. This behavior is clearly seen at an early stage in all the trajectories in Fig. 1. Nevertheless, in the cases shown in Fig. 1a, d, the treadmilling organism moves away from the wall after it has come close to the edge of the wall.

We now focus on the large-time behavior of treadmillers. To study this, we first introduce five regions shown in Fig. 2a. Regions 1–4 are the usual quadrants of the plane, but with a neighborhood of size 0.2ϵ around the wall removed; this removed neighborhood is region 5. The regions 1–4 are rendered in different shades of gray, and region 5 is rendered in white[1] (red,

---

[1]See http://arXiv.org/abs/1104.0146 for color figures.

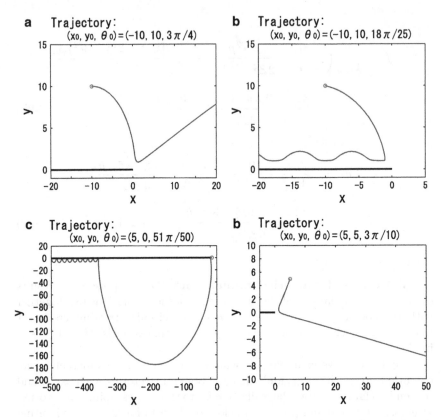

FIG. 1. *Examples of trajectories from the initial points* (x_{d0}, y_{d0}) *marked with circles, with an initial orientation* θ_0. *A treadmilling organism can escape from the wall, such as in (a) and (d), end up above the wall as in (b), or underneath the wall as in (c)*

blue, orange, green, and white online). From these regions, we can make a 'pie chart' around each point in the plane, an example of which is shown in Fig. 2b. The sectors of the pie chart correspond to a range initial angle θ_0; the shading (color online) of a sector describes which region the treadmiller ends up in for large times (here $t = 1{,}500$). Region 5 corresponds to 'crashing' trajectories. Thus, for a given position, the treadmiller may end up in different regions, and the relationship between angle and final region is not simple.

Figure 3 shows pie charts for many initial points (x_{d0}, y_{d0}). The most notable feature is the complexity of the pie charts for initial points near the wall and with large negative x coordinates. This reflects the fact that the treadmilling organism changes its heading direction and the quality of its trajectory significantly when it has come to the vicinity of the edge of the wall, $x = y = 0$ depending upon its (x_d, y_d, θ) at the time, and so even a very small difference in initial condition can result in a huge difference

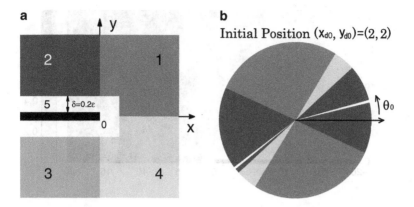

FIG. 2. *(a) The x-y plane divided into five regions. The thick black line corresponds to the wall. (b) An example of a pie chart. The different shadings (colors online) as a function of initial orientation θ_0 indicate the region of (a) where the treadmilling organism ends up after a sufficiently large time*

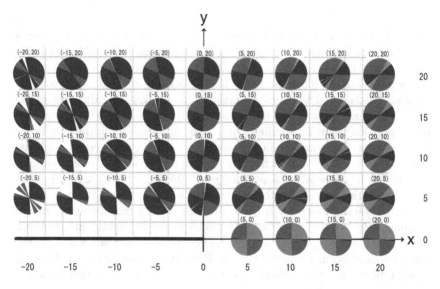

FIG. 3. *Pie charts at $t = 1{,}500$ for several initial conditions. Each pie chart is centered on the initial condition it corresponds to. The thick black line represents the wall. To help distinguish the shadings [colors online], note that region 3 only appears as very thin slivers in the pie charts along the x axis*

to its trajectory. (This is an example of chaotic scattering.) However, pie charts tend to become rather simple for larger y_{d0} for any fixed x_{d0}, since the treadmiller is then less influenced by the wall.

Now let us consider the *escape probability*, $P_E(x_{d0}, y_{d0})$, the probability that the treadmilling organism can escape from the wall region. We define

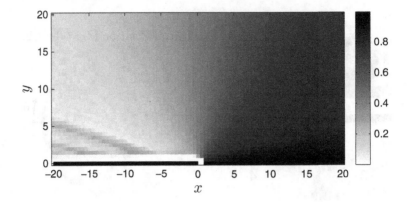

FIG. 4. *Escape probability P_E at $t = 1,500$ for different initial points. Each point corresponds to a given initial condition (x_{d0}, y_{d0}), integrated over all possible starting angles θ_0. A trajectory escapes if it ends up in regions 1 or 4, as defined in Fig. 2. The excluded white area near the wall corresponds to initial conditions where the swimmer would be partially inside the wall*

this as the probability that the treadmilling organism be in region 1 or 4 at sufficiently large time. We assume that the initial angle θ_0 is uniformly distributed. The escape probability $P_E(x_{d0}, y_{d0})$ then corresponds to the fraction of angles in a pie chart that lead to region 1 or 4, for a given initial position (x_{d0}, y_{d0}). The escape probability for $t = 1,500$ for different initial conditions is shown in Fig. 4, where each (x, y) coordinate corresponds to an initial position of the treadmiller. Note that for almost all initial conditions there is some finite probability of escaping or being trapped, though the escaping probability approaches unity along the positive x axis, and goes to zero for large negative x and large y. Observe the strong 'wedge' of trajectories that are likely to escape ($P_E > 0.6$), though there is also a backwards-facing wedge of trajectories that have a reasonable change of escaping ($P_E > 0.3$). This suggests that a probe or pipette inserted in a medium is likely to 'push away' treadmillers to some degree. Note also the 'tongues' of abnormally-high ($P_E \simeq 0.4$) escape probability near the wall, which reflects the complicated structure of the pie charts in that region.

4. Discussion and Conclusions. Following [4, 5], we derived the governing evolution equations for a treadmilling microorganism near a semi-infinite no-slip wall. This can also be regarded as a two-dimensional model of a probe or pipette inserted in the fluid. We then numerically calculated the trajectories of the organism for different initial conditions. The treadmilling organism was usually attracted to the wall for early times, but for later times often escaped the wall region. Typically this happened when the organism came close to the tip of the wall. To investigate this behavior further, we looked at 'pie chart' diagrams, where the sectors indicate which region a swimmer eventually ends up in as a function of its initial angular orientation. Initial points with larger negative x coordinates have more complex pie charts compared to those with smaller negative x coordinates, because they are sensitive to 'chaotic scattering' off the tip of the wall.

We then examined the escape probability P_E, the probability that the treadmilling organism can escape to the right of the wall region, assuming that it is initially randomly oriented. There is an evident 'wedge' of initial conditions that is likely to escape the wall. This suggests that a probe or pipette inserted could 'push' the organisms out of the way (with the wedge replaced by a cone in three dimensions), though since the organisms slow down as they get further away from the wall the effect might not be very pronounced.

Acknowledgments. The authors are grateful for the hospitality of the Geophysical Fluid Dynamics Program at the Woods Hole Oceanographic Institution (supported by NSF), and thank Matthew D. Finn for his helpful advice and suggestions. Some of the numerical calculations for this project were performed at the Institute for Information Management and Communication of Kyoto University.

REFERENCES

[1] Avron JE, Kenneth O, Oakmin DH (2005) Pushmepullyou: an efficient microswimmer. New J Phys 7:234

[2] Berke AP, Turner L, Berg HC, Lauga E (2008) Hydrodynamic attraction of swimming microorganisms by surfaces. Phys Rev E 101:038102

[3] Cosson J, Huitorel P, Gagnon C (2003) How spermatozoa come to be confined to surfaces. Cell Motil Cytoskel 54:56–63

[4] Crowdy DG, Or Y (2010) Two-dimensional point singularity model of a low-Reynolds-number swimmer near a wall. Phys Rev E 81:036313

[5] Crowdy DG, Samson O (2011) Hydrodynamic bound states of a low-Reynolds-number swimmer near a gap in a wall. J Fluid Mech 667:309–335

[6] Drescher K, Leptos K, Tuval I, Ishikawa T, Pedley TJ, Goldstein RE (2009) Dancing volvox: hydrodynamic bound states of swimming algae. Phys Rev Lett 102:168101

[7] Hernandez-Ortiz JP, Dtolz CG, Graham MD (2005) Transport and collective dynamics in suspensions of confined swimming particles. Phys Rev Lett 95:204501

[8] Lauga E, DiLuzio WR, Whitesides GM, Stone HA (2006) Swimming in circles: motion of bacteria near solid boundaries. Biophys J 90:400–412

[9] Lauga E, Powers TR (2009) The hydrodynamics of swimming micro-organisms. Rep Prog Phys 72:096601

[10] Leshansky AM, Kenneth O, Gat O, Avron JE (2007) A frictionless microswimmer. New J Phys 9:145

[11] Najafi A, Golestanian R (2004) Simple swimmer at low Reynolds number: three linked spheres. Phys Rev E 69:062901

[12] Obuse K (2010) Trajectories of a low Reynolds number treadmilling organism near a half-infinite no-slip wall. In: Proceedings of the 2010 summer program in geophysical fluid dynamics, Woods Hole Oceanographic Institute, Woods Hole

[13] Or Y, Murray RM (2009) Dynamics and stability of a class of low Reynolds number swimmers near a wall. Phys Rev E 79:045302

[14] Pedley TJ, Kessler JO (1992) Hydrodynamic phenomena in suspensions of swim-
       ming microorganisms. Annu Rev Fluid Mech 24:313–358
[15] Ramia M, Tullock DL, Phan-Thien N (1993) The role of hydrodynamics interaction
       in the locomotion of microorganism. Biophys J 65:755–778
[16] Rothschild AJ (1963) Non-random distribution of bull spermatozoa in a drop of
       sperm suspension. Nature 198:1221–1222
[17] Shapere A, Wilczek F (1989) Geometry of self-propulsion at low Reynolds numbers.
       J Fluid Mech 198:557–585
[18] Winet H, Bernstein GS, Head J (1984) Observations on the response of human
       spermatozoa to gravity, boundaries and fluid shear. J Reprod Fert 70:511–523
[19] Zhang S, Or Y, Murray RM (2010) Experimental demonstration of the dynam-
       ics and stability of a low Reynolds number swimmer near a plane wall. In:
       Proceedings of american control conference, Baltimore, pp 4205–4210

CILIA INDUCED BENDING OF *PARAMECIUM* IN MICROCHANNELS

SAIKAT JANA[†], JUNIL KIM[‡], SUNG YANG[§], AND SUNGHWAN JUNG(✉)[†]

Abstract. Most living organisms in nature have a preferential gait and direction along which they locomote, presumably derived from the evolutionary/mechanical advantage provided by the gaits. However under the influence of constrained geometries, organisms often exhibit peculiar locomotory characteristics. A *Paramecium* in its natural state preferentially swims in a helical path in the anterior direction. When introduced into channels with dimensions smaller than its length, a posterior swimming *Paramecium* bends its flexible body, executes a flip, and swims in the anterior direction again. We study the deformation of the body shape caused by forces generated by beating cilia, which are assumed to be acting at the tip of the organism. This method may lead to a non-invasive method of measuring the forces exerted during bending by self propelling organisms having high aspect ratio.

Key words. Low Reynolds numbers, swimming, cell bending

AMS(MOS) subject classifications. Primary 1234, 5678, 9101112

1. Introduction. Microorganisms move using propulsive organelles, which presumably provide them with efficient strategies for locomotion in their natural habitat. Often the motion of microorganisms is envisioned in unbounded fluids, a perspective which has led to an improved understanding of the basic principles of propulsion at small scales [11]. However, the natural habitat of microorganisms generally overlaps with many other organisms and substrates which can significantly modify their motion. Investigations of hydrodynamic coupling due to boundaries have revealed a variety of peculiar characteristics: *E. Coli* swimming in circles near boundaries [12] or on a particular side of the channel [2], and *Spermatozoa* forming vortex patterns [14]. All of these examples reveal a wide range of interactive forces and external cues in the extracellular environment that may influence motion at micro-scales.

Cells swimming in micro-channels can be viewed as soft-bodied organisms which may change their shape or locomotory behavior in response to various cues arising from the immediate environment in which they are swimming/crawling/locomoting. Morphological changes in the shape of microorganisms or other flexible bodies due to constricted geometries are an active area of interest [13]. For example, red blood cells in channels actively

[†]Department of Engineering Science and Mechanics, Virginia Tech, Blacksburg, VA 24061, USA. sunnyjsh@vt.edu

[‡]Department of Medical System Engineering, Gwangju Institute of Science and Technology, 123 Cheomdan-gwagiro (Oryong-dong), Buk-gu, Gwangju, 500-712, Republic of Korea

[§]School of Mechatronics, Department of Medical System Engineering, Department of Nanobio Materials and Electronics, Gwangju Institute of Science and Technology, 123 Cheomdan-gwagiro (Oryong-dong), Buk-gu, Gwangju, 500-712, Republic of Korea

S. Childress et al. (eds.), *Natural Locomotion in Fluids and on Surfaces*, IMA 155, DOI 10.1007/978-1-4614-3997-4_16,
© Springer Science+Business Media New York 2012

deform their shape in response to varying shear rates [1]. Under varying stress conditions cells have exhibited variable mechanical properties, transitioning from soft cells to glassy polymers [16]. Numerous investigations into the elasticity of cells have been undertaken, but unfortunately all of them involve subjecting the cell to invasive forces [7]. For example, atomic force microscopy (AFM) can be used to measure cell elasticity using the contact or semi-contact mode which can cause damage to the cell [10]. Similarly, optical tweezers have been used to measure the elasticity of cells by subjecting them to invasive forces [15, 21].

In addition to cell elasticity, propulsive appendages of cells are of interest since they provide the force required to propel the organism through viscous environments at small scales. Investigations into the force generated by organelles like cilia using AFM coupled with opto-electric system have revealed a broad range of values from $40 \sim 160\,$pN in frog epithelial ciliary hairs [8] to $500 \sim 3{,}000\,$pN for *Mytilus edulis* [17].

Most animals have a natural proclivity to locomote in a particular direction. For example humans walk and run using two feet in the forward direction under the normal conditions or even when they are shepherded in most crowded conditions [5]; birds fly by flapping their wings and aligning their body to maximize aerodynamic efficiency [18, 20]. Similarly in the micro-world most microorganisms have a preferential direction of swimming and a particular gait that they use for moving in their natural habitat. Under threat from predators, *Dinoflagellates* exhibit variable pitch and radius while swimming in a complex environment, but do not change their natural direction [19].

In this article, we investigate the bending of a *Paramecium* when introduced into a channel having lateral dimension smaller than its length. A posterior swimming *Paramecium* executes a somersault within the channel and reverses its swimming direction in order to navigate the channel along its preferred anterior direction. Actively beating cilia generate sufficient force to allow the organism to hinge about a point and bend the cell in the small channel. Our method exploits the preferential swimming direction of microorganisms under static fluid conditions. In Sect. 2, we describe the experimental techniques associated with fabricating the required microfluidic devices, introducing the organisms, and recording the motion of the bent body. In Sect. 3, beam bending theory is employed to rationalize the changing body shapes driven by the varying force exerted at the tip of the beam by the ciliary hairs. We compare the experimental body shapes with the ones predicted by theory. Finally, we discuss challenges and future directions.

2. Experiments.

2.1. Materials and Methods. The initial culture of *Paramecium multimicronucleatum* was obtained from Carolina Biological Supply and placed in double wheat medium at a temperature of 22°C. Subcultures were placed every 11 days when they reached their peak population. To increase

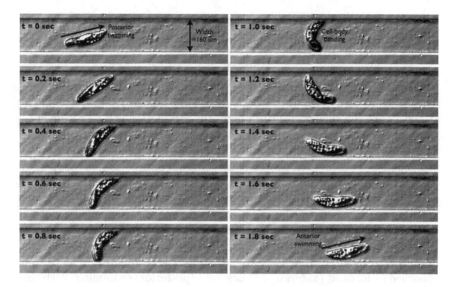

FIG. 1. *Paramecium swimming and tumbling in a PDMS channel; observed at 5X using DIC optics*

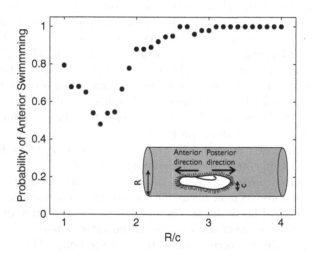

FIG. 2. *Plot of probability of anterior swimming in Paramecia versus non dimensional channel width in glass capillaries*

the cell density and remove debris, the cultures were centrifuged and washed twice in a Buffer solution consisting of 9 mM $CaCl_2$, 3 mM KCl and 5 mM $Tris - HCl$ (pH 7.2). The clean cells were allowed to equilibrate for 2 h and then observed under the Leica DMI 3000 microscope at 5X, 20X or 40X with DIC optics and their motion was recorded using a

IDT MotionXtra N3 camera. The cells in the buffer solution under normal experimental conditions were measured to be between $214\pm12\,\mu$m in length and $57\pm5\,\mu$m in diameter.

Cultures of the concentrated cells were pushed into $160\,\mu$m-wide rectangular polydimethylsiloxane (PDMS) channels. The microfluidic channel was fabricated using a conventional PDMS replica molding technique [4, 22]. Since the channel is $60\,\mu$m high, the cell motion can be assumed to be confined in a quasi-two-dimensional space. The pressure at both ends was allowed to equilibrate for $10\sim15$ min so that there was no flow in the channel. This allowed us to neglect external flow effects and helped in investigating the true swimming characteristics.

Ciliary coordination is often controlled by a complex collection of external cues that causes the organism to change the frequency or other parameters of wave propagation [6]. However the difference in direction of the propagating metachronal wave and swimming direction causes the organism to move in a helical path. Broadly the locomotory gait can be classified as forward (anterior) or backward (posterior) swimming; with the forward swimming exhibiting different helical modes [3].

2.2. Experimental Observation. Under these experimental conditions, most cells swam along their anterior direction from one end of the reservoir to the other end. This is the natural gait of the organism and is commonly observed in large capillary tubes. The swimming *Paramecia* traced sinusoidal trajectories with velocities varying from 557 to 943 µm/s. The wavelength of the sinusoidal path was roughly $350\,\mu$m. These trajectories have been reported previously and are commonly observed in swimming *Paramecia* [3]. From our data of swimming in a glass capillary we see that the maximal tendency of *Paramecia* to swim in the posterior direction occurs for $R/c\sim1.6$ as shown in Fig. 1.

In certain cases, as shown in Fig. 2, a *Paramecium* swimming in the posterior direction enters the channel and presumably struggles to swim in its non-preferred direction. When the channel width is smaller than the length of the *Paramecium*, it cannot execute the somersault while maintaining its original body shape. Instead, the posterior portion of the body hinges at one of the walls and the cilia of the anterior portion beat in synchrony to provide the force required to bend the body. The *Paramecium* then successively deforms its shape in time, eventually switching to anterior swimming direction. The novelty of this investigation lies in the fact that the cell body deforms its shape on its own and hence we can estimate the forces exerted by the cell on the substrate; without inducing any external stimuli or forces which may change the behavior of the cell significantly.

3. Model and Results. The bending of soft-bodied organisms like *Paramecia* can provide insight into the flexibility of organisms in their natural swimming state. Since the aspect ratio of the organism is $3.5\sim4$, these organisms can be modeled as cylindrical bars with a force acting

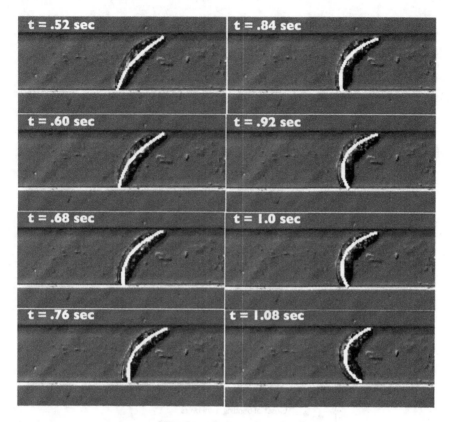

FIG. 3. *Comparison between experiments and shape predicted from our theoretical model (white line)*

on the tip. The approximation of a bending beam is compared with the experimental observations of a *Paramecium* that bends its body in the channel by generating forces using the anterior cilia.

The bending energy of a curved beam can be approximated as $\mathcal{E}_B = \int \frac{EI}{2} \kappa^2 ds$ where E is the Young's modulus, κ the curvature, I the area moment of inertia, and s the arc length along the cell's centerline. This bending energy is minimized under the constraint that the channel has a finite width. Corresponding to the coordinate system introduced in the Fig. 4, the channel width can be expressed as the integral of local angles, $W = \int_0^L \cos\theta(s) ds$ where θ denotes the angle between the centerline and the x-axis. Using the Lagrange multiplier λ, the new free energy becomes,

$$\mathcal{E} = \mathcal{E}_B + \lambda W = \int_0^L \frac{1}{2} \left[EI \left(\frac{\partial\theta}{\partial s} \right)^2 + \lambda \cos\theta \right] ds \qquad (1)$$

which can be minimized subject to a fixed channel width. To find the equilibrium profile of a cell's deformation, the variation of total free energy becomes:

$$\delta\mathcal{E} = -\int_0^L \left[\frac{\partial}{\partial s} \left(EI\frac{\partial\theta}{\partial s} \right) + \lambda\sin\theta \right] \delta\theta ds. \tag{2}$$

Finally, this minimization of the free energy leads to the equation describing the static shape of the cell in the constrained channel as:

$$\frac{d}{ds} \left(EI\frac{d\theta}{ds} \right) = -\lambda\sin\theta \tag{3}$$

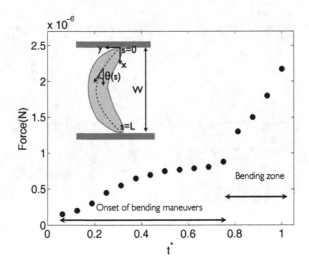

FIG. 4. *Top left corner shows the schematic of our theoretical model. The plot show the force exerted by the ends of Paramecium on the substrate at each different time-steps starting from the onset of bending. At initial time steps it is observed that the force exerted is very small but increases rapidly as the organism bends into a crescent shape. Also* $t^* = t/T$ *where* t^* *denotes the non-dimensional time as a ratio of time interval at each bending step(t) to the total time required to sense the wall and bend into the crescent shape(T)*

where λ represents the force applied at both ends normal to the surface and is estimated from the force generated by the number of cilia covering the surface at point of contact of the of cell body. In this context we would like to draw attention to the physical world and the fact that at higher viscosities the metachronal waves breaks down over longer distances but they exist for only a small distance. Since the tip is very close to wall it experiences viscous forces; and metachronal waves exist for small distance. We assume that this wave generates the force required to bend the cell body. Also there are metachronal waves at the sides of the body but they are disconnected from the waves at the tip. Knowing the cilia density all over the body of *Paramecium* one can do a conservative estimate of the number of cilia at the tip and hence the force. Owing to the slender shape, we model this organism as cylinder with a constant radius equal to the mean width of the cell. We consider the simple case of small angles $\theta \ll 1$ and a constant area moment of inertia $I_0 = \frac{\pi R^4}{4}$. This gives us a linear ODE of the form:

$$\frac{d^2\theta}{ds^2} = -\frac{\lambda}{EI_0}\theta \tag{4}$$

for which the solution can be written as $\theta(s) = \theta_0 \sin\left(-\frac{(s-L/2)}{L_{el}}\right)$ where $L_{el} = \sqrt{\frac{EI_0}{\lambda}}$ is the characteristic bending length. If the body length L is larger than L_{el} it represents the bent state of the *Paramecium* however for the opposite case we have the organism touching the walls of the channel without bending. The nominal moment of inertia obtained from the mean width of the organism is $I_0 = \pi R_0^4/4$ where $R_0 = 27.5\,\mu$m from our images. The other parameters $E \sim 10^5\,\text{N/m}^2$ and $\lambda \sim 10^{-6}\,\text{N}$ are estimated using values available in the literature [9, 10]. These values can then be used to evaluate equation (4).

As shown in Fig. 2, *Paramecium* usually hinges its body on the upper wall, and hence we assume zero bending moment at this end. This leads to the boundary condition, $\frac{d\theta}{ds}\big|_{s=0} = 0$. Also, the length of the *Paramecium* within the channel should be constant which is a physical constraint obtained from the shape of the organism. We use Runge Kutta Method in MATLAB to solve the beam equation and impose the condition of constant length on the *Paramecium*.

In order to get the bending states as shown in Fig. 3, different tip forces λ is assumed to be generated by the cilia at each time step, t. Total time T corresponds to the time required for the organism to bend from the straight line shape while touching both walls to the half crescent shape. The variation of the tip forces is reflected in the bending length ($L_{el} = \sqrt{EI_0/\lambda}$), which is chosen to match the experimental body shape at different time step. Figure 4 shows the variation of forces that is exerted by the cell on the substrate when it is bending. Initially we see that the forces are small which represents the time steps during which the organism is sensing the wall. However at around $t^* \sim 0.7$ we observe a spike in the force which corresponds to the state when the organism is bending its body and changing its swimming direction.

4. Conclusion. We have experimentally studied bending of *Paramecium* induced by the preferred swimming and theoretically shown that the organism's shape is consistent with the shapes of elastic bending rods in a confined channel. In this study only a single PDMS channel with a width of $160\,\mu$m was used, which caused similar bending patterns. In future experiments, by varying the channel width, we can determine the maximum bending and curvature that a freely swimming body can generate. This method could help us to determine the cell elasticity using the preferential gait direction. The cell body in our experiments is mostly hinged about the posterior end, and the anterior portion moves to cause the deformation. Also, during the bending of the body, certain cilia beat with a higher frequency as compared to others; visualizing the motion of cilia over the cell surface would help us in predicting the weakest point about which the bending occurs.

REFERENCES

[1] Abkarian M and Viallat A (2008) Vesicles and red blood cells in shear flow. Soft Matter 4(4):653–657

[2] DiLuzio WR et al (2005) Escherichia coli swim on the right-hand side. Nature 435(7046):1271–1274

[3] Dryl S, Grebecki A (1966) Progress in the study of excitation and response in ciliates. Protoplasma 62(2):255–284

[4] Duffy D et al (1998) Rapid prototyping of microfluidic systems in poly (dimethylsiloxane). Anal Chem 70(23): 4974–4984

[5] Dyer J et al (2008) Consensus decision making in human crowds. Anim Behav 75(2):461–470

[6] Gheber L, Korngreen A, Priel Z (1998) Effect of viscosity on metachrony in mucus propelling cilia. Cell Motil Cytoskeleton 39(1):9–20

[7] Guck J et al (2000) Optical deformability of soft biological dielectrics. Phys Rev Lett 84(23):5451

[8] Hill DB et al (2010) Force generation and dynamics of individual cilia under external loading. Biophys J 98(1):57–66

[9] Janmey P, McCulloch C (2007) Cell mechanics: integrating cell responses to mechanical stimuli. Annu Rev Biomed Eng 9:1–34

[10] Kuznetsova TG et al (2007) Atomic force microscopy probing of cell elasticity. Micron 38(8):824–833

[11] Lauga E, Powers TR (2009) The hydrodynamics of swimming microorganisms. Rep Prog Phys 72(9):096601

[12] Lauga E et al (2006) Swimming in circles: motion of Bacteria near solid boundaries. Biophys J 90(2):400–412

[13] Mannik J et al (2009) Bacterial growth and motility in sub-micron constrictions. Proc Natl Acad Sci 106(35):14861–14866

[14] Riedel I, Kruse K, Howard J (2005) A self-organized vortex array of hydrodynamically entrained sperm cells. Science 309:300

[15] Sleep J et al (1999) Elasticity of the red cell membrane and its relation to hemolytic disorders: an optical tweezers study. Biophys J 77(6):3085–3095

[16] Stamenovic D (2006) Two regimes, maybe three. Nat Mater 5:5978

[17] Teff Z, Priel Z, Gheber LA (2007) Forces applied by cilia measured on explants from mucociliary tissue. Biophys J 92(5):1813–1823

[18] Taylor G, Nudds R, Thomas A (2003) Flying and swimming animals cruise at a strouhal number tuned for high power efficiency. Nature 425:707–11

[19] Sheng J et al (2007) Digital holographic microscopy reveals prey-induced changes in swimming behavior of predatory dinoflagellates. Proc Natl Acad Sci 104(44):17512

[20] Spear L, Ainley D (1997) Flight behaviour of seabirds in relation to wind direction and wing morphology. Ibis 139(2):221–233

[21] Zhang H and Liu K (2008) Optical tweezers for single cells. J R Soc Interface 5(24):671
[22] Zhao X, Xia Y, Whitesides G (1997) Soft lithographic methods for nano-fabrication. J Mater Chem 7(7):1069–1074

RHEOLOGY OF SHEARED BACTERIAL SUSPENSIONS

ZHENLU CUI(✉)* AND XIAO-MING ZENG†

Abstract. A continuum hydrodynamic model for flowing active liquid crystals has been used to characterize active particle systems such as bacterial suspensions. The behavior of such systems subjected to a weak steady shear is analyzed. We explore the steady states and perform their stability analysis. We predict the rheology of active systems including an activity thickening or thinning behavior of the apparent viscosity and a negative apparent viscosity depending on the particle type, flow alignment and the anchoring conditions, which can be tested on bacterial suspensions. We find remarkable dualities which show that flow-aligning rodlike pullers (pushers) particles are dynamically and rheologically equivalent to flow-aligning discoid pushers (pullers) particles for both tangential and homeotropic anchoring conditions.

Key words. Bacterial suspensions, active liquid crystals, stability, rheology

AMS(MOS) subject classifications. 76A15, 82D30, 92C05.

1. Introduction. Bacterial suspensions and living cells are examples of active particle systems composed of interacting units that absorb energy and generate motion. Due to their anisotropic shapes, active particles can exhibit orientational order and form nematic phases, characterized by a macroscopic axis of mean orientation identified by a unit vector **n** and global symmetry for **n** → −**n**, likened to "living liquid crystals" or active liquid crystals [4]. Properties of these active systems are of fundamental interest to potential technological applications [8]. They are also interesting from a more fundamental perspective as their dynamic phenomenon such as bioconvection [3] and the spontaneous flow [4] are both physically fascinating and potentially of great biological significance.

According to the forces they exert on the surrounding fluid, active particles can be classified as pushers and pullers. A pusher such as most bacteria (e.g. *E. Coli*) is a swimming particle whose motion is actuated along the posterior of the body, while for a puller such as Chlamydomonas Reinhardtii, motion is actuated along the anterior. These two actuation modes result in oppositely signed active stress contributions to the fluid. We note that this distinction is by no means exclusive, as certain highly symmetric organisms such as spherical multicellular algae (e.g., Volvox) may fall between this pusher-puller distinction.

---

*Department of Mathematics and Computer Science; The Center for Defense and Homeland Security, Fayetteville State University, Fayetteville, NC 28301, USA, zcui@uncfsu.edu. The work is supported in part by Fayetteville State University Faculty Research Grants.

†School of Mathematical Sciences, Xiamen University, Xiamen 361005, China. The work is supported in part by the National Science Foundation of China (Grant No. 61170324).

S. Childress et al. (eds.), *Natural Locomotion in Fluids and on Surfaces*, IMA 155, DOI 10.1007/978-1-4614-3997-4_17, © Springer Science+Business Media New York 2012

Theoretical efforts [2, 4, 6, 7, 10] have been made in understanding the hydrodynamics and rheology of active particle suspensions. Hatwalne et al. [7] first generalized the de Gennes theory [5] for liquid crystals to model the rheology of active suspensions and pointed out that activity lowers the viscosity of a pusher system, while it enhances the viscosity of a puller system. Giomi et al. [6] and Saintillan [10] studied the rheological behavior of an active system under shear and confirmed previous predictions of Hatwalne et al. [7]. The author [2] extended these studies to a larger parameter spaces. Edwards et al. [4] used Leslie-Ericken-Parodi continuum theory to model bacterial suspensions. They numerically explored the range of possible steady states and distinguished six spontaneous flow states and described their response to an applied shear. However, they did not consider the rheology and the effects of the boundary conditions. In this work, we will focus on these issues. The paper is organized as follows. The model is presented in Sect. 2. Section 3 investigates the sheared steady states, conducts their stability analysis and predict steady rheology. Section 4 gives conclusions.

2. Model Formulation. We adopt the Erickson-Leslie-Parodi model [4], in which the nematic order parameter is a fixed-magnitude unit vector field $\tilde{n}$ which evolves according to

$$\frac{\partial \tilde{n}}{\partial t} + \tilde{\mathbf{v}} \cdot \nabla \tilde{n} = \lambda \mathbf{D} \cdot \tilde{n} - \mathbf{\Omega} \cdot \tilde{n} + \Gamma \tilde{h} \tag{1}$$

where $\tilde{\mathbf{v}}$ is the velocity field of the solvent; λ is the flow alignment parameter, Γ is a rotational viscosity; $\tilde{h} = K \Delta \tilde{n}$ is the molecular field and K is the Frank elasticity constant (single constant approximation), $\Omega = \frac{1}{2}(\nabla \mathbf{v} - \nabla \mathbf{v}^T)$ and $\mathbf{D} = \frac{1}{2}(\nabla \mathbf{v} + \nabla \mathbf{v}^T)$ are the rate-of-vorticity and the rate-of-strain tensors, respectively

The first two terms on the right-hand side of (1) describe alignment (or tumbling) of the director field by local shear flow. The third term accounts for the tendency of the ordered nematic to resist distortions, and arises ultimately from excluded volume interactions between individual particles. The flow field $\tilde{\mathbf{v}}$ obeys the Navier-Stokes equation

$$\rho(\frac{\partial \tilde{\mathbf{v}}}{\partial t} + \tilde{\mathbf{v}} \cdot \nabla \tilde{\mathbf{v}}) = \nabla \cdot (\vec{\tau} + 2\eta \mathbf{D}) \tag{2}$$

with the continuity equation $\nabla \cdot \tilde{\mathbf{v}} = 0$ to guarantee incompressibility, the stress tensor given by the passive and active contributions, $\tau = \tau^p + \tau^a$.

$$\tau^p = -p\mathbf{I} - \frac{\lambda}{2}[\tilde{n}\tilde{h} + (\tilde{n}\tilde{h})^T] + \frac{1}{2}[\tilde{n}\tilde{h} - (\tilde{n}\tilde{h})^T] \tag{3}$$

where ρ is the fluid density and η is the viscosity, p is the pressure and λ is the flow-alignment parameter. The magnitude of λ controls how the director field responds to a shear flow. $|\lambda| > 1$ corresponds to flow aligning regime in which the director tends to align to the flow direction at the Leslie alignment angle $\theta_L = \frac{1}{2}\cos^{-1}\frac{1}{\lambda}$ while $|\lambda| < 1$ corresponds to flow tumbling regime in which the director continuously rotates under shear.

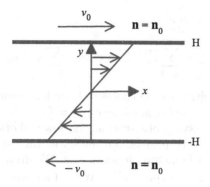

FIG. 1. *Plane shear flow geometry. Nonslip boundary conditions for the velocity and boundary anchoring for the orientation tensor is assumed to equal to its quiescent nematic equilibrium value*

The value of λ is mainly determined by the shape of the active particles. $\lambda > 0$ corresponds to a rod-shaped particle, $\lambda < 0$ for a disc-shaped particle and $\lambda = 0$ for a spherical particle.

The active contribution can be expressed as [4, 6],

$$\tau^a = \delta\tilde{\mathbf{n}}\tilde{\mathbf{n}} \tag{4}$$

The sign of δ determines whether the particles are pushers ($\delta < 0$) or pullers ($\delta > 0$).

3. Weak Steady Shear Flows. We consider shear flow between two parallel plates located at $y = \pm H$ and moving with corresponding velocity

$$\mathbf{v} = (\pm v_0, 0, 0) \ . \tag{5}$$

We assume strong particle anchoring at the plates given by

$$\mathbf{n_0} = (\cos\varphi_0, \sin\varphi_0, 0), \tag{6}$$

where $\mathbf{n_0}$ is the initial director across the plates in absence of shear flow. Figure 1 depicts the cross section of the shear flow on the (x,y) plane. Variations in the direction of flow (x) and primary vorticity direction (z), and transport in the vertical (y) direction are suppressed.

We seek asymptotic solutions of the governing system of equations with the boundary conditions given by (5) and (6). We consider the in-plane orientation of active suspensions confined to the shearing plane (x, y).

The director $\mathbf{n}$ confined to the shearing plane (x, y) and parameterized by the director angle φ,

$$\mathbf{n} = (\cos\varphi, \sin\varphi, 0). \tag{7}$$

We propose the solution ansatz

$$v_x = \sum_{k=1}^{\infty} v_x^{(k)} v_0^k, \varphi = \varphi_0 + \sum_{k=1}^{\infty} \varphi^{(k)} v_0^k \tag{8}$$

The solution is sensitive to the choice of boundary conditions, we restrict to tangential ($\varphi_0 = 0$) and homeotropic ($\varphi_0 = \frac{\pi}{2}$) anchoring conditions. At the first order of the asymptotic scheme, the flow and orientational director dynamics dominate. We present the results at tangential anchoring and comment on those at homeotropic anchoring. We drop the subscript on φ for brevity and use v_x to express $v_x^{(1)}$. At the first order $O(v_0)$, the system reduces to

$$\frac{\partial \varphi}{\partial t} = A \frac{\partial^2 \varphi}{\partial y^2} + B \frac{\partial v_x}{\partial y},$$

$$\rho \frac{\partial v_x}{\partial t} = \frac{\partial \tau_{xy}}{\partial y}, \tag{9}$$

$$\tau_{xy} = C \frac{\partial^2 \varphi}{\partial y^2} + D \frac{\partial v_x}{\partial y} + \delta \varphi,$$

where

$$A = \Gamma K, B = \frac{\lambda-1}{2}, C = -KB, D = \eta. \tag{10}$$

The steady state of the system (9) is given

$$v_x = \frac{\sinh ry}{\sinh rH}, \varphi = -\frac{B \coth rH}{Ar} \left(\frac{\cosh ry}{\cosh rH} - 1 \right) \tag{11}$$

where $r = \sqrt{\frac{(\lambda-1)\delta}{2(AD-BC)}}$. It is real (imaginary) if $(\lambda - 1)\delta > 0 \ (< 0)$. This implies a remarkable duality for the steady state structures: *A puller (pusher) system with $\lambda > 1$ is the same as a pusher (puller) system with $\lambda < 1$.* One will see that this duality also holds true for the linear rheology (see (16) below). This duality is also predicted in other theoretical models [2, 4, 6].

One can see (11) is hyperbolic (sinusoidal) if $(\lambda - 1)\delta > 0 \ (< 0)$. However at high activity, the story will be totally changed: when $(\lambda-1)\delta > 0$, the flow velocity is zero and the director angle is a constant away from the plates, which is akin to permeation in passive cholesteric liquid crystals [1].

The transient solution for v_x and φ (the difference between the time-dependent solution and the steady state) obeys the same homogeneous linear partial differential equations but satisfies a zero boundary condition. Its behavior dictates the stability of the steady state within the asymptotic balance model: the steady state is asymptotically stable if the transient solution vanishes as $t \to \infty$. By the energy method, we can prove the following theorem.

THEOREM 3.1. *The steady state is stable for a tangential anchoring if $(\lambda - 1)\delta > 0$. If $(\lambda - 1)\delta < 0$, then the steady state is stable for $|\delta| << 1$, i.e., in a low activity regime.*

Proof. We first note that $A > 0$, $BC < 0$ and $D = \eta > 0$. For simplicity, in the following proof, we use $\varphi_y, \varphi_{yy}, \varphi_{yyy}, v_{x,y}$ and $v_{x,yy}$ to express $\frac{d\varphi}{dy}, \frac{d^2\varphi}{dy^2}, \frac{d^3\varphi}{dy^3}, \frac{dv_x}{dy}$ and $\frac{d^2v_x}{dy^2}$, respectively.

Extending $(9)_1$ to the boundary and accounting for the boundary condition $\varphi(-H, t) = \varphi(H, t) = 0$, we have

$$(A\tfrac{\partial^2 \varphi}{\partial y^2} + B\tfrac{\partial v_x}{\partial y})|_{y=\pm 1} = 0. \tag{12}$$

We introduce a Liapunov function

$$I(t) = \int_{-H}^{H} [\gamma_1 \varphi_y^2 + \gamma_2 v_x^2 + \gamma_3 \varphi^2] dy \tag{13}$$

with $\gamma_1 > 0$, $\gamma_2 > 0$ and $\gamma_3 > 0$.

If $(\lambda - 1)\delta > 0$, then B and δ have same signs. We choose $\gamma_1 = |C|$, $\gamma_2 = |B|\rho$ and $\gamma_3 = |B|\frac{\delta}{B}$ and integrating by parts, the time derivative of the nonnegative functional can be estimated:

$$\frac{dI(t)}{dt} = -2\int_{-H}^{H} [\gamma_1 A\varphi_{yy}^2 + (\gamma_1 B + \gamma_2 C)\varphi_{yy}v_{x,y}$$

$$+ \gamma_2 D v_{x,y}^2 + \gamma_3 A\varphi_y^2 + (\gamma_3 B - \gamma_2 \delta)\varphi_y v_x] dy \tag{14}$$

$$= -2\int_{-H}^{H} [|C|A\varphi_{yy}^2 + |B|Dv_{x,y}^2 + A|B|\frac{\delta}{B}\varphi_y^2] dy < 0.$$

This shows that the steady solution of the system is stable.

If $(\lambda - 1)\delta < 0$, then B and δ have opposite signs. We choose $\gamma_1 = |C|$, $\gamma_2 = |B|\rho$ and $\gamma_3 = 0$, then

$$\frac{dI(t)}{dt} = -2\int_{-H}^{H} [|C|A\varphi_{yy}^2 + |B|Dv_{x,y}^2] dy + 2\delta|B| \int_{-H}^{H} \varphi_y v_x dy. \tag{15}$$

If $|\delta| << 1$, i.e., in a low activity regime, then the first integral of the right-hand side of (15) is dominant, and since it is negative, $\frac{dI(t)}{dt} < 0$. Hence the system is stable. The proof is complete.

□

The rheological property of interest is the apparent viscosity defined by $\eta_{app} = \frac{\tau_{xy}}{2V}$ [2] where $V = \int_0^H v_x(y) dy$ is the flow rate per unit length and $\tau_{xy} = \frac{B\delta \coth rH}{Ar}$ is the shear stress, which is a constant at this order across the shear cell. The resulting apparent viscosity is given by

$$\eta_{app} = \frac{B\delta \cosh rH}{2A(\cosh rH - 1)} = \frac{(\lambda-1)\delta \cosh rH}{4\Gamma K(\cosh rH - 1)} \tag{16}$$

Figure 2 shows the apparent viscosity versus the activity for flowing aligning and tumbling rodlike swimmers. The apparent viscosity of a flow-aligning rodlike puller system or a tumbling rodlike pusher system is thickened by the activity while the apparent viscosity of a flow-aligning rodlike pusher or a tumbling rodlike puller system is thinned by the low activity. A negative apparent viscosity is found for a tumbling rodlike puller system in a high activity regime. Common swimming bacteria, such as Bacillus Subtilis, *E. Coli*, and many others, rodlike swimmers (length about $\sim$5 μm, diameter of the order of $\sim$1 μm) are flow-aligning rodlike pushers, corresponding to $\lambda_L > 1$. The apparent viscosity of these systems is thinned by the activity for tangential anchoring. In contrast, Chlamydomonas Reinhardtii, spheroidal in shape or egg-shaped, are tumbling pullers, corresponding to $\mid \lambda_L \mid < 1$. The apparent viscosity of such systems is thinned for a tangential anchoring.

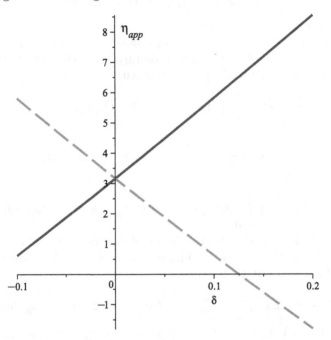

FIG. 2. *The apparent viscosity versus the activity for rodlike swimmers for tangential anchoring condition at selected values of $\lambda = 1.5$ (flow-aligning, solid line) and $\lambda = 0.5$ (tumbling, dashed line). The values of other parameters are taken from Ref. [4]: $\Gamma = 0.2, K = 0.04$ and $\eta = 1.27$. We also set $H = 1$*

The apparent viscosity for other systems can be obtained by the duality and we summarize in Fig. 3. These results are in agreement with the predictions by a kinetic model in [2].

4. Conclusions. We have derived the governing equations for flowing active suspensions based on Leslie-Ericken-Parodi continuum model. We establish the asymptotic formulas of the steady boundary-value problem subject to a steady weak shear and give stability analysis of the steady structures. We find remarkable dualities which show that pullers (pushers) with $\lambda > 1$ are dynamically and rheologically equivalent to pushers

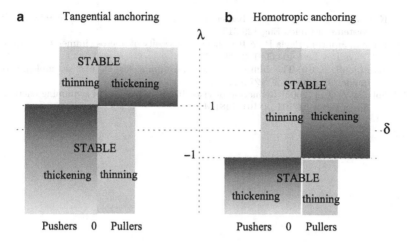

FIG. 3. *Phase diagram in the (δ, λ_L) plane. Steady state stable regions and apparent viscosity thinning/thickening regions. Steady spontaneous oscillatory states arises in unstable regions (unshaded)*

(pullers) with $\lambda < 1$ for a tangential anchoring while pullers (pushers) with $\lambda > -1$ are dynamically and rheologically equivalent to pushers (pullers) with $\lambda < -1$ for a homeotropic anchoring. These results demonstrate a remarkable richness in the steady-state hydrodynamic and rheological behaviors of nematic active suspensions to an external forcing. They are consistent with the previous results on rodlike particles [2, 4, 6, 9–11] and discoid particles [2, 4, 6], experimentally and theoretically. Clearly, more dedicated controlled experiments with active suspensions of different shapes and further generalizations of the obtained results to more general flow geometries are needed and we look forward to tests of these predictions in experiments on active systems.

REFERENCES

[1] Cui Z (2010) Small amplitude oscillatory shear permeation flow of cholesteric liquid crystal polymers. Commun Math Sci 8:943–963.

[2] Cui Z (2011) Weakly sheared active suspensions: hydrodynamics, stability and rheology. Phys Rev E 83:031911

[3] Dombrowski C et al (2004) Self-concentration and large-scale coherence in bacterial dynamics. Phys Rev Lett 93:098103–098106

[4] Edwards SA, Yeomans JM (2009) Spontaneous flow states in active nematics: a unified picture. Europhys Lett 85:18008

[5] de Gennes PG, Prost J (1993) The physics of liquid crystals. Oxford University Press, London

[6] Giomi L et al (2010) Sheared active fluids: thickening, thinning, and vanishing viscosity. Phys Rev E 81:051908

[7] Hatwalne Y et al (2004) Rheology of active-particle suspensions. Phys Rev Lett 92:118191–118194

[8] Kim MJ et al (2007) Use of bacterial carpets to enhance mixing in microfluidic systems. J Fluids Eng 129:319–324

[9] Rafaï S, Jibuti L, Peyla P (2010) Effective viscosity of microswimmer suspensions. Phys Rev Lett 104:098102–098105

[10] Saintillan D (2010) The dilute rheology of swimming suspensions: a simple kinetic model. Exp Mech 50:1275–1281

[11] Sokolov A et al (2009) Reduction of viscosity in suspension of swimming bacteria. Phys Rev Lett 103:148101–148104

A USER-FRIENDLY FORMULATION OF THE NEWTONIAN DYNAMICS FOR THE COUPLED WING-BODY SYSTEM IN INSECT FLIGHT

SHENG XU(✉)*

Abstract. We establish a user-friendly matrix formulation of the Newtonian dynamics for the free flight of an insect. In our formulation, we can easily change the number of insect wings, prescribe either the kinematics of each wing relative to the body or the torque on each wing exerted by the body, and allow for the dependence of fluid force and torque on acceleration. The implementation of the formulation is straightforward.

Key words. Matrix formulation, Newtonian dynamics, insect flight

AMS(MOS) subject classifications. Primary 70E55, 76Z99

1. Introduction. A flying insect is a multibody system consisting of a body and multiple wings. The physical connections between its body and wings are physical constrains in the system. To study the stability and maneuverability of free flight, we need to couple the Newtonian dynamics and aerodynamics of this multibody system. General formulations for writing general codes are available for the Newtonian dynamics of multibody systems [1]. Equations specific for the Newtonian dynamics for insect flight were also presented in [2, 3]. However, the implementation of these formulations and equations is not straightforward and can be very messy.

The goal of this paper is to establish a user-friendly formulation of the Newtonian dynamics for insect flight. In this formulation, we can easily change the number of insect wings, switch between the case of prescribed wing kinematics and the case of prescribed body-to-wing torque, and allow for dependence of fluid force and torque on acceleration. The formulation is neat and concise. The implementation is simple and straightforward. The basic idea to derive this formulation is to mathematically manipulate the dynamical equations and physical constraints of a wing-body system in suitable reference frames such that we can obtain the translational and angular accelerations of the body and the wings just by inverting an explicitly given matrix, which only depends on the shapes, positions and mass distributions of the insect body and wings.

2. Reference Frames. As shown in Fig. 1, we look at the positions and orientations of the body and wings of an insect in the following different reference frames

- $\vec{x}^l = (x^l, y^l, z^l)$: the static lab frame,

---

*Department of Mathematics, Southern Methodist University, Dallas, TX 75275–0156, USA. sxu@smu.edu. The work of the author is supported by NSF grant DMS-0915237.

S. Childress et al. (eds.), *Natural Locomotion in Fluids and on Surfaces*, IMA 155, DOI 10.1007/978-1-4614-3997-4_18,
© Springer Science+Business Media New York 2012

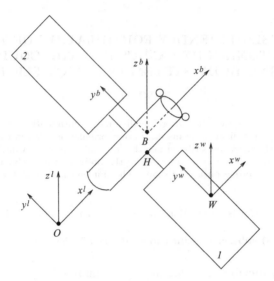

FIG. 1. *Multiple reference frames for an insect*

- $\vec{x}^b = (x^b, y^b, z^b)$: the moving body frame attached to the insect body,
- $\vec{x}^w = (x^w, y^w, z^w)$: a moving wing frame attached to each insect wing.

We set the origin B of the body frame to be the center of mass (CM) of the body and the x^b, y^b, and z^b axes as the principle axes of the moments of inertia I_B of the body. We set the origin W and the x^w, y^w, and z^w axes of each wing frame similarly. These settings help simplify dynamical equations for the body and each wing.

Hereafter, we use the lowercase superscripts l, b, and w to specify the lab, the body, and a wing frame, respectively, the capital subscripts B, W, H to specify the body, a wing, and a body-wing hinge, respectively, the superscript T to denote the transpose of a matrix, and one dot and two dots atop a symbol to denote first- and second-order time derivatives, respectively.

Let $\vec{x}$ be a position vector and $\vec{q}$ be a velocity, angular velocity, force, or torque vector. We can transform $\vec{x}$ and $\vec{q}$ between different reference frames as follows

$$\vec{x}^b = R_{lb}(\vec{x}^l - \vec{x}^l_B), \quad \vec{q}^b = R_{lb}\vec{q}^l, \tag{1}$$

$$\vec{x}^w = R_{bw}(\vec{x}^b - \vec{x}^b_W), \quad \vec{q}^w = R_{bw}\vec{q}^b, \tag{2}$$

$$\vec{x}^w = R_{lw}(\vec{x}^l - \vec{x}^l_W), \quad \vec{q}^w = R_{lw}\vec{q}^l, \tag{3}$$

where R_{lb}, R_{bw}, and R_{lw} are the lab-to-body, body-to-wing, and lab-to-wing transformation matrices, respectively. They are *orthogonal* matrices (see Appendix).

We denote the angular velocity of the body relative to the lab frame as $\vec{\Omega}_B$ and the angular velocity of a wing relative to the body as $\vec{\Pi}_W$. In the corresponding wing frame, the angular velocity of the wing relative to the lab frame is given by

$$\vec{\Omega}_W^w = \vec{\Omega}_B^w + \vec{\Pi}_W^w = R_{bw}\vec{\Omega}_B^b + \vec{\Pi}_W^w, \tag{4}$$

which can be differentiated with respect to time to obtain the following relation between angular accelerations

$$\dot{\vec{\Omega}}_W^w = -R_{bw}((R_{bw}^T\vec{\Pi}_W^w) \times \vec{\Omega}_B^b) + R_{bw}\dot{\vec{\Omega}}_B^b + \dot{\vec{\Pi}}_W^w. \tag{5}$$

This relation will be used to compose our final matrix formulation. The first term at the right hand side of this relation is derived from

$$R_{bw}R_{bw}^T = E \Rightarrow \dot{R}_{bw}R_{bw}^T = -R_{bw}\dot{R}_{bw}^T,$$

where E is the 3 by 3 identity matrix, and

$$\dot{R}_{bw}^T R_{bw}\vec{q}^b = \vec{\Pi}_W^b \times \vec{q}^b.$$

In specific, it is derived as follows

$$\dot{R}_{bw}\vec{\Omega}_B^b = \dot{R}_{bw}R_{bw}^T R_{bw}\vec{\Omega}_B^b = -R_{bw}\dot{R}_{bw}^T R_{bw}\vec{\Omega}_B^b$$
$$= -R_{bw}\vec{\Pi}_W^b \times \vec{\Omega}_B^b = -R_{bw}((R_{bw}^T\vec{\Pi}_W^w) \times \vec{\Omega}_B^b).$$

3. Body-Wing Constraints.

By Eqs. 1 and 3, we have

$$\vec{x}_H^l - \vec{x}_B^l = R_{lb}^T\vec{x}_H^b, \tag{6}$$
$$\vec{x}_H^l - \vec{x}_W^l = R_{lw}^T\vec{x}_H^w, \tag{7}$$

where $\vec{x}_H^b$ and $\vec{x}_H^w$, the coordinates of a body-wing hinge in the body and the corresponding wing frame, are *time-independent* if both the body and the wing are rigid. The following geometric body-wing constraint is obtained by subtracting the above two equations

$$\vec{x}_W^l - \vec{x}_B^l = R_{lb}^T\vec{x}_H^b - R_{lw}^T\vec{x}_H^w. \tag{8}$$

By differentiating the geometric constraint with respect to time twice, we can obtain a relation between the translational and angular accelerations of the body and the wing, which will be used to compose our final matrix formulation.

Differentiating Eq. 8 once, we obtain the kinematic constraint

$$\dot{\vec{x}}_W^l - \dot{\vec{x}}_B^l = R_{lb}^T(\vec{\Omega}_B^b \times \vec{x}_H^b) - R_{lw}^T(\vec{\Omega}_W^w \times \vec{x}_H^w), \tag{9}$$

where the first term at the right hand side is derived from (noticing that $\vec{x}_H^b$ is time-independent)

$$\dot{R}_{lb}^T\vec{x}_H^b = (\dot{R}_{lb}^T R_{lb})(R_{lb}^T\vec{x}_H^b) = \vec{\Omega}_B^l \times (R_{lb}^T\vec{x}_H^b)$$
$$= (R_{lb}^T\vec{\Omega}_B^b) \times (R_{lb}^T\vec{x}_H^b) = R_{lb}^T(\vec{\Omega}_B^b \times \vec{x}_H^b),$$

and the second term can be derived similarly. The following general formulas are used in the derivation:

$$\dot{R}_{lb}^T R_{lb} \vec{q}^l = \vec{\Omega}_B^l \times \vec{q}^l,$$
$$(R\vec{\beta}) \times (R\vec{\gamma}) = R(\vec{\beta} \times \vec{\gamma}),$$

where R is a 3×3 orthogonal matrix and $\vec{\beta}$ and $\vec{\gamma}$ are any column vectors.

After differentiating Eq. 9, we then obtain the following relation for the accelerations

$$\ddot{\vec{x}}_W^l - \ddot{\vec{x}}_B^l = \vec{d}_W^l - R_{lb}^T H_b \dot{\vec{\Omega}}_B^b + R_{lw}^T H_w \dot{\vec{\Omega}}_W^w, \qquad (10)$$

where $\vec{d}_W^l$ can be shown to be

$$\vec{d}_W^l = -R_{lb}^T(\|\vec{\Omega}_B^b\|_2^2 \vec{x}_H^b - (\vec{\Omega}_B^b \cdot \vec{x}_H^b)\vec{\Omega}_B^b)$$
$$+R_{lw}^T(\|\vec{\Omega}_W^w\|_2^2 \vec{x}_H^w - (\vec{\Omega}_W^w \cdot \vec{x}_H^w)\vec{\Omega}_W^w), \qquad (11)$$

and the matrices H_b and H_w are formed from $\vec{x}_H^b$ and $\vec{x}_H^w$, respectively, to convert a cross product to a matrix-vector product according to the general formula

$$\begin{pmatrix} \beta_1 \\ \beta_2 \\ \beta_3 \end{pmatrix} \times \vec{\gamma} = \begin{pmatrix} 0 & -\beta_3 & \beta_2 \\ \beta_3 & 0 & -\beta_1 \\ -\beta_2 & \beta_1 & 0 \end{pmatrix} \vec{\gamma}.$$

4. External Force and Torque. The fluid force and torque on the body in the lab frame are denoted as $\vec{f}_{fB}^l$ and $\vec{\tau}_{fB}^l$, respectively, and the fluid force and torque on a wing in the lab frame are denoted as $\vec{f}_{fW}^l$ and $\vec{\tau}_{fW}^l$, respectively. To be general, we allow fluid force and torque to explicitly depend on acceleration as follows

$$\vec{f}_{fB}^l = \vec{f}_{0B}^l + A_{fB} R_{lb}^T \dot{\vec{\Omega}}_B^b + m_{aB} \ddot{\vec{x}}_B^l, \qquad (12)$$

$$\vec{f}_{fW}^l = \vec{f}_{0W}^l + A_{fW} R_{lw}^T \dot{\vec{\Omega}}_W^w + m_{aW} \ddot{\vec{x}}_W^l, \qquad (13)$$

$$\vec{\tau}_{fB}^l = \vec{\tau}_{0B}^l + A_{\tau B} R_{lb}^T \dot{\vec{\Omega}}_B^b, \qquad (14)$$

$$\vec{\tau}_{fW}^l = \vec{\tau}_{0W}^l + A_{\tau W} R_{lw}^T \dot{\vec{\Omega}}_W^w, \qquad (15)$$

where angular acceleration in the lab frame is transformed to the corresponding body or wing frame, to be consistent with the choice of frames for rotation dynamics in Sect. 6.

In the lab frame, the gravitational force (due to weight and buoyancy) on the body and a wing are given by

$$\vec{f}_{gB}^l = (m_B - m_{fB})\vec{g}^l, \qquad (16)$$

$$\vec{f}_{gW}^l = (m_W - m_{fW})\vec{g}^l, \qquad (17)$$

where m_B and m_W are the mass of the body and the wing, respectively, m_{fB} and m_{fW} are the mass of the fluid displaced by the body and the wing, respectively, and $\vec{g}^l$ is the gravitational acceleration. The gravitational force on the whole insect is

$$\vec{f}_{gB}^l = \vec{f}_{gB}^l + \sum_W \vec{f}_{gW}^l, \tag{18}$$

where $\sum_W$ means summation over all wings.

5. Translational Dynamics. In the lab frame, the Newton's equation for the CM of the whole free-flying insect reads

$$m_B \ddot{\vec{x}}_B^l + \sum_W m_W \ddot{\vec{x}}_W^l = \vec{f}_{gS}^l + \vec{f}_{fB}^l + \sum_W \vec{f}_{fW}^l. \tag{19}$$

In the lab frame, the Newton's equation for the CM of a wing reads

$$m_W \ddot{\vec{x}}_W^l = \vec{f}_{gW}^l + \vec{f}_{fW}^l + \vec{f}_{bW}^l, \tag{20}$$

where $\vec{f}_{bW}^l$ is the force exerted by the body on the wing at the hinge. By Newton's third law, the force exerted by this wing on the body at the hinge is given by

$$\vec{f}_{wB}^l = -\vec{f}_{bW}^l = \vec{f}_{gW}^l + \vec{f}_{fW}^l - m_W \ddot{\vec{x}}_W^l. \tag{21}$$

6. Rotational Dynamics. In the body frame, the Euler's equation for the body reads

$$I_B \dot{\vec{\Omega}}_B^b + \vec{\Omega}_B^b \times (I_B \vec{\Omega}_B^b) = \vec{\tau}_{fB}^b + \sum_W \vec{\tau}_{wB}^b, \tag{22}$$

where $\vec{\tau}_{wB}$ is the torque on the body received from a body-wing hinge. Using Eq. 21, we have

$$\begin{aligned}
\vec{\tau}_{wB}^b &= \vec{c}_{wB}^b + \vec{x}_H^b \times \vec{f}_{wB}^b, \\
&= -\vec{c}_{bW}^b + H_b R_{lb}(\vec{f}_{gW}^l + \vec{f}_{fW}^l - m_W \ddot{\vec{x}}_W^l),
\end{aligned} \tag{23}$$

where $\vec{c}_{wB} = -\vec{c}_{bW}$ and $\vec{c}_{bW}$ is the muscle torque actively exerted by the body on this wing.

In the corresponding wing frame, the Euler's equation for a wing reads

$$I_W \dot{\vec{\Omega}}_W^w + \vec{\Omega}_W^w \times (I_W \vec{\Omega}_W^w) = \vec{\tau}_{fW}^w + \vec{\tau}_{bW}^w, \tag{24}$$

where $\vec{\tau}_{bW}$ is the torque on the wing received from the body-wing hinge. Using Eq. 21, we have

$$\begin{aligned}
\vec{\tau}_{bW}^w &= \vec{c}_{bW}^w + \vec{x}_H^w \times \vec{f}_{bW}^w \\
&= R_{bw}\vec{c}_{bW}^b + H_w R_{lw}(m_W \ddot{\vec{x}}_W^l - \vec{f}_{gW}^l - \vec{f}_{fW}^l).
\end{aligned} \tag{25}$$

7. Matrix Formulation. Finally, we combine the results in previous sections to establish the matrix formulation of the Newtonian dynamics for insect flight. To unify the case of prescribed wing kinematics and the case of prescribed body-to-wing torque in the formulation, we introduce

$$\vec{a}_W = \delta_{i2}\dot{\vec{\Pi}}^w_W + \delta_{i1}\vec{c}^b_{bW}, \tag{26}$$

$$\vec{b}_W = \delta_{i1}\dot{\vec{\Pi}}^w_W + \delta_{i2}\vec{c}^b_{bW}, \tag{27}$$

where δ_{ij} is the Kronecker delta. If $i = 1$, the kinematics of each wing relative to the body is prescribed ($\vec{b}_W = \dot{\vec{\Pi}}^w_W$ is known and $\vec{a}_W = \vec{c}^b_{bW}$ is unknown). If $i = 2$, the torque exerted by the body on each wing is prescribed ($\vec{b}_W = \vec{c}^b_{bW}$ is known and $\vec{a}_W = \dot{\vec{\Pi}}^w_W$ is unknown). In either case, $\vec{a}_W$ is unknown, and $\vec{b}_W$ is known. We then choose the following quantities as unknowns in our matrix formulation: $\ddot{\vec{x}}^l_B$ and $\dot{\vec{\Omega}}^b_B$ of the body, and $\ddot{\vec{x}}^l_W$, $\dot{\vec{\Omega}}^w_W$, $\vec{f}^l_{fW}$, and $\vec{a}_W$ of each wing.

After we substitute $\dot{\vec{\Pi}}^w_W$ and $\vec{c}^b_{bW}$ by

$$\dot{\vec{\Pi}}^w_W = \delta_{i2}\vec{a}_W + \delta_{i1}\vec{b}_W, \tag{28}$$

$$\vec{c}^b_{bW} = \delta_{i1}\vec{a}_W + \delta_{i2}\vec{b}_W, \tag{29}$$

and fluid force and torque by Eqs. 12, 14 and 15, we can put together Eqs. 5, 10, 13, 19, 22, and 24, to form the final matrix formulation. If there is only one wing, the matrix formulation reads

$$C_1 \mathbf{w}_1 = \mathbf{d}_1. \tag{30}$$

The coefficient matrix C_1 is

$$
C_1 =
\begin{pmatrix}
m^*_B E & -A_{fB}R^T_{lb} & m_W E & 0 & -E & 0 \\
0 & I^*_B & m_W H_b R_{lb} & 0 & -H_b R_{lb} & \delta_{i1} E \\
0 & 0 & -m_W H_w R_{lw} & I^*_W & H_w R_{lw} & -\delta_{i1} R_{bw} \\
-E & R^T_{lb} H_b & E & -R^T_{lw} H_w & 0 & 0 \\
0 & -R_{bw} & 0 & E & 0 & -\delta_{i2} E \\
0 & 0 & -m_{aW} E & -A_{fW} R^T_{lw} & E & 0
\end{pmatrix}
$$

where $m^*_B = m_B - m_{aB}$, $I^*_B = I_B - R_{lb}A_{\tau B}R^T_{lb}$ and $I^*_W = I_W - R_{lw}A_{\tau W}R^T_{lw}$. The vector of unknowns $\mathbf{w}_1$ is

$$\mathbf{w}_1 = [\ddot{\vec{x}}^l_B, \dot{\vec{\Omega}}^b_B, \ddot{\vec{x}}^l_W, \dot{\vec{\Omega}}^w_W, \vec{f}^l_{fW}, \vec{a}_W]^T.$$

The right-hand-side vector $\mathbf{d}_1$ is

$$\mathbf{d}_1 = [\vec{d}^1, \vec{d}^2, \vec{d}_W^3, \vec{d}_W^4, \vec{d}_W^5, \vec{d}_W^6]^T =$$

$$
\begin{pmatrix}
\vec{f}_{gS}^l + \vec{f}_{0B}^l \\
-\vec{\Omega}_B^b \times (I_B \vec{\Omega}_B^b) + R_{lb}\vec{\tau}_{0B}^l + \sum_W(-\delta_{i2}\vec{b}_W + H_b R_{lb}\vec{f}_{gW}^l) \\
-\vec{\Omega}_W^w \times (I_W \vec{\Omega}_W^w) + R_{lw}\vec{\tau}_{0W}^l + \delta_{i2}R_{bw}\vec{b}_W - H_W R_{lw}\vec{f}_{gW}^l \\
-R_{lb}^T(\|\vec{\Omega}_B^b\|_2^2 \vec{x}_H^b - (\vec{\Omega}_B^b \cdot \vec{x}_H^b)\vec{\Omega}_B^b) + R_{lw}^T(\|\vec{\Omega}_W^w\|_2^2 \vec{x}_H^w - (\vec{\Omega}_W^w \cdot \vec{x}_H^w)\vec{\Omega}_W^w) \\
\delta_{i1}\vec{b}_W - R_{bw}((R_{bw}^T\vec{\Pi}_W^w) \times \vec{\Omega}_B^b) \\
\vec{f}_{0W}^l
\end{pmatrix}.
$$

Obviously, the coefficient matrix only depends on the shapes, positions, and mass distributions of the wing and the body, and the right hand side additionally depends on kinematics of the wing and the body.

The position and kinematics of the body and the wing is fully described by

$$\mathbf{p} = [\vec{x}_B^l, \dot{\vec{x}}_B^l, \vec{\alpha}_{lb}, \vec{\Omega}_B^b, \vec{x}_W^l, \dot{\vec{x}}_W^l, \vec{\alpha}_{bw}, \vec{\Pi}_W^w]^T, \tag{31}$$

where $\vec{\alpha}_{lb}$ and $\vec{\alpha}_{bw}$ are the vectors formed by Tait-Bryan angles between the lab and body frames and between the body and wing frames, respectively. Definitions of these angles are given in Appendix. After we invert C_1 in Eq. 30 to solve for $\mathbf{w}_1$ including all the accelerations, we can numerically integrate in time the following first-order ODEs to update the position and kinematics of the body and the wing

$$\dot{\mathbf{p}} = \left[\dot{\vec{x}}_B^l, \ddot{\vec{x}}_B^l, K_B^{-1}\vec{\Omega}_B^b, \dot{\vec{\Omega}}_B^b, \dot{\vec{x}}_W^l, \ddot{\vec{x}}_W^l, K_W^{-1}\vec{\Pi}_W^w, \dot{\vec{\Pi}}_W^w\right]^T, \tag{32}$$

where the right-hand side becomes known and the matrices K_B and K_W are given in Appendix.

We can easily include multiple wings in the matrix formulation. For example, if we add a second wing W_2, we have

$$
\begin{pmatrix}
C_1 & C_2^{12} \\
C_2^{21} & C_2^{22}
\end{pmatrix}
\begin{pmatrix}
\mathbf{w}_1 \\
\ddot{\vec{x}}_{W_2}^l \\
\dot{\vec{\Omega}}_{W_2}^w \\
\vec{f}_{fW}^l \\
\vec{a}_{W_2}
\end{pmatrix}
=
\begin{pmatrix}
\mathbf{d}_1^* \\
\vec{d}_{W_2}^3 \\
\vec{d}_{W_2}^4 \\
\vec{d}_{W_2}^5 \\
\vec{d}_{W_2}^6
\end{pmatrix}, \tag{33}
$$

where C_1 and $\mathbf{w}_1$ are the coefficient matrix and right hand side of Eq. 30 for the first wing, and $\mathbf{d}_1^*$ is the same as $\mathbf{d}_1$ in Eq. 30 except that the summation in the second entry $\vec{d}^{2*}$ now involves the second wing, and the sub-matrices C_2^{12}, C_2^{21}, and C_2^{22} associated with the second wing are given by

$$C_2^{12} = \begin{pmatrix} m_W E & 0 & -E & 0 \\ m_W H_b R_{lb} & 0 & -H_b R_{lb} & \delta_{i1} E \\ 0 & 0 & 0 & 0 \\ 0 & 0 & 0 & 0 \\ 0 & 0 & 0 & 0 \\ 0 & 0 & 0 & 0 \end{pmatrix},$$

$$C_2^{21} = \begin{pmatrix} 0 & 0 & 0 & 0 & 0 & 0 \\ -E & R_{lb}^T H_b & 0 & 0 & 0 & 0 \\ 0 & -R_{bw} & 0 & 0 & 0 & 0 \\ 0 & 0 & 0 & 0 & 0 & 0 \end{pmatrix},$$

$$C_2^{22} = \begin{pmatrix} -m_W H_w R_{lw} & I_W^* & H_w R_{lw} & -\delta_{i1} R_{bw} \\ E & -R_{lw}^T H_w & 0 & 0 \\ 0 & E & 0 & -\delta_{i2} E \\ -m_{aW} W & -A_{fW} R_{lw}^T & E & 0 \end{pmatrix}.$$

8. Tests. We perform two tests of our formulation using a two-wing insect model. We recommend that only the orientation of the wings be updated from numerical integration and the position of the wings be obtained directly from the geometric constraint in Eq. 8, which avoids the breakdown of the constraint due to numerical errors. Note that the initial conditions for integrating the system of ODEs in Eq. 32 must be consistent with the geometric and kinematic constraints given in Eqs. 8 and 9.

In the first test, we examine conservation of the angular momentum of a torque-free insect model. In this case, two identical wings are fixed relative to the body with reflective symmetry through the $x^b - z^b$ plane, and the CM of the model falls on the x^b axis. If the initial angular velocity of the model has only the rolling component, then the model keeps rolling about the rolling x^b axis with the same initial rate. The simulation verifies this conservation, as shown in Fig. 2. In the second test, we examine Newton's second law for an insect model subject to an external force which is constant in the lab frame. Even thought the two wings of the model flap with respect to the body under the action of prescribed body-to-wing torque, the CM of the model follows a known trajectory according to Newton's second law, which is also verified in the results shown in Fig. 2.

9. Conclusions. We derive a user-friendly matrix formulation of the Newtonian dynamics for studying the stability and maneuverability of insect flight. The formulation is neat, concise, simple and convenient. It includes multiple wings easily, unifies the case with prescribed wing kinematics and the case with prescribed body-to-wing torque, and allows for

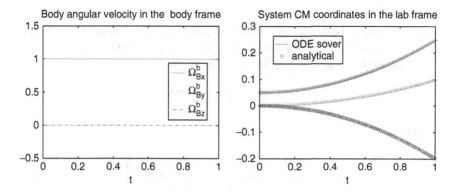

FIG. 2. *Left: Conservation of the rolling angular momentum of an insect model; Right: Comparison of the CM position of an insect model between the numerical and exact results*

the dependence of fluid force and torque on acceleration. A straightforward Matlab implementation of the formulation is available on http://faculty.smu.edu/sxu. In the future, we will extend the formulation to insect flight with flexible wings.

APPENDIX

To write the explicit forms of the frame-to-frame transformation matrices, we use the following Tait-Bryan angles (angles of roll, pitch, and yaw) to describe the orientation of one frame relative to another

- ϕ_{lb}, θ_{lb}, and ψ_{lb}: the roll (around the x^b axis), pitch (around the y^b axis), and yaw (around the z^b axis) angles of the body relative to the lab frame, respectively;
- ϕ_{bw}, θ_{bw}, and ψ_{bw}: the roll (around the x^w axis), pitch (around the y^w axis), and yaw (around the z^w axis) angles of a wing relative to the body, respectively.

We then define $\vec{\alpha}_{lb} = [\phi_{lb}, \theta_{lb}, \psi_{lb}]^T$ and $\vec{\alpha}_{bw} = [\phi_{bw}, \theta_{bw}, \psi_{bw}]^T$.

Applying the rotations in the order of yaw, pitch, and roll, we have

$$R_{lb} = R_{lb}^{roll} R_{lb}^{pitch} R_{lb}^{yaw},$$
$$R_{bw} = R_{bw}^{roll} R_{bw}^{pitch} R_{bw}^{yaw}, \qquad (34)$$
$$R_{lw} = R_{bw} R_{lb}.$$

The lab-to-body transformation matrices are

$$R_{lb}^{roll} = \begin{pmatrix} 1 & 0 & 0 \\ 0 & \cos\phi_{lb} & \sin\phi_{lb} \\ 0 & -\sin\phi_{lb} & \cos\phi_{lb} \end{pmatrix}, \quad R_{lb}^{pitch} = \begin{pmatrix} \cos\theta_{lb} & 0 & -\sin\theta_{lb} \\ 0 & 1 & 0 \\ \sin\theta_{lb} & 0 & \cos\theta_{lb} \end{pmatrix},$$

$$R_{lb}^{yaw} = \begin{pmatrix} \cos\psi_{lb} & \sin\psi_{lb} & 0 \\ -\sin\psi_{lb} & \cos\psi_{lb} & 0 \\ 0 & 0 & 1 \end{pmatrix}. \qquad (35)$$

We can obtain body-to-wing transformation matrices similarly. Every transformation matrix is orthogonal, so its inverse is its transpose.

The angular velocity $\vec{\Omega}_B$ in the body frame is

$$\vec{\Omega}_B^b = K_B \dot{\vec{\alpha}}_{lb}, \tag{36}$$

where

$$K_B = \begin{pmatrix} 1 & 0 & -\sin\theta_{lb} \\ 0 & \cos\phi_{lb} & \sin\phi_{lb}\cos\theta_{lb} \\ 0 & -\sin\phi_{lb} & \cos\phi_{lb}\cos\theta_{lb} \end{pmatrix}. \tag{37}$$

Similarly the angular velocity $\vec{\Pi}_W$ in the wing frame can be written as $\vec{\Pi}_W^w = K_W \dot{\vec{\alpha}}_{bw}$. If the Tait-Bryan angles of the wing relative to the body are prescribed, then $\dot{\vec{\Pi}}_W^w = \dot{K}_W \dot{\vec{\alpha}}_{bw} + K_W \ddot{\vec{\alpha}}_{bw}$ is known.

REFERENCES

[1] Roberson RE, Schwertassek R (1998) Dynamics of multibody systems. Springer, Berlin/New York
[2] Gebert G, Gallmeier P, Evers J (2002) Equations of motion for flapping flight. AIAA paper, AIAA, pp 2002–4872
[3] Sun M, Wang J, Xiong Y (2007) Dynamic flight stability of hovering insects. Acta Mech Sin 23:231–246

EFFICIENT FLAPPING FLIGHT USING FLEXIBLE WINGS OSCILLATING AT RESONANCE

HASSAN MASOUD* AND ALEXANDER ALEXEEV(⊠)*

Abstract. We use fully-coupled three-dimensional computer simulations to examine aerodynamics of elastic wings oscillating at resonance. Wings are modeled as planar elastic plates plunging sinusoidally at a low Reynolds number. The wings are tilted from horizontal, thereby generating asymmetric flow patterns and non-zero net aerodynamic forces. Our simulations reveal that resonance oscillations of elastic wings drastically enhance aerodynamic lift, thrust, and efficiency. We show that flexible wings driven at resonance by a simple harmonic stroke generate lift comparable to that of small insects that employ a significantly more complicated stroke kinematics. The results of our simulations point to the feasibility of using flexible resonant wings with a simple stroke for designing efficient microscale flying vehicles.

Key words. Low Reynolds number, flapping flight, flexible wing, resonance, lattice Boltzmann model, MAV

AMS(MOS) subject classifications. Primary 76Z10, 74F10, 76M28

1. Introduction. Winged insects seem to be the perfect prototype for designing microscale flying machines. After many millions of years of evolution, insects have developed a unique ability to fly, hover, and maneuver in complex, confined environments. Insects are remarkably efficient in producing lift and thrust by flapping their tiny, flexible wings, which are extremely light and account for only a few percent of body mass [1]. Insect wings have span lengths from millimeters to several centimeters and beat with frequencies in the range from a few hertz to hundreds of hertz [1]. These small scale flapping wings typically operate at a low Reynolds number and effectively generate unsteady vortical flows that keep insects aloft [2–6].

For many years, researchers have been interested in designing microscale air vehicles (MAV) that replicate flapping flight of winged insects [5, 7–9]. However, flapping techniques used by winged insects are rather sophisticated and typically involve a combination of pitching and plunging motions [10], making the design of a flapping MAV that precisely mimics an insect-like stroke especially challenging. From a practical point of view, it is advantageous to use MAV wings with a simple kinematic pattern, such as a sinusoidal plunging motion. While this stroke can be more readily implemented in MAVs, experiments with rigid wings show that sinusoidal oscillations are relatively inefficient in generating lift [11].

Herein, we examine how wing flexibility can be harnessed to improve the aerodynamic performance of flapping wings driven by a simple

---

*George W. Woodruff School of Mechanical Engineering, Georgia Institute of Technology, Atlanta, Georgia 30332, USA, alexander.alexeev@me.gatech.edu and Hassan.Masoud@gatech.edu. The work was supported in part by the NSF through TeraGrid computational resources.

harmonic stroke. Recent studies indicate that resonance oscillations may improve wing performance and, therefore, could be useful for designing flapping MAVs [12–15]. Vanella et al. [12] employed a 2D, two-link model to study the hovering of flexible wings at low Re. They considered frequencies in the range $1/2$–$1/6$ of the wing's undamped resonance frequency, and showed that the aerodynamic performance is enhanced when the wing oscillates at $1/3$ of its natural frequency. Michelin and Smith [13] examined the resonance of 2D heaving wings in inviscid fluids. Using a reduced-order model, they showed that resonance maximizes magnitudes of the mean thrust and power input. Spagnolie et al. [15] studied the effect of wing flexibility on the locomotion of freely moving, passively pitching flapping wings. Using experiment and 2D modeling, they demonstrated that wing horizontal velocity is maximum when the flapping frequency is near the resonance frequency of wings in fluid.

While it seems obvious that wing flexibility may play a key role in flapping flight [12, 13, 15–18], until now there have been only a few studies that deal with the fully-coupled 3D computational analysis of aerodynamics of flexible flapping wings. Ho et al. [19] developed a 3D numerical model of a flapping elastic wing at a relatively high Reynolds number to probe the effectiveness of active control of elastic wings. A feedback control algorithm was integrated with a fluid solver, to optimize the internal structure of a flexible wing. It was shown that elastic wings have better lift production and power generation than similar rigid wings. Zhu [20] employed a thin plate theory coupled with the boundary element method for inviscid fluids to analyze the effect of spanwise and chordwise flexibility on the thrust and propulsion efficiency of periodically oscillating wings. His results suggest that wing flexibility has a significant influence on the effective locomotion of aerial and aquatic animals. Liu and Aono [21] investigated the size effects on insect hovering over a wide range of Reynolds numbers. They coupled a CFD-based flight simulator with a FEM-based wing model and found a distinct dissimilarity in the vortex dynamics and structure among four different insects hovering at different Re. Recently, Kweon and Choi [22] explored the behavior of sectional lift coefficient of a fruit-fly wing in hovering. They showed that the time averaged lift is proportional to the spanwise distance from the rotation center. Using the lattice Boltzmann flexible particle method, Qi et al. [23] probed the hovering aerodynamics of plunging flexible wing at Reynolds number Re = 136. They showed that lift and drag initially increase and subsequently decrease with increasing the wing's bending rigidity in the spanwise direction. Their results also revealed that the downwash flows induced by wing tip and trailing vortices in the wake area are larger for flexible wings than for rigid wings and, therefore, result in a greater lift force.

The effect of wing elasticity is even more important when flapping wings operate close to their fundamental resonance frequency [14]. To examine the resonance aerodynamics at low Reynolds number, we employ

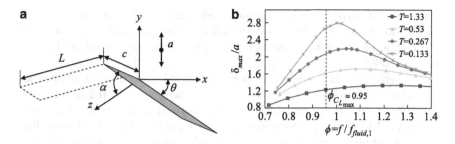

FIG. 1. (a) Schematic of oscillating wings. The dotted line indicates the wing located beyond the symmetry plane. (b) Maximum wing-tip defection $\delta_{\max}$ versus dimensionless frequency ϕ for different values of added mass parameter T. The tilt and incidence angles are respectively $\theta = 40°$, $\alpha = 0$ and Re = 100. The vertical dashed line in (b) shows the maximum lift frequency $\phi_{C_{L\max}} \approx 0.95$

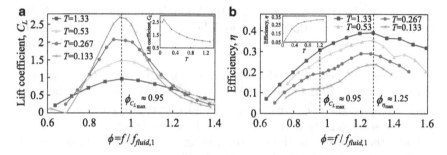

FIG. 2. (a) Lift coefficient C_L and (b) flapping efficiency η versus dimensionless frequency ϕ for different values of added mass parameter T. The tilt and incidence angles are respectively $\theta = 40°$, $\alpha = 0$ and Re = 100. Insets in (a) and (b) show, respectively, the maximum lift coefficient C_L and maximum efficiency η as a function of added mass parameter T

a fully-coupled three-dimensional computational model of flexible flapping wings. We focus on elastic wings driven by sinusoidal oscillations and probe how wing resonance affects aerodynamic forces. We first study hovering and then analyze how resonance flapping can be harnessed to generate steady thrust. Our study reveals that plunging wings at resonance can generate lift comparable to that of winged insects, thereby, indicating the feasibility of using flexible wings with simple stroke kinematics for designing efficient flapping MAVs.

2. Methodology. We model flexible wings as elastic rectangular plates plunged vertically according to a sinusoidal law (Fig. 1a). The wings of span L are tilted from the horizontal in the spanwise direction at an angle θ and in the cordwise direction at an angle α.

To capture the dynamic interactions between elastic, flapping wings and a viscous fluid, we employ a hybrid computational approach [14, 24–29] that integrates the lattice Boltzmann model (LBM) for the dynamics of incompressible viscous fluids and the lattice spring model (LSM) for the mechanics of elastic solids. The two models are coupled through appropriate boundary conditions at the movable solid-fluid interface [26, 30].

Briefly, the LBM is a lattice method that is based on the time integration of a discretized Boltzmann equation for particle distribution functions [31]. In three dimensions, LBM is characterized by a distribution function on a 19-velocity lattice describing the mass density of fluid particles. The hydrodynamic quantities are calculated as moments of the distribution function [31].

The elastic wings are modeled using the LSM. We use a two-dimensional triangular lattice with stretching and bending springs characterized by spring constants k_s and k_b, respectively. This spring arrangement yields a material with the Poisson's ratio $\nu = 1/3$ and the bending modulus $EI = \frac{3\sqrt{3}}{4}k_b c(1 - \nu^2)$, where E is the Young's modulus and $I = cb^3/12$ is the moment of inertia with c and b being the chord and thickness of the wing, respectively. In our simulations, we vary k_b to alter the wing elasticity.

We use the velocity Verlet algorithm to integrate Newton's equation of motion for the lattice nodes, $\mathbf{F}(\mathbf{r}_i) = m(d^2\mathbf{r}_i/dt^2)$, where $\mathbf{F}$ is the total force acting on the node with mass $m = \frac{\sqrt{3}}{2}\rho_s b\Delta x_{LS}^2$ at position $\mathbf{r}_i$. Here, ρ_s is the solid density and Δx_{LS} is the lattice spacing. The total force includes the force due to the interconnecting springs and the force exerted by the fluid at the solid-fluid interface [25, 26].

Our computational domain has sizes $4L \times 6L \times 6L$ in the x, y, and z directions, respectively (Fig. 1a). We use grid refinement [32] with lattice spacing $\Delta x_{LB} = 1$ and dimensions $2L \times 3L \times 3L$ to resolve the flow near the oscillating wing.[1] The coarse grid fills the rest of the computational domain with lattice spacing $\Delta x_{LB_c} = 2$.

The flexible wing is formed from 23×11 LSM nodes with $\Delta x_{LS} = 2.48$, yielding the wing sizes $L = 50$ and $c = 0.4L$ (Fig. 1a). Two rows of nodes beyond the symmetry boundary at $x = 0$ serve to impose the wing vertical oscillations with an amplitude $a = 0.2L$. At the rest of the outer boundaries, we apply the no-flow condition. Thus, we effectively model a pair of simultaneously oscillating symmetrically tilted wings with a symmetry plane at $x = 0$. We set the flapping frequency f such that wing oscillations yield a Reynolds number $\mathrm{Re} = 2\pi f a c/\nu$ equal to 100. For the fluid properties, we set the density $\rho = 1$ and the kinematic viscosity $\nu = 2.5 \times 10^{-3}$.

3. Model Validation. We have previously validated our model in the limit of $\mathrm{Re} \ll 1$ [25, 26, 33]. We further tested the model by simulating three-dimensional flows around rectangular wings at $\mathrm{Re} = 100$. When

---

[1] Unless stated otherwise, all dimensional values are given in lattice Boltzmann units.

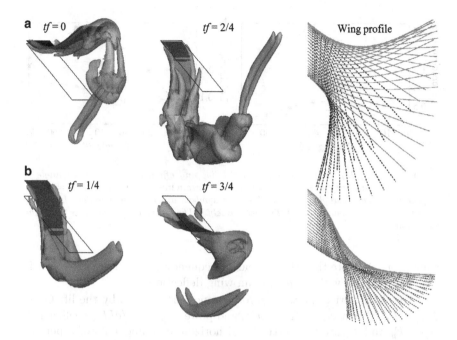

FIG. 3. *Snapshots illustrating vorticial flows and wing profiles for wings oscillating at (a)* $\phi_{C_{L\max}} \approx 0.95$ *and (b)* $\phi_{\eta_{\max}} \approx 1.25$. *The cyan and green surfaces indicate isovorticial surfaces with the vorticity magnitude* $\omega = 37.5f$ *and* $\omega = 12.5f$, *respectively. The solid black contours show the initial position of an undeformed wing. Wing profiles for downstroke are shown by the solid green lines, whereas the upstroke profiles are shown by the dashed blue lines. The wing parameters are* $\theta = 40°$, $\alpha = 0$, $T = 0.133$ *and* Re $= 100$

compared with the experimental data [34], we found good agreement with respect to drag and lift coefficients with errors less than 5%. We also calculated the drag coefficient on a plate oscillating at low Re ranging from 100 to 750. These results were found in good agreement with experiments [35, 36]. To assess the grid quality, we applied a fine grid for the entire computational domain and doubled the domain size. In both tests, the difference in the drag and lift coefficients between the simulations did not exceed 3%. We also verified that the LSM grid is sufficiently accurate to capture the dynamic wing deformations.

4. Problem Parameters. In addition to Re, we introduce the dimensionless oscillation frequency $\phi = f/f_{fluid,1}$ and the parameter $T = \rho c/\rho_s b$ that characterizes the ratio between added and apparent masses. Here, $f_{fluid,1}$ is the fundamental resonance frequency in viscous fluid, which is a function of Re, T, and EI, and calculated based on the linear theory of small amplitude oscillations of high-aspect-ratio elastic beams [37]. At resonance, the wing tip moves with a $\pi/2$ phase shift relative to the

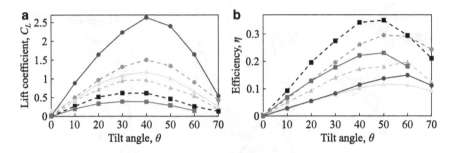

FIG. 4. *(a) Lift coefficient C_L and (b) flapping efficiency η versus tilt angle θ. The solid and dashed lines are for added mass parameter $T = 0.133$ and $T = 0.533$, respectively. The triangle, circle and square symbols are for dimensionless frequency $\phi = 0.8$, $\phi = 0.95$ and $\phi = 1.25$, respectively. The incidence angle is $\alpha = 0$ and Re $= 100$*

wing root. We note that the resonance frequency $f_{fluid,1}$ may not coincide with the frequency of the maximum wing deflection.

We characterize the wing aerodynamic performance by the lift $C_L = 2\bar{F}_y/cLU_0^2$, thrust $C_T = 2\bar{F}_z/cLU_0^2$, and power $C_p = 2P/cLU_0^3$ coefficients. Here, $\bar{F}_y$ and $\bar{F}_z$ are the vertical and horizontal components of a period-averaged aerodynamic force on the oscillating wing, P is the power required for flapping the wing during one period, and $U_0 = 2\pi a f$ is the characteristic velocity. We also introduce the wing efficiency $\eta = (C_L^2 + C_T^2)^{1/2}/C_p$ that represents the ratio between the total force generated by the wings and the power required for its production. Typically, aerodynamic lift and drag are defined with respect to the direction of incoming fluid flow [38, 39]. In the case of hovering in quiescent air, however, we define aerodynamic forces relative to the direction of wing root oscillations.

5. Results and Discussion. In Fig. 1b, we plot the maximum wing-tip deflection $\delta_{\max}$ as a function of ϕ for different values of T for wings tilted at $\theta = 40°$ and $\alpha = 0$. We find that at resonance ($\phi = 1$) the wing deflection is significantly increased for smaller T, whereas for larger T the change of $\delta_{\max}$ is relatively small.

One can expect that an increase in wing bending will enhance the lift force generated by flapping wings. Indeed, we find that C_L increases at the near resonance frequencies and maximizes at $\phi_{C_{L\max}} \approx 0.95$ (Fig. 2a), which is a frequency that differs from that of the maximum wing-tip deflection (Fig. 1b). Interestingly, C_L has a maximum at $\phi_{C_{L\max}}$ even for $T = 1.33$, for which $\delta_{\max}$ is greater for the post-resonance frequencies.

We find that $C_{L\max}$ increases for wings with smaller T for which the effect of wing inertia is more pronounced (see inset in Fig. 2a). In the case of $T = 0.133$, plunging flexible wing yields $C_{L\max} \approx 2.7$, which is comparable to the lift generated by small insects that employ a significantly more complex stroke kinematic [1, 5, 10, 40]. We find that even for light

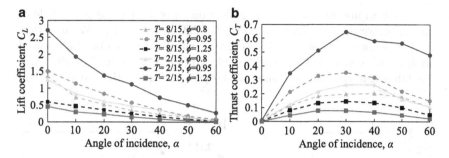

FIG. 5. *(a) Lift coefficient C_L and (b) thrust coefficient C_T versus incidence angle α for different values of dimensionless frequency ϕ and added mass parameter T. The tilt angle is $\theta = 40°$ and Re = 100*

wings with $T = 1.33$, $C_{L\text{max}}$ is about 100 times greater than that produced by rigid wings with identical geometry and stroke. It indicates that wing elasticity and dynamic deformations are critical for enhancing aerodynamic forces at resonance.

We also find that aerodynamic efficiency of flapping wings is significantly enhanced at resonance frequencies (Fig. 2b). In contrast to $C_{L\text{max}}$, however, the maximum efficiency η_{max} increases with increasing T (see inset in Fig. 2b). For all T, η_{max} occurs at $\phi_{\eta_{\text{max}}} \approx 1.25$ and, therefore, $\phi_{\eta_{\text{max}}}/\phi_{C_{L\text{max}}} \approx 1.3$. Thus, by changing the flapping frequency, the wing oscillations can be changed between flapping regimes leading to the maximum lift and efficiency.

Figure 3 illustrates vortex formation and wing bending at frequencies equal to $\phi_{C_{L\text{max}}}$ and $\phi_{\eta_{\text{max}}}$. At $\phi = \phi_{C_{L\text{max}}}$ (Fig. 3a), the wing generates vortices along the longitudinal edges, which form a horse-shoe pattern. When wing moves downwards, its surface is roughly normal to the direction of the root motion and the wing generates significant drag, whereas during the upstroke the wing bends towards the vertical axis and slides over the vortices with low resistance.

At resonance, the asymmetry between the downstroke and upstroke is much more pronounced compared to that of a rigid wing, in which case the asymmetry is only due to the wing initial tilt with respect to the horizontal. This enhanced asymmetry of elastic wing oscillations gives rise to an almost 100-fold increase in the net lift produced by resonant wings comparing to rigid tilted wings with similar geometry.

At $\phi = \phi_{\eta_{\text{max}}}$, the wing exhibits a different oscillation pattern. In this regime, the wing tip and root move out-of-phase (Fig. 3b) and the wing effectively rotates around a nodal line separating the inner and outer wing sections. Hence, when the inner section moves downwards, the outer section moves upwards, and vice versa. These out-of-phase oscillations reduce vortex shedding which seems to contribute to a higher efficiency at this oscillation regime.

To examine the effect of wing tilt θ on wing flapping at zero angle of incidence, we plot in Fig. 4 C_L and η for different T and ϕ. We find that C_L has a maximum at $\theta = 40°$ (Fig. 4a), which seems to be the optimal angle for lift generation by tilted flexible wings at resonance. The efficiency η exhibits a maximum for somewhat larger θ that lie in the range between $50°$ and $70°$.

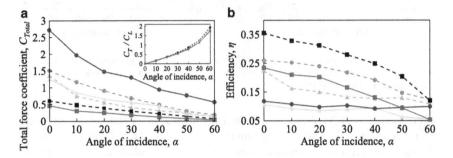

FIG. 6. *(a) Total force coefficient $C_{Total} = (C_L^2 + C_T^2)^{1/2}$ and (b) flapping efficiency η versus incidence angle α. The solid and dashed lines are for added mass parameter $T = 0.133$ and $T = 0.533$, respectively. The triangle, circle and square symbols are for dimensionless frequency $\phi = 0.8$, $\phi = 0.95$ and $\phi = 1.25$, respectively. The inset in (a) shows thrust to lift ratio C_T/C_L as a function of incidence angle α. The tilt angle is $\theta = 40°$ and $Re = 100$*

When the angle of incidence α is non-zero, flapping wings generate a thrust force which is directed in the positive z direction (Fig. 1a). Figure 5 shows how lift and thrust coefficients change with α for a range of T and ϕ. We find that C_L monotonically decreases with increasing α (Fig. 5a). Thus, the maximum lift is generated at $\alpha = 0$. Thrust coefficient C_T, on the other hand, increases with α and reaches a maximum at an angel of incidence about $30°$. For larger α, C_T monotonically decreases. Hence, $\alpha \approx 30°$ is optimal for the thrust generation by flapping resonant wings.

The summation of horizontal (thrust) and vertical (lift) components generated by the wings is presented in Fig. 6a. This figure shows that the total force reduces with increasing α. We can relate it to a decrease in the apparent wing area normal to the direction of imposed wing oscillations. Interestingly, we find that the ratio between C_T and C_L is only a function of α and does not depend on T and ϕ (see inset in Fig. 6a). Thus, redistribution of the net force generated by wings between the horizontal and vertical components does not depend on the flapping frequency.

Finally, we examine the flapping efficiency for wings with a non-zero angle of incidence. Figure 6b shows that η decreases with increasing α. This decrease, however, is not as significant as that of the total force (Fig. 6a). It indicates that C_p is also decreasing with α due to a lower power required to oscillate wings with non-zero angles of incidence.

6. Experimental Parameters. The dimensionless parameters can be readily used to specify physical properties of wings. Our simulations show that the added mass parameter $T \approx 1$ yields the maximum flapping efficiency at resonance. For Re = 100 and $T = 1$, resonance oscillations in the air can be obtained with a pair of flat plates with $L \approx 5\,\text{mm}$, $b \approx 2\,\mu\text{m}$, $\rho_s \approx 1{,}000\,\text{kg/m}^3$, $E \approx 75\,\text{GPa}$ that oscillate at $f \approx 125\,\text{Hz}$. Wings with similar mechanical properties have been recently manufactured using MEMS technology [41]. Additional control over wing flexibility can be obtained using wings with structured surfaces.

7. Summary. We used three-dimensional computational modeling to examine the flapping aerodynamics of flexible wings plunging at Re = 100. Our simulations revealed that resonant wings driven by a simple harmonic stroke can generate lift comparable to that of winged insects that typically use rather complicated stroke kinematics integrating a series of plunging and pitching motions [1, 5, 10, 40]. We also showed that, at resonance, flexible wings can generate a lift force that exceeds by two orders of magnitude the force produced by otherwise identical rigid wings. This demonstrates the dramatic effect that wing elasticity can play in the aerodynamics of flapping wings.

Our results suggest that wing elasticity in combination with resonance oscillations can be harnessed to effectively replace the complex stroke kinematics that is needed for rigid wings to generate adequate lift and thrust. Thus, our findings demonstrate a new approach for designing flapping wings driven by a simple harmonic stroke that can be readily implemented in MAVs, thereby yielding mechanically simple, lightweight, and robust flying machines.

REFERENCES

[1] Ellington CP (1984) The aerodynamics of hovering insect flight. Part 2. Morphological parameters. Philos Trans R Soc B 305(1122):17–40
[2] Dickinson MH, Lehmann FO, Sane SP (1999) Wing rotation and the aerodynamic basis of insect flight. Science 284(5422):1954–1960
[3] Sane SP (2003) The aerodynamics of insect flight. J Exp Biol 206(23):4191–4208
[4] Lehmann FO (2008) When wings touch wakes: understanding locomotor force control by wake-wing interference in insect wings. J Exp Biol 211(2):224–233
[5] Shyy W, Lian Y, Tang J, Liu H, Trizila P, Stanford B, Bernal L, Cesnik C, Friedmann P, Ifju P (2008) Computational aerodynamics of low Reynolds number plunging, pitching and flexible wings for MAV applications. Acta Mech Sin 24(4):351–373
[6] Pesavento U, Wang ZJ (2009) Flapping wing flight can save aerodynamic power compared to steady flight. Phys Rev Lett 103(11):118102–118104
[7] Ansari SA, Zbikowski R, Knowles K (2006) Aerodynamic modelling of insect-like flapping flight for micro air vehicles. Prog Aerosp Sci 42(2):129–172
[8] Zbikowski R (2002) On aerodynamic modelling of an insect-like flapping wring in hover for micro air vehicles. Philos Trans R Soc A 360(1791):273–290
[9] Wood RJ (2008) The first takeoff of a biologically inspired at-scale robotic insect. IEEE Trans Robot 24(2):341–347

[10] Ellington CP (1984) The aerodynamics of hovering insect flight. Part 3. Kinematics. Philos Trans R Soc B 305(1122):41–78

[11] Watman D, Furukawa T (2008) A system for motion control and analysis of high-speed passively twisting flapping wings. In: Proceedings of the IEEE international conference robotics, Pasadena, pp 1576–1581

[12] Vanella M, Fitzgerald T, Preidikman S, Balaras E, Balachandran B (2009) Influence of flexibility on the aerodynamic performance of a hovering wing. J Exp Biol 212(1):95–105

[13] Michelin S, Smith SGL (2009) Resonance and propulsion performance of a heaving flexible wing. Phys Fluids 21(7):071902

[14] Masoud H, Alexeev A (2010) Resonance of flexible flapping wings at low Reynolds number. Phys Rev E 81(5):056304

[15] Spagnolie SE, Moret L, Shelley MJ, Zhang J (2010) Surprising behaviors in flapping locomotion with passive pitching. Phys Fluids 22(4):041903

[16] Liu L, Fang Z, He Z (2008) Optimization design of flapping mechanism and wings for flapping-wing MAVs. Intell Robot Appl 5314:245–255

[17] Thiria B, Godoy-Diana R (2010) How wing compliance drives the efficiency of self-propelled flapping flyers. Phys Rev E 82(1):015303

[18] Yin B, Luo H (2010) Effect of wing inertia on hovering performance of flexible flapping wings. Phys Fluids 22(11):111902

[19] Ho S, Nassef H, Pornsinsirirak N, Tai YC, Ho CM (2003) Unsteady aerodynamics and flow control for flapping wing flyers. Prog Aerosp Sci 39(8):635–681

[20] Zhu Q (2007) Numerical simulation of a flapping foil with chordwise or spanwise flexibility. AIAA J 45(10):2448–2457

[21] Liu H, Aono H (2009) Size effects on insect hovering aerodynamics: an integrated computational study. Bioinspir Biomim 4(1):015002

[22] Kweon J, Choi H (2010) Sectional lift coefficient of a flapping wing in hovering motion. Phys Fluids 22(7):071703

[23] Qi DW, Liu YM, Shyy W, Aono H (2010) Simulations of dynamics of plunge and pitch of a three-dimensional flexible wing in a low Reynolds number flow. Phys Fluid 22(9):091901

[24] Masoud H, Alexeev A (2010) Modeling magnetic microcapsules that crawl in microchannels. Soft Matter 6(4):794–799

[25] Alexeev A, Verber R, Balazs AC (2006) Designing compliant substrates to regulate the motion of vesicles. Phys Rev Lett 96(14):148103

[26] Alexeev A, Verberg R, Balazs AC (2005) Modeling the motion of microcapsules on compliant polymeric surfaces. Macromolecules 38(24):10244–10260

[27] Alexeev A, Yeomans JM, Balazs AC (2008) Designing synthetic, pumping cilia that switch the flow direction in microchannels. Langmuir 24(21):12102–12106

[28] Alexeev A, Balazs AC (2007) Designing smart systems to selectively entrap and burst microcapsules. Soft Matter 3(12):1500–1505

[29] Smith KA, Alexeev A, Verberg R, Balazs AC (2006) Designing a simple ratcheting system to sort microcapsules by mechanical properties. Langmuir 22(16):6739–6742

[30] Bouzidi M, Firdaouss M, Lallemand P (2001) Momentum transfer of a Boltzmann-lattice fluid with boundaries. Phys Fluid 13(11):3452–3459

[31] Succi S (2001) The lattice Boltzmann equation for fluid dynamics and beyond. Oxford University Press, Oxford

[32] Chen H, Filippova O, Hoch J, Molvig K, Shock R, Teixeira C, Zhang R (2006) Grid refinement in lattice Boltzmann methods based on volumetric formulation. Phys A 362(1):158–167

[33] Zhu G, Alexeev A, Balazs AC (2007) Designing constricted microchannels to selectively entrap soft particles. Macromolecules 40(14):5176–5181

[34] Taira K, Colonius T (2009) Three-dimensional flows around low-aspect-ratio flat-plate wings at low Reynolds numbers. J Fluid Mech 623:187–207

[35] Shih CC, Buchanan HJ (1971) Drag on oscillating flat plates in liquids at low Reynolds numbers. J. Fluid Mech 48(2):229–239

[36] Keulegan GH, Carpenter LH (1958) Forces on cylinders and plates in an oscillating fluid. J Res Natl Bur Stand 60(5):423–440

[37] Van Eysden CA, Sader JE (2007) Frequency response of cantilever beams immersed in viscous fluids with applications to the atomic force microscope: arbitrary mode order. J Appl Phys 101(4):044908

[38] Berman GJ, Wang ZJ (2007) Energy-minimizing kinematics in hovering insect flight. J Fluid Mech 582:153–168

[39] Wang ZJ (2004) The role of drag in insect hovering. J Exp Biol 207(23):4147–4155

[40] Wang ZJ (2005) Dissecting insect flight. Annu Rev Fluid Mech 37:183–210

[41] Bronson JR, Pulskamp JS, Polcawich RG, Kroninger CM, Wetzel ED (2009) PZT MEMS actuated flapping wings for insect-inspired robotics. In: Proceedings of the IEEE 22nd international conference micro electro mechanical systems, Sorrento, pp 1047–1050

STABILITY OF PASSIVE LOCOMOTION
IN PERIODICALLY-GENERATED VORTEX WAKES

BABAK G. OSKOUEI* AND EVA KANSO(✉)*

Abstract. The passive locomotion of a body placed in a thrust wake is examined. This work is motivated by a common belief that live and inanimate objects may extract energy from unsteady flows for locomotory advantages. We propose idealized wake models using periodically-generated point vortices to emulate shedding of vortices from an un-modeled moving (thrust) object. We investigate the two-way coupled dynamics of a submerged rigid body with such thrust wakes. In particular, we seek and obtain periodic trajectories where a circular body 'swims' passively against the flow of a thrust wake by extracting energy from the ambient vortices. These periodic trajectories are unstable, as indicated by the associated Floquet multipliers. The instabilities are particularly strong for elliptic bodies where rotational effects are at play.

Key words. Body-vortex interactions, passive locomotion, stability

AMS(MOS) subject classifications. 76Z10s, 76B47, 37N25

1. Introduction. It is widely believed that fish exploit the unsteadiness in the flow to their advantage. For example, dolphins are known to surf in the bow waves of ships [6] while whales gain significant thrust benefits from swimming near surface waves [4]. River fish entrain in wakes and low-speed regions adjacent to shear layers partly to optimize their net energy expense [8, 15].

Harvesting energy from vortical flows and its beneficial use for locomotion was examined in recent laboratory experiments conducted by [11] and [3] on live and dead rainbow trout placed in the wake of an oscillating cylinder. The live trout was reported to alter its flapping motion to synchronize in both frequency and phase with the incoming wake, suggesting that the trout is taking advantage of the energy in the wake of the upstream obstacle. The body of the anesthetized fish – still elastic – was also observed to 'swim' upstream. The generation of thrust force responsible for this unquestionably passive motion is attributed to the flapping motion of the elastic body induced by the vortical wake. Motivated by these results, [7] conducted high-fidelity numerical simulations of a planar three-link articulated body in the wake of an oscillating cylinder and showed that, for certain ratios of body length to cylinder diameter, the fish-like body – whether deformable or rigidly-locked – is passively propelled upstream, suggesting that body elasticity is not essential for this passive motion. The authors attributed this locomotion to the fact that the leading-edge suction at the body's head overcomes the stream-wise friction on the aft portion of the body. They also noted that the presence of the fish-like body in the cylinder's wake extends greatly the suction region behind the cylinder.

*Aerospace and Mechanical Engineering, University of Southern California, Los Angeles, CA 90089, USA, kanso@usc.edu

S. Childress et al. (eds.), *Natural Locomotion in Fluids and on Surfaces*, IMA 155, DOI 10.1007/978-1-4614-3997-4_20,
© Springer Science+Business Media New York 2012

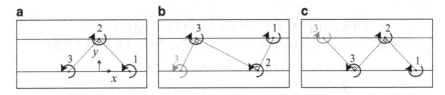

FIG. 1. *Periodically-generated vortices are used to emulate the wake of a body located and swimming to the right : (a) three vortices at t = 0, (b) at t = T/2, the left-most vortex is removed and a new vortex is introduced to the right, (c) at t = T, the process is repeated such that the final vortex configuration is identical to that at t = 0*

In this work, we consider the motion of rigid bodies placed in thrust wakes as opposed to drag wakes as in the experiments of [11] and [3]. The motion of finite-sized objects placed in such wakes received little or no attention though it may provide insights into the passive locomotion of smaller fish and nutrients in the wakes of large fish. We present an idealized mathematical model of a rigid body interacting freely with periodically-generated point vortices that emulate vortical wakes without accounting explicitly for the upstream object responsible for generating these wakes, thus excluding the effects of a suction region. This modeling approach is used in [10] for a class of symmetric wakes and is reminiscent to the classical work of [16] and the more recent work of [1]. These works focus on investigating oscillations and deformations of a 'pinned' body that are optimal for thrust production and extraction of energy from the flow. In our models, we allow the body to freely move in the plane and study the two-way coupled dynamics between the body's motion and the wake's evolution. We make no assumptions on smallness of motion, we do however assume the fluid to be inviscid and incompressible and the flow to be irrotational everywhere except at the locations of the point vortices. Under these idealized conditions, we show the existence of periodic solutions where the rigid body 'swims' in a thrust wake in the direction opposite to the motion of the flow at no energy cost, whereas passive tracers move with the flow. The body itself does not generate any force and its motion is entirely due to energy exploited from the presence of the vortices. We comment on the stability of these periodic trajectories.

2. Wake Model. We propose an idealized wake model inspired by the von Kármán street, but unlike the infinite von Kármán street, we describe the wake using a finite number of point vortices. Namely, we periodically introduce point vortices into a chosen region in the fluid domain to emulate vortex shedding from an external source and periodically remove point vortices as they move out of this region. The removal of point vortices can be justified on the ground that vorticity further downstream from the region of interest has diminishing effects on the dynamics of an object immersed in that region. We first describe the dynamics and stability of these periodically-generated wakes. In Sect. 3, we examine the dynamics of a passive body interacting freely with these wakes.

Let (x, y) denote the coordinates in the plane of motion measured from an orthonormal inertial frame $(\mathbf{e}_1, \mathbf{e}_2)$. It is more convenient for what follows to introduce the complex coordinate $z = x + iy$ (where $i = \sqrt{-1}$) and its complex conjugate $\bar{z} = x - iy$. For concreteness, consider three point vortices in a staggered configuration initially placed at $z_1(0) = L$, $z_2(0) = iL$ and $z_3(0) = -L$ and let the vortex at $y = L$ have circulation $-\Gamma$ and those at $y = 0$ have circulation $\Gamma > 0$. Note that the wake model we propose may be generalized to any number of vortices and choice of vortex spacing but this particular choice allows one to conveniently obtain explicit expressions for the time evolution of the three vortices (see [2, 12]). Namely, the vortices move parallel to the x-axis such that, at any instant in time t, their positions define the vertices of a right triangle,

$$z_1(t) = -\frac{t}{2} + \sqrt{\frac{t^2}{4} + 1}, \quad z_2(t) = -t + i, \quad z_3(t) = -\frac{t}{2} - \sqrt{\frac{t^2}{4} + 1}. \quad (1)$$

Here, we non-dimensionalized the problem using the length scale L and the time scale $2\pi L^2/\Gamma$. The periodically-generated wake (with period T) is obtained by letting the vortices evolve according to (1) from $t = 0$ to $t = T^-/2$. At $t = T/2$, the vortex $z_3(t = T^-/2)$ is removed by hand, so-to-speak, and a new vortex is introduced at the location $z_{T/2}$ such that $z_{T/2}$ is diagonally opposite to the location of the removed vortex and $z_{T/2}$, $z_1(t = T^-/2)$ and $z_2(t = T^-/2)$ define the vertices of a right triangle, see Fig. 1b. These vortices are then relabeled z_1, z_2 and z_3, respectively. Their evolution from $t = T^+/2$ to $t = T^-$ has a similar form to the expressions in (1) (with mirror symmetry about $T/2$) with the understanding that at $t = T/2$, one has a jump in the total impulse of the fluid $p = \sum_i \Gamma_i z_i$, ($i = 1, 2, 3$). At $t = T^-$, the process is repeated. Namely, the vortex $z_3(t = T^-)$ is removed by hand and a new vortex is introduced at the location z_T such that z_T, $z_1(t = T^-)$ and $z_2(t = T^-)$ define the vertices of a right triangle, see Fig. 1c. These vortices are relabeled z_1, z_2 and z_3, respectively, and one has yet another jump in the total impulse p of the fluid at $t = T$. One can readily verify that if the period is set to $T = 3$, at the end of each period, the vortices return to their initial location in space. If $T < 3$, at the end of each period, the vortices are shifted along the positive x-axis by an amount $\Delta x = 1 - 3T/4 + \sqrt{(T/4)^2 + 1}$. These wakes can be expressed in the form of a discrete map [12] whose dynamics admit a fixed point corresponding to the periodic trajectory described here. Note that, while the fluid is inviscid, the periodically-generated vortex wake is not a Hamiltonian system. Indeed, one can view it as a periodically forced system. The stability of this periodic trajectory can be assessed by studying the associated Floquet multipliers, i.e., the eigenvalues of the discrete map.

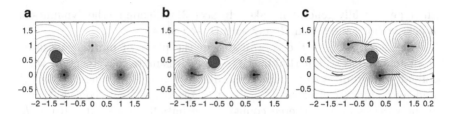

FIG. 2. *Periodic trajectory of body-fluid system depicted at three snapshots (a)* $t = 0$, *(b)* $t = T/2$ *and (c)* $t = T$ *showing the trajectories of the cylinder and the point vortices. Streamlines of the fluid motion are superimposed at these three instants. The parameter values are* $r_o = 0.2$, $T = 1$, $z_{T/2} = 1.56 + i0.96$, $z_T = 2.29$ *and the initial conditions are* $z_1(0) = -1, z_2(0) = i1, z_3(0) = 1$, *while* $z_{cyl}(0) = -1.27 + i0.75$ *and* $v_{cyl}(0) = 0.79 + i0.035$

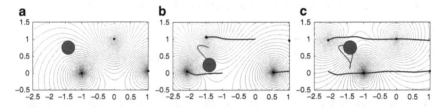

FIG. 3. *Periodic trajectory of body-fluid system depicted at three snapshots (a)* $t = 0$, *(b)* $t = T/2$ *and (c)* $t = T$ *showing the trajectories of the cylinder and the point vortices. Streamlines of the fluid motion are superimposed at these three instants. The parameter values are* $r_o = 0.2$, $T = 3$, $z_{T/2} = 1.03 + i0.95$ *and* $z_T = 1.03 + i0.058$. *The initial conditions are* $z_1(0) = 1.03 + i0.058, z_2(0) = i1, z_3(0) = -1$ *while* $z_{cyl}(0) = -1.41 + i0.75$, $v_{cyl}(0) = -0.72 + i0.32$

One finds that for $T < 3$, the periodically-generated vortex wake is linearly stable (all eigenvalues lie within the unit circle), it is marginally stable for $T = 3$ and unstable otherwise.

3. Passive Locomotion in Periodically-Generated Wakes. We now examine the dynamics of a rigid body interacting freely with these periodically-generated vortex wakes. The body is considered to be neutrally-buoyant, that is, its density (normalized to 1) is equal to the fluid density. At any instant in time apart from $t = kT$ and $t = kT + T/2$ (k integer), that is to say, apart from the times when a vortex is removed and a new vortex is introduced, the system corresponds to a rigid body dynamically interacting with three point vortices of non-zero total strength whereas the net circulation around the rigid body is set to zero at all times. The equations of motion governing the dynamics of such systems is discussed in [5, 10, 13, 14]. We use the equations derived in the latter and omitted here for brevity. At the time instants $t = kT$ and $t = kT + T/2$, one has a jump in the total impulse of the body-fluid system causing the cylinder's velocity to be C^0 continuous.

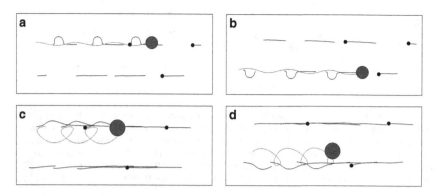

FIG. 4. *Periodic trajectories for four sets of initial conditions and parameter values. In all cases, the total integration time is* $t = 3T$, $r_o = 0.2$, $z_1(0) = 1, z_2(0) = i1, z_3(0) = -1$. *In (a)* $T = 1$, $z_{cyl}(0) = -1.35 + i1.08$, $v_{cyl}(0) = 0.48 - i0.027$, $z_{T/2} = 1.56 + i0.99$, $z_T = 2.28 - i0.0067$. *In (b)* $T = 1$, $z_{cyl}(0) = -1.53 + i0.059$, $v_{cyl}(0) = 1.52 + i0.42$, $z_{T/2} = 1.63 + i0.95$, $z_T = 2.36 - i0.26$. *In (c)* $T = 2$, $z_{cyl}(0) = -1.26 + i0.98$, $v_{cyl}(0) = -0.61 - i0.043$, $z_{T/2} = 1.22 + i0.98$, $z_T = 1.61 - i0.011$. *In (d)* $T = 2$, $z_{cyl}(0) = -1.47 + i0.38$, $v_{cyl}(0) = -0.61 - i0.043$, $z_{T/2} = 1.30 + i0.91$, $z_T = 1.67 - i0.027$

In the case of a circular cylinder of radius r_o, one obtains a 10-dimensional dynamical system with discontinuities. Namely, one has 2 degrees of freedom to determine the position of the cylinder and 2 additional variables to determine its velocity as well as 6 degrees of freedom to determine the location of the point vortices. We look for periodic trajectories of the 10-dimensional body-fluid system for a given period T. To this end, we use a random initial guess and a shooting method to solve for initial conditions (position z_{cyl} and velocity v_{cyl} of cylinder and positions z_1, z_2, z_3 of point vortices) and location $z_{T/2}$ of the introduced vortex such that the solution is periodic with period T.

In Fig. 2, we show a periodic trajectory of body-fluid system at three snapshots (a) $t = 0$, (b) $t = T/2$ and (c) $t = T$. The cylinder radius is set to 0.2 and the period to $T = 1$. Each period, the cylinder undergoes a net horizontal translation of $\Delta x = 1.29$ and its velocity at $t = T$ returns to its value at $t = 0$. That is to say, while the vortices and passive tracers are advected to the left, the rigid cylinder moves to the right in the direction opposite to the flow by extracting energy from the surrounding fluid since we only have hydrodynamic forces applied on the cylinder. Figure 3 depicts another periodic trajectory with $T = 3$ and no net horizontal translation. In other words, the cylinder 'holds station' in one region while the vortices are advected to the left. In Fig. 4, we show a family of periodic trajectories for various periods, initial conditions and parameter values. The cylinder in all these cases moves in the direction opposite to the flow by extracting energy from the surrounding fluid. These periodic trajectories are relatively easy to find using our shooting method suggesting that the dynamics is not

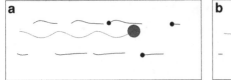 

FIG. 5. *Periodic trajectories for (a) circular cylinder of radius $r_o = 0.2$ and (b) an elliptic cylinder of semi-major axis 0.55 and semi-minor axis 0.073, i.e., both cylinders have the same area 0.04π. The trajectory in (a) corresponds to the same conditions as that in Fig. 2 but here shown for a total time of 3T. The trajectory in (b) has period $T = 0.4$, the vortices of strength $+0.2, -0.2, +0.2$ are located at $0.69, 0.23 + i0.96, -0.99-i0.028$ while $z_{cyl}(0) = -0.96+i0.55$, $\theta(0) = -0.0859$, $v_{cyl}(0) = 2.28-i0.22$, $\dot{\theta}(0) = -1.16$*

very sensitive to small perturbations in initial conditions. We computed the Floquet multipliers of the associated 10-dimensional discrete map and found the system to be unstable with eigenvalues lying outside but near the unit circle.

In the case of an elliptic cylinder, one has two additional dimensions associated with the orientation θ and angular velocity $\dot{\theta}$ of the ellipse (where the dot is used to denote time derivative). Periodic trajectories of this 12-dimensional body-fluid system for a given period T are extremely difficult to find by hand using our naive shooting method. The reason for this lies in the fact that rotational effects easily destabilize the motion of the rigid elliptic cylinder, making small variations in the initial guess result in large variations at $t = T$. Figure 5 shows a periodic trajectory for the circular cylinder (left) and a periodic trajectory for elliptic body (right). The largest Floquet multiplier for the elliptic body is of the order of 1,000 indicating that indeed a small variations in initial conditions is amplified 1,000 times over one period, hence the difficulty in locating such trajectories.

4. Conclusions. We used periodically-generated point vortices to emulate vortex shedding from an un-modeled moving object and examined the dynamics of a free rigid body placed in such thrust wakes. We identified several periodic trajectories of the body-fluid system where the body moves passively, i.e., via hydrodynamic forces only, in the opposite direction to the fluid. These trajectories are found to be unstable to small perturbations in initial conditions. In the case of an elliptic cylinder, the instability (the largest Floquet multiplier) is so large that it is difficult to find periodic trajectories using a naive shooting method. This is consistent with the intuition that rotational effects destabilize the passive motion of elongated bodies in inviscid fluids. Yet, the laboratory experiments with the anesthetized fish [3] suggest that its passive locomotion in unsteady vortex wakes is stable. Although the experiments are concerned with drag wakes as opposed to the thrust wakes considered here, we speculate that body elasticity (see [9]) and fluid viscosity – both effects not accounted for in the present model – play an important role in stabilizing this passive motion.

Acknowledgement. This work is supported by the NSF CAREER award CMMI 06-44925 and the grant CCF08-11480.

REFERENCES

[1] Alben S (2010) Passive and active bodies in vortex-street wakes. J Fluid Mech 642:95–125

[2] Aref H, Rott N, Thomann H (1992) Gröblis solution of the three-vortex problem. Annu Rev Fluid Mech 24:1–20

[3] Beal DN, Hover FS, Triantafyllou MS, Liao JC, Lauder GV (2006) Passive propulsion in vortex wakes. J Fluid Mech 549:385–402

[4] Bose N, Lien J (1990) Energy absorption from ocean waves: a free ride on cetaceans. Proc R Soc Lond 240:591–605

[5] Borisov AV, Mamaev IS, Ramodanov SM (2007) Dynamic interaction of point vortices and a two-dimensional cylinder. J Math Phys 48:1–9

[6] Hayes WD (1953) Wave riding of dolphins. Nature 172:1060

[7] Eldredge JD, Pisani D (2008) Passive locomotion of a simple articulated fish-like system in the wake of an obstacle. J Fluid Mech 607:279–288

[8] Fausch KD (1993) Experimental analysis of microhabitat selection by juvenile steel head (Oncorhynchus mykiss) and coho salmon (O. kisutch) in a British Columbia stream. Can J Fish Aquat Sci 50:1198–1207

[9] Jing F, Kanso E (2012) Effects of body elasticity on the stability of underwater motion, J Fluid Mech 690:461–473

[10] Kanso E, Oskouei BG (2008) Stability of a coupled body-vortex system. J Fluid Mech 800:77–94

[11] Liao JC, Beal DN, Lauder GV, Triantafyllou MS (2003) Fish exploiting vortices decrease muscle activity. Science 302:1566–1569

[12] Oskouei BG (2011) Body-vortex dynamics: application to aquatic locomotion. Ph.D. dissertation, University of Southern California

[13] Roenby J, Aref H (2010) Chaos in body-vortex interactions. Proc R Soc Lond A 466:1871–1891

[14] Shashikanth BN, Marsden JE, Burdick JW, Kelly SD (2002) The Hamiltonian structure of a 2D rigid circular cylinder interacting dynamically with N point vortices. Phys Fluid 14:1214–1227

[15] Webb PW (1998) Entrainment by river chub Nocomis micropogon and smallmouth bass micropterus dolomieu on cylinders. J Exp Biol 201:2403–2412

[16] Wu TY, Chwang AT (1974) Extraction of flow energy by fish and birds in a wavy stream. Proceeding of the symposium on swimming and flying in nature, California Institute of Technology, Pasadena, pp 687–702

SIMULATING VORTEX WAKES OF FLAPPING PLATES

J.X. SHENG[*], A. YSASI[†], D. KOLOMENSKIY[‡], E. KANSO[†],
M. NITSCHE(✉)[§], AND K. SCHNEIDER[¶]

Abstract. We compare different models to simulate two-dimensional vortex wakes behind oscillating plates. In particular, we compare solutions using a vortex sheet model and the simpler Brown–Michael model to solutions of the full Navier–Stokes equations obtained using a penalization method. The goal is to determine whether simpler models can be used to obtain good approximations to the form of the wake and the induced forces on the body.

Key words. Separated shear flows, vortex sheets, Brown–Michael, penalization method

AMS(MOS) subject classifications. Primary 76B47

1. Introduction. A key component to better understand the locomotion and efficiency of swimming fish is the study of vortex separation, its subsequent evolution and its interaction with the moving body. Several studies have focused on the problem of flow past flapping plates. Of interest is, for example, the shape of the vortex wake as a function of the plate's oscillation profile, as well as the resulting forces on the plate.

Recent experimental work includes soap film experiments by Schnipper et al. [22], who found a variety of wake types as a function of the generating parameters, and particle image velocimetry measurements by Godoy-Diana et al. [9]. They observed the transition between drag and thrust producing motions. The associated wake signatures, often referred to as von Karman vs. reversed von Karman wakes, have been widely studied (see e.g. [25]). Numerical simulations have followed several approaches. Examples of full Navier-Stokes simulations of flow past flapping bodies, using either compact finite difference, mixed Fourier/finite difference, or viscous vortex particle methods, are given in [2, 7, 12, 27]. Inviscid vortex sheet separation models [1, 15, 16, 18, 23, 24] are less costly since the problem is reduced to a lower-dimensional one. They are meant to be a good approximation of the outer flow away from viscous boundary layers and have been used extensively to study, for example, flapping flags in a crossflow, the motion of falling cards,

---

[*]M2P2 CNRS, Aix-Marseille Université, 13453 Marseille, France
ETH Zurich, 8092 Zurich, Switzerland shengj@ethz.ch

[†]University of Southern California, Los Angeles, CA 90089, USA

[‡]M2P2 CNRS, Aix-Marseille Université, 13453 Marseille, France
Centre Européen de Recherche et de Formation Avancée en Calcul Scientifique (CERFACS), Toulouse, France

[§]Corresponding author. Department of Mathematics and Statistics, University of New Mexico, Albuquerque, NM 87131, USA

[¶]M2P2 CNRS, Aix-Marseille Université, 13453 Marseille, France

S. Childress et al. (eds.), *Natural Locomotion in Fluids and on Surfaces*, IMA 155, DOI 10.1007/978-1-4614-3997-4_21,
© Springer Science+Business Media New York 2012

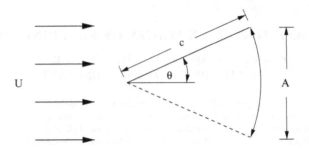

FIG. 1. *Sketch illustrating the plate of length c hinged at one point, and flapping with maximal amplitude A in a background flow of magnitude U*

and wakes of plates and flexible membranes. The most computationally efficient wake models are point vortex models [4, 6, 8, 17, 28].

In this paper we compare full Navier-Stokes simulations to the computationally simpler vortex sheet and point vortex approximations. The goal is to determine to what extent the wakes and resulting body forces are recovered by the simpler models. Similar comparisons of viscous and inviscid point vortex or vortex sheet approximations are also presented in [8, 19]. Such studies are needed to determine whether the simpler models can be used to improve understanding of how fish swim and steer themselves by manipulating the surrounding fluid [10, 11].

2. Problem Description. Figure 1 illustrates the problem considered here. An idealized rigid plate of zero thickness immersed in a fluid of kinematic viscosity ν is hinged at one point in an oncoming parallel flow with speed U. The plate has length c and oscillates with angle θ about a line parallel to the oncoming flow, with maximal tip displacement A. The plate motion is described by $\theta(t) = \theta_m \sin(2\pi t/\tau)$, where $\theta_m = \sin^{-1}(A/(2c))$. The governing dimensionless parameters are

$$Re = \frac{cU}{\nu}, \qquad St = \frac{fA}{U}, \qquad A_c = \frac{A}{c}. \qquad (1)$$

Throughout this paper we consider, as an illustration, the values $St = 0.4$ with $A_c = 0.8$. Biologists have found that larger swimming fish, namely cetaceans such as dolphins and wales, swim in the range $0.25 \leq St \leq 0.35$, even though associated Reynolds numbers vary by a factor of 10 [20, 26]. The value chosen here is slightly above the observed range. We note that, for fixed f and A, the larger the Strouhal number is, the slower is the background velocity U. This causes the vorticity shed from the plate to remain near and interact longer with the plate, which increases the computational difficulty in resolving the flow. The results below are thus in the more computationally difficult regime of the observed values.

3. Numerical Methods. We simulate the fluid flow using three numerical models of increasing simplicity. In the first model, the plate is replaced by one of finite thickness, and the flow approximated by the penalized Navier–Stokes equations [3, 21],

$$\partial_t \mathbf{u} + \mathbf{u} \cdot \nabla \mathbf{u} + \nabla p - \nu \Delta \mathbf{u} + \frac{1}{\eta} \chi_\Omega (\mathbf{u} - \mathbf{u}_p) = 0 \; , \tag{2}$$

where $\nabla \cdot \mathbf{u} = 0$ and the density has been normalized to 1. Here, χ_Ω is a mask function which is 1 inside the region Ω occupied by the plate, and 0 elsewhere, and $\mathbf{u}_p$ is the plate velocity. As $\eta \to 0$, the solution of (2) converges to the solution of the Navier–Stokes equations in the complement of Ω [5]. Following [14], (2) is solved on a periodic domain using a classical Fourier pseudo-spectral method for the spatial discretization, and an adaptive second order Adams–Bashforth method for the time discretization. The results presented below were obtained for fixed $Re = 1{,}000$, plate thickness $1/(16c)$, and $\eta = 0.001$, using $N_x \times N_y = 8,192 \times 4,096$ grid points on the computational domain $[0, 24] \times [-8, 4]$, where the plate is placed at $y = 0$.

In the vortex sheet model the fluid is treated as a purely inviscid one. The plate is modeled as a bound vortex sheet that satisfies zero normal flow through the plate. A point vortex is released at each time step from the trailing edge, and the shed vorticity is modeled as a regularized free sheet [13]. No separation is allowed at the leading edge. A key component is the algorithm used to determine the shed circulation $\Gamma(t)$. Here, we follow [18] and impose the Kutta condition

$$\frac{d\Gamma}{dt} = -\frac{1}{2}(u_+^2 - u_-^2) \; , \tag{3}$$

where $u_\pm$ are the tangential velocities above and below the plate, at the trailing edge. An alternative method introduced by Jones [15] is based on representing the flow in the complex plane (see also [1, 16, 23, 24]). We confirmed that the two methods give identical results for an example presented in [15], even though the implementation details differ significantly. The vortex sheet model depends on the regularization parameter for the free sheet, which in the results below is set to $\delta = 0.04c$.

The simplest model we consider is the Brown–Michael point vortex model [6, 8, 17, 28]. Here, a single point vortex is shed from the edge of the plate at the beginning of the motion. Its circulation and position changes in time so as to satisfy the Kutta condition, that is, flow tangency at the edge. At the instant the rate of change of the vortex circulation vanishes it is released: it moves with the fluid velocity and its circulation remains constant. At the same time another vortex is shed from the edge and the process repeats. The advantage of this method is that it is extremely fast.

4. Numerical Results. Figures 2–4 present the numerical results computed with the three methods, using $St = 0.4$, $A_c = 0.8$, normalized by c, at $t/\tau = 1, 2, 5$. Figure 2 plots vorticity contours computed with the

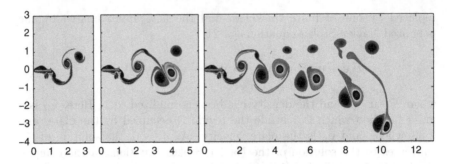

FIG. 2. *Vorticity contours of the solution to the penalized Navier-Stokes equations, for Re = 1,000, at $t/\tau = 1, 2, 5$*

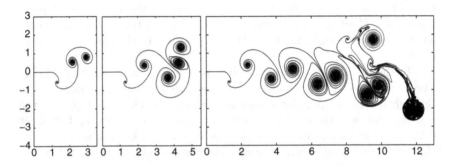

FIG. 3. *Vortex sheet position at $t/\tau = 1, 2, 5$*

Navier–Stokes penalization method, for $Re = 1,000$. Darkly colored vortex regions denote vortices of negative vorticity, regions colored in light grey with a white ring and dark interior denote vortices of positive vorticity. The vorticity values range from -50 to 50.

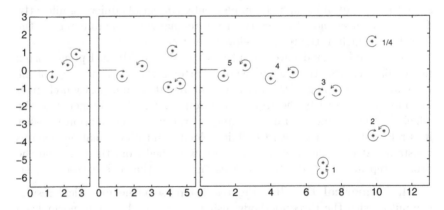

FIG. 4. *Point vortex positions computed using Brown-Michael model, at $t/\tau = 1, 2, 5$*

The plate is impulsively started from the horizontal position $\theta = 0$ with positive velocity. In the first 1/4 period upward sweep, negative vorticity separates from the upper side of the plate, forming a vortex of negative circulation. This is the leading vortex in the first frame in Fig. 2, centered around $x = 3$. As can be seen from the second and third frame, this lead vortex travels downstream in time but remains centered above $y = 0$, by itself. With each consecutive down- and upward sweeps of the plate, two vortices of opposite sign are shed. Unlike the first, these two vortices pair up and travel diagonally downwards. For example, at $t/\tau = 2$, one well-developed pair is observed, with a second nascent one near the edge of the plate. At $t/\tau = 5$, four vortex pairs are observed, with a fifth nascent one near the edge. The leading pair at $t/\tau = 5$ is beginning to move backwards.

Note also that vorticity is generated along the walls of the plate which, at the times shown, separates from the bottom wall. Furthermore, the shear layer connecting consecutive vortex pairs begins to roll up, forming smaller secondary vortices lined up above the vortex pairs. We have performed further computations which indicate that the strength of these secondary vortices depends on the plate thickness. This is consistent with experimental results by Schnipper et al. [22]. Their Fig. 3d, e, f, all with the same value of the Strouhal number as defined in (1), indicate that by increasing the width of the leading edge in their case, the number and strength of secondary vortices per flapping cycle increases.

Figure 3 plots the position of the sheet computed with the inviscid vortex sheet model. The wall vorticity is absent, as well as the secondary vortices, but the location of the primary vortices is in good agreement with the viscous simulations. Some differences are noticeable at later times. For example, at $t/\tau = 5$ the vortex pairs travel at an angle that is less inclined, and do not travel as far downward as in the viscous case. This may be attributed to the fact that, as the viscous vortices evolve, their strength decays by diffusion changing the relative circulation between them. As a result, the leading viscous vortices are relatively weaker and may be convected more strongly by vortices behind them.

Figure 4 plots the position of the point vortices computed with the Brown–Michael model. This simple model captures the pairing of shed vortices and approximate travel direction very well. The position of the first vortex, denoted by "1/4", is in good agreement with the other methods. Differences can be observed at $t/\tau = 5$. For example, the first vortex pair, denoted by "1", has travelled down and backwards further than in Figs. 2 or 3. Also the second and third pairs, denoted by "2" and "3", have travelled down further than in Fig. 2. This contrasts the vortex sheet results, in which these vortices have travelled down less than in Fig. 2.

Figures 5a, b compare the normalized shed circulation $\Gamma(t)/(Uc)$ and drag force $F_x/(U^2 c)$ for the three methods. In the vortex sheet case, the force is computed following [1, 15, 24]; for the Brown–Michael model, the formulation for the force is derived following the outline given in [17].

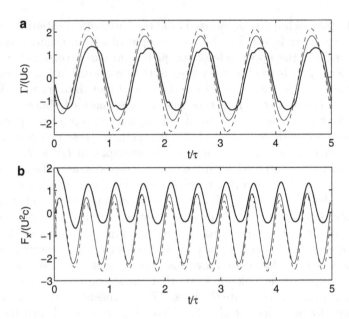

FIG. 5. *(a)* *Normalized shed circulation* $\Gamma(t)/(Uc)$, *and (b)* *normalized drag force* $C_d = F_x/(U^2c)$. *Each plot shows results of the viscous simulation (thick), the vortex sheet simulation (thin) and the Brown-Michael model (dashed)*

Both circulation and drag synchronize closely with the oscillation of the plate. In all models, the circulation oscillates about approximately zero mean value. The oscillation amplitudes are about 1.4–1.7 times larger in both inviscid models than in the viscous one. It is interesting that even though the circulation in the two inviscid models is quite similar, the corresponding vortex positions differ in comparison with the viscous ones. Whether this depends on the distribution of vorticity between vortices or the initial vortex placement remains to be understood. It is also curious that the viscous values in Fig. 5 lag a small time behind the other two.

In the case of the drag force in Fig. 5b, the oscillation amplitude is about 1.8 times larger in the inviscid models than in the viscous one. More significantly, however, is that the mean values of the oscillation disagree. In the inviscid models it is negative, predicting a net thrust of -0.72 (vortex sheet) to -0.95 (Brown–Michael). In the viscous model the mean drag is positive, predicting a net drag of about 0.45. Further investigation of the pressure component and the viscous component of the force, and of simpler cases for which theoretical results are available, may help elucidate which one is responsible for this discrepancy.

5. Summary. We compared simulations of the vortex wake behind a flapping plate using three models: the viscous penalized Navier–Stokes equations, a vortex sheet model, and a point vortex model. We find that the wake structure is qualitatively similar in all three models, and the shed circulation and drag coefficient are of the same order of magnitude. There are differences in the vortex position that become more apparent at larger times. Also, in the inviscid models, the shed circulation values are larger, as is the drag oscillation amplitude. Most interestingly, the skin drag seems to dominate the horizontal force in the viscous case and not in the inviscid one. However, we have presented results for only one value of Re, plate thickness, and δ, and further study is necessary to determine the effect of these parameters on the observed differences.

Acknowledgement. The work of AY and EK is supported by the NSF CAREER award CMMI 06-44925 and the grant CCF08-11480.

REFERENCES

[1] Alben S (2010) Passive and active bodies in vortex-street wakes. J Fluid Mech 642:92–125

[2] Alben S, Shelley M (2005) Coherent locomotion as an attracting state for a free flapping body. PNAS 102:11163–11166

[3] Angot P, Bruneau C-H, Fabrie P (1999) A penalization method to take into account obstacles in incompressible viscous flows. Numer Math 81:497–520

[4] Aref H, Stremler M, Ponta F (2006) Exotic vortex wakes – point vortex solutions. J Fluids Struct 22:929–940

[5] Carbou G, Fabrie P (2003) Boundary layer for a penalization method for viscous incompressible flow. Adv Differ Equ 8:1453–1480

[6] Cortelezzi L, Leonard A (1993) Point vortex model of the undsteady separated flow past a semi-infinted plate with transverse motion. Fluid Dyn Res 11:264–295

[7] Eldredge JD (2005) Efficient tools for the simulation of flapping wing flows. In: AIAA 2005–0085, Reno, pp 1–11

[8] Eldredge JD, Wang C (2010) High-fidelity simulations and low-order modeling of a rapidly pitching plate. In: AIAA 2010–4281, Chicago, pp 1–19

[9] Godoy-Diana R, Aider J-L, Wesfreid JE (2008) Transitions in the wake of a flapping foil. Phys Rev E 77:016308

[10] Kanso E (2009) Swimming due to transverse shape deformations. J Fluid Mech 631:127–148

[11] Kanso E, Marsden JE, Rowley CW, Melli-Huber JB (2005) Locomotion of articulated bodies in a perfect fluid. J Nonlinear Sci 15:255–289

[12] Kern S, Koumoutsakos P (2006) Simulations of optimized anguilliform swimming. J Exp Biol 209:4841–4857

[13] Krasny R (1989) Desingularization of periodic vortex sheet roll-up. J Comput Phys 65:292–313

[14] Kolomenskiy D, Schneider K (2009) A Fourier spectral method for the Navier–Stokes equations with volume penalization for moving solid obstacles. J Comp Phys 228:5687–5709

[15] Jones MA (2003) The separated flow of an invisicd fluid around a moving flat plate. J Fluid Mech 496:405–441

[16] Jones MA, Shelley MJ (2005) Falling cards. J Fluid Mech 540:393–425

[17] Michelin S, Llewellyn Smith SG (2009) An unsteady point vortex method for coupled fluid-solid problems. Theor Comput Fluid Dyn 23:127–153

[18] Nitsche M, Krasny R (1994) A numerical study of vortex ring formation at the edge of a circular tube. J Fluid Mech 276:139–161

[19] Pullin DI, Wang ZJ (2004) Unsteady forces on an accelerating plate and applications to hovering insect flight. J Fluid Mech 509:1–21

[20] Rohr JJ, Fish FE (2004) Strouhal numbers and optimization of swimming by odontocete cetaceans. J Exp Biol 206:1633–1642

[21] Schneider K (2005) Numerical simulation of the transient flow behaviour in chemical reactors using a penalization method. Comput Fluid 34:1223–1238

[22] Schnipper T, Andersen A, Bohr T (2009) Vortex wakes of a flapping foil. J Fluid Mech 633:411–423

[23] Shelley M, Alben S (2008) Flapping states of a flag in an inviscid fluid: bistability and the transition to chaos. Phys Rev Lett 100:074301

[24] Shukla RK, Eldredge JD (2007) An inviscid model for vortex shedding from a deforming body. Theor Comput Fluid Dyn 21:343–368

[25] Triantafyllou MS, Triantafyllou GS, Gophalkrishnan R (1991) Wake mechanics for thrust generation in oscillation foils. Phys Fluid A 3(12):2835–2837

[26] Triantafyllou MS, Triantafyllou GS, Yue DKP (2000) Hydrodynamics of fishlike swimming. Annu Rev Fluid Mech 32:33–53

[27] Wang ZJ (2000) Vortex shedding and frequency selection in flapping flight. J Fluid Mech 410:323–341

[28] Ysasi A, Kanso E, Newton PK (2011) Wake structure of a deformable Joukowski airfoil. Physica D 240(20):1574–1582

A VELOCITY DECOMPOSITION APPROACH FOR SOLVING THE IMMERSED INTERFACE PROBLEM WITH DIRICHLET BOUNDARY CONDITIONS

ANITA T. LAYTON(✉)*

Abstract. In a previous study, we presented a second-order accurate method for computing the coupled motion of a viscous fluid and an elastic material interface with zero thickness (Beale and Layton, J Comput Phys 228:3358–3367, 2009). The fluid flow was described by the Navier-Stokes equations with periodic boundary conditions, and the deformation of the moving interface exerts a singular force onto the fluid. In this study, we extend that method to Dirichlet boundary conditions. We decompose the velocity into three parts: a "Stokes" part, a "regular" part, and a "boundary correction" part. The "Stokes" part is determined by the Stokes equations and the singular interfacial force. The Stokes solution is obtained using the immersed interface method, which gives second-order accurate values by incorporating known jumps for the solution and its derivatives into a finite difference method. The regular part of the velocity is given by the Navier-Stokes equations with a body force resulting from the Stokes part, and with periodic boundary conditions. The regular velocity is obtained using a time-stepping method that combines the semi-Lagrangian method with the backward difference formula. Because the body force is continuous, jump conditions are not necessary. The boundary correction solution is described by the unforced Navier-Stokes equations, with Dirichet boundary conditions given by the difference between the Dirichlet boundary conditions of the overall Navier-Stokes solution, and the boundary values of the Stokes and regular velocities. Because the boundary correction solution is sufficiently smooth, jump conditions are also not necessary. Numerical results exhibit approximately second-order accuracy in time and space.

1. Introduction. Perhaps the most notable advance in the simulation of moving boundaries within a viscous fluid domain is the immersed boundary method, proposed by Peskin [6]. The singular boundary forces are transferred onto the underlying fluid using approximate (smooth) Dirac delta functions typically with $\mathcal{O}(h)$ support. In general, the immersed boundary method computes approximations with first-order spatial accuracy.

The singular boundary forces induce jump discontinuities in the fluid solution (e.g., pressure). By approximating the Dirac delta function with a smooth approximation, the immersed boundary method does not capture those jumps at the immersed boundary, but rather approximates the solution or its derivative as a continuous function with a large gradient. An alternative approach that captures the jumps in the solution and its derivatives sharply, and that generates approximations with second-order accuracy, is the immersed interface method developed by LeVeque and Li [3, 4]. The key idea is the incorporation of known jumps in the solution or its derivatives into the finite difference schemes. This method was first applied to the Stokes equations [4], which describe viscous flows at the zero

---

*Department of Mathematics, Duke University, Durham, NC 27708, USA, alayton@math.duke.edu. This work was supported in part by NSF grant DMS-0715021.

S. Childress et al. (eds.), *Natural Locomotion in Fluids and on Surfaces*, IMA 155, DOI 10.1007/978-1-4614-3997-4_22,
© Springer Science+Business Media New York 2012

Reynolds number limit. The discretized Stokes equations form an elliptic system. The method can also be applied to the full Navier-Stokes equations, as was first done in [5] and [2], although, owing to the large number of cases and corrections needed, the implementation can be quite involved.

In a previous study [1], we proposed a method of velocity decomposition for the immersed interface problem. That method, like the immersed interface method, captures the jumps in the solution sharply and generates solutions that are second-order accurate in space and time. Compared to the immersed interface method, the method of velocity decomposition requires substantially fewer correction terms, and thus, in some sense, is simpler to implement. The method of velocity decomposition was motivated by the key observation that the jump conditions in the fluid variables for the problem with Navier-Stokes flow are the same as those for Stokes flow (see below). Given that consideration, the method of velocity decomposition splits the velocity and pressure into two parts, one part determined by the (steady) Stokes equations and the interfacial force on the immersed interface, and a second, more regular part which can be calculated on a regular grid without special treatment near the interface. As noted above, the decomposition allows the discontinuities in fluid variables to be accounted for accurately in a relatively simple manner. The method of velocity decomposition was originally developed for a two-dimensional domain with bi-periodic boundary conditions [1]. The goal of this study to extend that method to the Dirichlet boundary conditions.

2. Model Equations. We consider the coupled motion of a viscous fluid and an immersed boundary in a two-dimensional computational domain Ω. The fluid flow is described by the Navier-Stokes equations. Dirichlet boundary conditions are imposed at the domain boundary $\partial\Omega$. The immersed boundary Γ is assumed to be a simple closed curve that separates Ω into two subdomains, Ω^+ (exterior) and Ω^- (interior), so that $\Omega = \Omega^+ \cup \Gamma \cup \Omega^-$. (Generalization to multiple closed immersed boundaries is straightforward.) We assume that the fluid properties are the same in Ω^+ and Ω^-. Γ may exert a force on the fluid (see below), which induces discontinuities in the fluid pressure and velocity gradient at Γ.

The Navier-Stokes equations for the fluid motion are

$$\frac{\partial \mathbf{u}}{\partial t} + \mathbf{u} \cdot \nabla \mathbf{u} = -\nabla p + \mu \nabla^2 \mathbf{u} + \mathbf{f}, \quad \nabla \cdot \mathbf{u} = \mathbf{0}, \tag{1}$$

$$\mathbf{u}|_{\partial\Omega} = \mathbf{u}_b \tag{2}$$

where $\mathbf{u} = (u, v)$ denotes the fluid velocity; p is the pressure; μ is the fluid viscosity, assumed to be constant; and $\mathbf{f} = (f_1, f_2)$ is the interfacial force, supported entirely along Γ. The fluid density is set to 1.

The interfacial force $\mathbf{f}$ is frequently assumed to be a function of the configuration of the immersed interface. In some cases, Γ is assumed to be a membrane consisting of elastic material, so that when it is distorted from its rest state by stretching or relaxing, it exerts a restoring elastic tension force. The stretching is naturally expressed in terms of a material coordinate α on Γ, chosen to be arclength in the rest state. At time t the material point with label α has current position $\mathbf{X}(\alpha, t)$. We denote the arclength on Γ at the current time t by s and write the force $\mathbf{f} = (f_1, f_2)$ as

$$f_i(\mathbf{x}, t) = \int_0^L F_i(s, t)\delta(\mathbf{x} - \mathbf{X}(\alpha(s), t))ds, \qquad i = 1, 2. \tag{3}$$

where F_i is the force strength at point s and δ is the two-dimensional delta function. The tension force $\mathbf{F}$ is given by

$$\mathbf{F}(s, t) = \frac{\partial}{\partial s}(T(s, t)\boldsymbol{\tau}(s, t)), \tag{4}$$

where we assume the tension $T(s, t)$ is given by $T(s, t) = T_0 \left(\left| \frac{\partial \mathbf{X}}{\partial \alpha} \right| - 1 \right)$ when the material is stretched. The unit tangent vector $\boldsymbol{\tau}(s, t)$ to Γ is $\boldsymbol{\tau}(s, t) = \frac{\partial \mathbf{X}}{\partial s} = \frac{\partial \mathbf{X}/\partial \alpha}{|\partial \mathbf{X}/\partial \alpha|}$. Thus the force density can be computed directly from the location $\mathbf{X}(\alpha, t)$ of the boundary Γ. Note that in the relaxed state $|\partial \mathbf{X}/\partial \alpha| = 1$, and the tension vanishes.

Owing to the singularity in $\mathbf{f}$, the solution of (1) and (2) is not smooth across Γ. Its effect can be expressed in terms of jump conditions in p and $\mathbf{u}$ [4, 5, 7]):

$$[p] = \mathbf{f} \cdot \mathbf{n}, \quad \left[\frac{\partial p}{\partial n} \right] = \frac{\partial}{\partial s}(\mathbf{f} \cdot \boldsymbol{\tau}), \tag{5}$$

$$[\mathbf{u}] = 0, \quad \mu \left[\frac{\partial \mathbf{u}}{\partial \mathbf{n}} \right] = -(\mathbf{f} \cdot \boldsymbol{\tau})\boldsymbol{\tau}. \tag{6}$$

Unless special care is taken, these discontinuities tend to introduce substantial inaccuracy into the computed solution obtained by means of a standard finite difference method.

3. Numerical Method. Below we describe the method of velocity decomposition and discuss its formulation for Dirichlet boundary conditions. The method of velocity decomposition takes advantage of the fact that the jump conditions in the fluid variables for the problem with Navier-Stokes flow are the same as those for Stokes flow. To compute the solution of (1) and (2), we express the fluid velocity and pressure as the sum of a *Stokes* part, denoted by the subscript 's,' a *regular* part, denoted by the subscript 'r,' and a *boundary correction* part, denoted by the subscript 'bc':

$$\mathbf{u} = \mathbf{u}_s + \mathbf{u}_r + \mathbf{u}_{bc}, \quad p = p_s + p_r + p_{bc}. \tag{7}$$

The Stokes part of the solution is determined by the Stokes equations, including the boundary force:

$$\nabla p_s = \mu \nabla^2 \mathbf{u}_s + \mathbf{f}, \quad \nabla \cdot \mathbf{u}_s = 0 \tag{8}$$

The Stokes solution is assumed to satisfy the free-space boundary conditions, not the Dirichlet boundary conditions satisfied by the overall solution. As previously noted, the jump conditions for $\mathbf{u}_s$, p_s are the same as those for $\mathbf{u}$, p in (5) and (6). This follows from the continuity of $\mathbf{u}$ at Γ, which implies that the total derivative of $\mathbf{u}$ is continuous as well.

Taking the difference of (1) and (8), one obtains the equation for the regular and boundary correction parts of the solution

$$\frac{\partial}{\partial t}\left(\mathbf{u}_r + \mathbf{u}_{bc}\right) + \mathbf{u} \cdot \nabla \left(\mathbf{u}_r + \mathbf{u}_{bc}\right) = -\nabla \left(p_r + p_{bc}\right) + \mu\nabla^2 \left(\mathbf{u}_r + \mathbf{u}_{bc}\right) + \mathbf{f}_b,$$
$$\tag{9}$$

$$\nabla \cdot \left(\mathbf{u}_r + \mathbf{u}_{bc}\right) = 0, \tag{10}$$

where $\mathbf{f}_b$ is a body force given by the total derivative of the Stokes velocity: $\mathbf{f}_b = -\frac{\partial \mathbf{u}_s}{\partial t} - \mathbf{u} \cdot \nabla\mathbf{u}_s$. We let the regular solution satisfy the Navier-Stokes equations with the body force $\mathbf{f}_b$, i.e.,

$$\frac{\partial}{\partial t}\mathbf{u}_r + \mathbf{u} \cdot \nabla\mathbf{u}_r = -\nabla p_r + \mu\nabla^2\mathbf{u}_r + \mathbf{f}_b, \quad \nabla \cdot \mathbf{u}_r = 0. \tag{11}$$

The regular solution is assumed to satisfy the bi-periodic boundary conditions. Notice that the transport of both $\mathbf{u}_r$ on the left side of (9) and (11), and of $\mathbf{u}_s$ in $\mathbf{f}_b$ are with the full velocity $\mathbf{u}$. Unlike the interfacial force $\mathbf{f}$, the body force $\mathbf{f}_b$ is not singularly supported on Γ, nor does $\mathbf{f}_b$ have a jump discontinuity across Γ. $\mathbf{f}_b$ is a continuous function on Ω, since $\mathbf{u}_s$, and thus its total derivative, are continuous across Γ. However, the gradient of $\mathbf{f}_b$ has a jump discontinuity across Γ. Because the jump conditions for $\mathbf{u}_s$, p_s are the same as those for $\mathbf{u}$, p, the corresponding jumps for $\mathbf{u}_r$, p_r are zero. This fact and the continuity of $\mathbf{f}_b$ suggest that we can solve for $\mathbf{u}_r$, p_r on a regular grid accurately without jump terms at the interface.

At the domain boundary $\partial\Omega$, the velocity given by $\mathbf{u}_s + \mathbf{u}_r$ does not necessarily match the prescribed boundary velocity $\mathbf{u}_b$. Thus, to satisfy the Dirichlet boundary conditions (2), we solve for the boundary correction solution given by

$$\frac{\partial \mathbf{u}_{bc}}{\partial t} + \mathbf{u} \cdot \nabla\mathbf{u}_{bc} = -\nabla p_{bc} + \mu\nabla^2\mathbf{u}_{bc}, \quad \nabla \cdot \mathbf{u}_{bc} = 0, \tag{12}$$

$$\mathbf{u}_{bc}|_{\partial\Omega} = \mathbf{u}_b - \mathbf{u}_s|_{\partial\Omega} - \mathbf{u}_r|_{\partial\Omega} \tag{13}$$

To describe the discretized equations, we use the following notations. Let $\Delta t > 0$ be the time step, and let $t_n \equiv n\Delta t$ be the nth time level, for $n = 0, 1, \ldots$. For any time-dependent quantity ψ, we write ψ^n for $\psi(t_n)$. We use rectangular grids with grid interval h_x and h_y along the x- and y-axis. For notational simplicity, we assume $h_x = h_y \equiv h$. Given a computational domain $[x_a, x_b] \times [y_a, y_b]$, we compute values of fluid quantities at grid points $(x_a + ih, y_a + jh)$ where $i, j = 0, 1, 2, \ldots, N$. The position of the immersed boundary at time t is represented by a set of boundary markers

$\mathbf{X}_k(t) \equiv (X_k(t), Y_k(t))$, for $k = 0, 1, 2, \ldots, N_k$, where $\mathbf{X}_0(t) = \mathbf{X}_{N_k}(t)$, since the boundary is assumed to be a simple closed curve. The kth boundary marker approximates $\mathbf{X}(\alpha_k, t)$, where $\alpha_k = kL_0/N_k$, and L_0 is the length of Γ in the unstretched state; that is, the boundary markers are chosen so that they are equally spaced in the unstretched or relaxed state.

3.1. Computing the Stokes and Regular Solutions. We first compute the Stokes pressure p_s and velocity $\mathbf{u}_s \equiv (u_s, v_s)$ by solving (8). We follow the approach of LeVeque and Li [4] and reduce (8) to a sequence of Poisson problems, one for each unknown p_s, u_s, v_s; the resulting Poisson problems are solved together with the jump conditions for the unknowns. Once $\mathbf{u}_s$ is known, the body force $\mathbf{f}_b$, which is given by the total derivative of $\mathbf{u}_s$, can found at the current time. We advance $\mathbf{u}_r$ on a regular grid, by means of the semi-Lagrangian treatment of the total derivatives, to take advantage of the smoothness of those total derivatives along fluid trajectories. Implementation details can be found in [1].

3.2. Computing the Boundary Correction Solution. After $\mathbf{u}_s$ and $\mathbf{u}_r$ have been updated, we compute $\mathbf{u}_{bc}$ to ensure that the overall solution satisfies the Dirichlet boundary conditions (2). To that end, we first evaluate the Dirichlet boundary conditions $\mathbf{u}_{bc}^{n+1}|_{\partial\Omega} = \mathbf{u}_b^{n+1} - \mathbf{u}_s^{n+1}|_{\partial\Omega} - \mathbf{u}_r^{n+1}|_{\partial\Omega}$, using known boundary values of $\mathbf{u}_s^{n+1}$ and $\mathbf{u}_r^{n+1}$. Then a second-order projection method is used to solve (12) and (13) to yield $\mathbf{u}_{bc}^{n+1}$. Because the boundary correction solution is sufficiently smooth, no jump conditions are necessary. A second-order Eulerian time-stepping method is used, with which the time-discretized form of (12) is given by

$$\frac{\mathbf{u}_{bc}^{n+1} - \mathbf{u}_{bc}^n}{\Delta t} + \nabla p_{bc}^{n+\frac{1}{2}} = -\frac{1}{2}\left(\mathbf{u}_{bc}^{n+1}\cdot\nabla\mathbf{u}_{bc}^{n+1} + \mathbf{u}_{bc}^n\cdot\nabla\mathbf{u}_{bc}^n\right)$$
$$+ \frac{1}{2}\mu\nabla^2\left(\mathbf{u}_{bc}^{n+1} + \mathbf{u}_{bc}^n\right) \qquad (14)$$

3.3. Computing Boundary Motion. To advance the boundary configuration, we first evaluate the fluid velocity at the immersed interface, which we denote by $\mathbf{u}_\Gamma$. As in the case of the overall solution $\mathbf{u}$, the boundary velocity $\mathbf{u}_\Gamma$ can be expressed as the sum of a Stokes part and a regular part:

$$\mathbf{u}_\Gamma^n = \mathbf{u}_{s\Gamma}^n + \mathbf{u}_{r\Gamma}^n + \mathbf{u}_{bc\Gamma}^n. \qquad (15)$$

We first update $\mathbf{u}_{s\Gamma}^n$ by means of the representation of the Stokes velocity as a layer potential:

$$\mathbf{u}_{s\Gamma}^n = \int_\Gamma V(\mathbf{X}^n - \mathbf{y})\mathbf{f}^n(\mathbf{y})ds(\mathbf{y}), \qquad (16)$$

assuming free-space boundary conditions [8].

The values of $\mathbf{u}_{r\Gamma}$ and $\mathbf{u}_{bc\Gamma}$ can be approximated on the interface using spatial interpolations. These solutions are sufficiently smooth that no corrections are required in the interpolations. To advance the boundary configuration from t_n to t_{n+1}, we update the boundary markers using the two-step Adam-Bashforth method.

4. Numerical Results. We simulated the motion of a relaxing or oscillating ellipse, an example that has frequently been used to test numerical methods for immersed boundary problems [2, 4, 9]. The initial boundary is an ellipse with major and minor axes set to $a = 0.7513$ and $b = 0.4926$, respectively. The unstretched boundary was taken to be a circle with radius $r_0 = 0.5$. The tension coefficient T_0 was set to 0.2. We tested the method for a fluid that is relatively viscous, with the diffusion coefficient μ set to 0.1, and then for another fluid that is significantly less viscous with $\mu = 0.01$. In both cases, the computational domain was $[-1.2, 1.2] \times [-1.2, 1.2]$ with homogeneous Dirichlet boundary conditions. The fluid was initialized to be at rest, i.e., $\mathbf{u} = 0$ and $p = 0$ at $t = 0$.

After a sufficiently long simulation time, the interface is expected to converge to a circle with radius $r_e = \sqrt{ab} \approx 0.6084$, which is larger than the unstretched boundary but which has the same area as the initial ellipse, owing to the incompressibility of the enclosed fluid. At steady state, the fluid velocity vanishes everywhere; p attains constant values inside and outside the boundary, with a jump $[p] < 0$, because the boundary is initialized to a stretched state and because the limiting fluid velocity is zero. At steady state, the ellipse converged to a circle with a radius of ~ 0.61, which demonstrates that the method conserves area to a satisfactory extent.

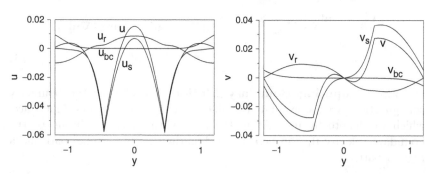

FIG. 1. *The fluid velocity at $x = 0.3$ and $t = 0.3$. u, v, full velocity; u_s, v_s, Stokes part; u_r, v_r, regular part; u_{bc}, v_{bc}, boundary correction part*

The model equations (1) and (2) were integrated using the method of velocity decomposition to nondimensional time $t = 10$. At the final time, in both cases, the boundary approaches its steady state, but, for $\mu = 0.01$, it oscillates at earlier times. Figure 1 shows the Stokes velocity (u_s and v_s), regular velocity (u_r and v_r), boundary correction velocity (u_{bc} and v_{bc}), and the overall fluid velocity (u and v), at $t = 0.3$ along

$x = 0.3$, for $\mu = 0.1$. The jump discontinuities in the normal derivatives of u_s, v_s, u, and v were captured sharply by the method. In contrast, the normal derivatives of u_r, v_r, u_{bc}, and v_{bc} are continuous across Γ. The magnitude of u_{bc} and v_{bc} is largest neighboring the domain boundary (i.e., at $y = \pm 1.2$).

TABLE 1

Relative errors in velocity components u and v at grid points, computed using N = 1,280 as the reference solution

N	$L_2(u)$	$L_\infty(u)$	$L_2(v)$	$L_\infty(v)$
80	1.904E−3	4.160E−2	9.825E−3	7.239E−2
160	6.720E−5	6.572E−3	7.698E−4	1.853E−2
320	6.936E−6	1.949E−3	1.160E−4	7.051E−3
640	1.280E−6	7.148E−4	6.296E−6	1.619E−3

Table 1 shows convergence results for $N = 80$, 160, 320, and 640. The number of boundary markers scales with N; specifically, we set $N_k = N$. The time-step Δt was set to 0.2 h. The errors in velocity at time $t = 0.3$ were obtained at the grid points in the computational domain and at boundary markers on the boundary Γ, in L^2 norm and L^∞ norm, using the reference solution computed on a fine grid with $N = 1,280$. These results exhibit approximately second-order accuracy in space-time. The value of the area A enclosed by Γ, computed at the approximate steady state at $t = 10$, is compared with the exact value $A = 0.7853981635$, which is known owing to incompressibility. Results exhibit second-order convergence at sufficiently large N values. Similar convergence results were also obtained for $\mu = 0.01$.

REFERENCES

[1] Beale JT, Layton AT (2009) A velocity decomposition approach for moving interfaces in viscous fluids. J Comput Phys 228:3358–3367

[2] Lee L, LeVeque RJ (2003) An immersed interface method for the incompressible Navier-Stokes equations. SIAM J Sci Comput 25:832–856

[3] LeVeque RJ, Li Z (1994) The immersed interface method for elliptic equations with discontinuous coefficients and singular sources. SIAM J Numer Anal 31:1019–1044

[4] LeVeque RJ, Li Z (1997) Immersed interface methods for Stokes flow with elastic boundaries or surface tension. SIAM J Sci Comput 18(3):709–735

[5] Li Z, Lai M-C (2001) The immersed interface method for the Navier-Stokes equations with singular forces. J Comput Phys 171:822–842

[6] Peskin CS (2002) The immersed boundary method. Acta Numer 11:479–517

[7] Peskin CS, Printz BF (1993) Improved volume conservation in the computation of flows with immersed elastic boundaries. J Comput Phys 105:33–46

[8] Pozrikidis C (1992) Boundary integral and singularity methods for linearized viscous flow. Cambridge University Press, Cambridge

[9] Tu C, Peskin CS (1992) Stability and instability in the computation of flows with moving immersed boundaries: a comparison of three methods. SIAM J Sci Stat Comput 13:1361–1376

ELLIPTIC REGULARIZATION AND THE SOLVABILITY OF SELF-PROPELLED LOCOMOTION PROBLEMS

ADAM BOUCHER(✉)*

Humans have been fascinated by the wonders of animal locomotion since antiquity. The effortless grace of birds in flight, and the silent glide of fish through water exemplify the elegance possible with control over the local fluid flow. While we have been free to observe the poetry in these motions for centuries, the ability to predict and mimic these motions with any degree of accuracy is a much more recent development. Advances in computing hardware and our numerical simulation capabilities have opened the door to the analysis and prediction of the motion of complex shapes in fluid environments. As we continue to develop more depth and sophistication in our ability to simulate the mathematical models of fluid mechanics, we must be sure to maintain an awareness of the existence and smoothness of the exact solutions to our mathematical models. In this paper we present a very brief overview of some techniques from partial differential equations which can be used to analyze physically relevant problems in animal locomotion. The ideas used here give important insight into situations where control systems for ordinary differential equations may be used to control the trajectories of vehicles through fluid environments.

As we pass from description to prediction using numerical simulation, it is important to clarify our understanding of the underlying mathematical models as much as possible. Numerical simulations approximate mathematical models, and mathematical models approximate the behavior of materials in the physical world, thus the information from a numerical simulation is further removed from the physical world than the theoretical model from which it is derived. Understanding the solvability properties: existence, uniqueness and regularity (smoothness) of the solutions to the underlying mathematical models is necessary to inform and guide the construction and evaluation of numerical methods. In the study of conservation laws, for example, shock waves are a real feature of the solutions to the mathematical models, and numerical methods must be specially designed to account for their appearance and behavior [5]. When weak or discontinuous solutions are present, our expectations of accuracy and convergence must be modified appropriately.

The study of the solvability of self-propelled locomotion is a relatively new field of study. In [4] Galdi studies the solvability and long time dynamics of a body which is assumed to propel itself using tangential stresses generated along the fluid-body interface (as in the envelope model of ciliary propulsion) and Silvestre [14] establishes the existence of strong solutions

---
*Department of Mathematics and Statistics, University of New Hampshire, Durham, NH 03824, USA adam.boucher@unh.edu

S. Childress et al. (eds.), *Natural Locomotion in Fluids and on Surfaces*, IMA 155, DOI 10.1007/978-1-4614-3997-4_23, © Springer Science+Business Media New York 2012

to a similar problem using the Stokes equations as a fluid model. More recently J.S. Martin et al. have presented an analysis of the two dimensional flow generated by the self-propelled motion of a body undergoing prescribed body deformations and have proven the existence of strong solutions to the self-propelling problem under suitable assumptions [7, 8]. In the work by Galdi and Silvestre, the problems are analyzed on an exterior domain in body fixed coordinates. In the work on fish swimming by J.S. Martin et al., the fluid problem is lifted to a fixed reference domain, and the resulting problem is studied using the theory of nonlinear semi-groups using an extension of the ideas from Takahashi [15]. Here, I demonstrate how elliptic regularization provides a natural environment for the study of the solvability properties of locomotion problems. I give the fundamental existence result for the ambient flow problem, and give a short outline of how this result might be used to conduct a more thorough analysis of the three dimensional self-propelling problem as posed in [7].

The self-propelling problem as posed in [7] should be of considerable interest to mathematicians and engineers interested in the design and control of shape changing vehicles. The self-propelling assumptions central to this problem permit the researcher to work directly with the shape of the immersed body. This allows one to look at the effect of shape variables such as flapping kinematics to study locomotion without referencing any of the internal mechanics driving the deformation. In an ideal setting an engineer could design a locomotive gait using deformation as the driving force for locomotion, and an engineer, who may be completely ignorant of the underlying control problem could design a mechanical system which acheives the desired deformations.

1. The Ambient Flow Problem. One particularly fruitful and popular method for the study of animal locomotion consists of experimentally measuring the kinematics of the body of interest, then computationally simulating those kinematics to recover the ambient flow field around the body. Such methods allow the determination and analysis of the kinematic parameters used by animals in locomotion, and may give researchers insight into frequency selection, or optimal attack angles for various flight maneuvers [17, 18]. If we imagine that the body kinematics (including trajectory through space) are all completely prescribed, then we might take the surrounding flow field to be governed by the incompressible Navier-Stokes equations. To analyze the solvability of this situation we must deal with a non-homogeneous initial value problem defined on a non-cylindrical domain.

Let the location of a moving body be prescribed by the moving set $\mathcal{B}(t)$ on the time interval $[0, T]$. If we consider the flow in a bounded region, $D \subset \mathbb{R}^3$, surrounding the body, then by defining $\Omega(t) \equiv D - \mathcal{B}(t)$ the velocity and pressure for the fluid surrounding the body must satisfy:

$$\mathbf{u}_t - \nu \Delta \mathbf{u} + (\mathbf{u} \cdot \nabla) \mathbf{u} + \frac{1}{\rho} \nabla p = \mathbf{g}, \text{ in } \Omega(t), \text{ for each } t \in [0, T], \tag{1}$$

$$\nabla \cdot \mathbf{u} = 0 \text{ in } \Omega(t) \text{ for each } t \in [0, T], \tag{2}$$

$$\mathbf{u} = U(\mathbf{x}, t), \text{ on } \delta\mathcal{B}(t), \text{ for each } t \in [0, T], \tag{3}$$

$$\mathbf{u} = 0 \text{ on } \delta D \text{ for each } t \in [0, T], \tag{4}$$

$$\mathbf{u}(0) = \mathbf{u}_0, \text{ on } D - \mathcal{B}(0) \text{ at } t = 0. \tag{5}$$

Here ν is the viscosity of the fluid, ρ is the density of the fluid, $\mathbf{g}$ represents any external body forces on the fluid, and $U(\mathbf{x}, t)$ represents the velocity field for the material points along the surface of the body.

Using the technique of elliptic regularization originally created by J. Lions for the study of degenerate parabolic operators, and subsequently adapted by R. Salvi for the analysis of the Navier-Stokes equations, we can easily establish the existence of weak solutions to this problem. Below we include definitions of the relevant function spaces, the variational equation satisfied by the weak solution, the technical result and a brief outline of the estimates necessary for the proof.

To study the solvability of unsteady problems on moving domains one typically introduces a set of time dependent function spaces of the form, $L^2[0, T, H^m(\Omega(t))]$. These spaces can be understood as mappings from the interval $[0, T]$ into Hilbert spaces which depend upon the value of t. These spaces are Banach spaces with the following norms:

$$\|f\|_{L^2[0,T,H^m(\Omega(t))]} \equiv \left(\int_0^T \|f\|_{H^m(\Omega(t))}^2 \, dt \right)^{\frac{1}{2}}.$$

For each t the spaces $H^m(\Omega(t))$ are classical Sobolev spaces, with elements possessing weak derivatives up to and including order m in $L^2(\Omega(t))$. A rough overview of these spaces can be found in Evans [3]. The precise notation used here is explained in the work of Rodolfo Salvi [10–12]. We use a subscript of 0 to denote the zero trace subspace, and we use a subscript of σ to denote the divergence free subspace (e.g. $H_{0,\sigma}^1(\Omega(t))$ is the space of divergence free functions with weak first derivatives, and zero trace along the boundary of $\Omega(t)$). We do not distinguish between scalar and vector versions of these function spaces. Briefly we define:

$$(\mathbf{u}, \mathbf{v})_{\Omega(t)} = \int_{\Omega(t)} \sum_i u_i v_i d\mathbf{x}, \qquad |\mathbf{u}|_{\Omega(t)}^2 = (\mathbf{u}, \mathbf{u})_{\Omega(t)},$$

$$((\mathbf{u}, \mathbf{v}))_{\Omega(t)} = \int_{\Omega(t)} \sum_i \nabla u_i \cdot \nabla v_i d\mathbf{x}, \qquad \|\mathbf{u}\|_{\Omega(t)}^2 = ((\mathbf{u}, \mathbf{u}))_{\Omega(t)}.$$

We make the following assumptions about the boundary velocity:

(A1) The boundary velocity $U(\mathbf{x}, t)$ possesses a divergence free extension belonging to at least $C^1[0, T; H^1_{0,\sigma}(\Omega(t))]$, with the gradient ∇U belonging to the space $C^0[0, T; H^1(\Omega(t))]$. For brevity, we use the same symbol U to denote the extended boundary velocity.

(REMARK. Using this assumption we change variables to render the boundary conditions homogeneous. This change of variables leads to the weak variational equation analyzed in Theorem 1.1.)

(A2) The boundary velocity U obeys a 'smallness' condition to ensure strong positivity in Theorem 1.1.

Specifically we require that:

$$\sup_{t \in [0,T]} \|U(t)\|_{\Omega(t)} \leq \alpha$$

where α is chosen sufficiently small to ensure the form $a_m(\boldsymbol{\psi}, \boldsymbol{\psi})$ (defined below) is strongly positive.

Taking the momentum equation, multiplying by a smooth, divergence free test function and integrating by parts in space and time we can obtain the following weak, variational form of the problem defined in (1)–(5).

Find a distribution $\mathbf{u} \in L^\infty[0, T, L^2(\Omega(t))] \cap L^2[0, T, H^1_{0,\sigma}(\Omega(t))]$ satisfying the following variational equation:

$$\int_0^T \left[-(\mathbf{u}, \partial_t \boldsymbol{\phi})_{\Omega(t)} + \nu((\mathbf{u}, \boldsymbol{\phi}))_{\Omega(t)} + ((\mathbf{u} - U) \cdot \nabla \mathbf{u}, \boldsymbol{\phi})_{\Omega(t)} \right.$$

$$\left. + (\mathbf{u} \cdot \nabla U, \boldsymbol{\phi})_{\Omega(t)} - (\mathbf{f}, \boldsymbol{\phi})_{\Omega(t)} \right] dt \qquad (6)$$

$$= (\mathbf{u}_0, \boldsymbol{\phi}(0))_{\Omega(0)} - (\mathbf{u}(T), \boldsymbol{\phi}(T))_{\Omega(T)},$$

for any test function $\boldsymbol{\phi} \in \mathcal{D}_\sigma(\Omega_T)$.

The existence of a weak solution to this problem is granted by Theorem 1.1

THEOREM 1.1. *Assuming the domain $\Omega(t)$ is uniformly C^3 and (A1) and (A2) are satisfied, there exists at least one weak solution, $\mathbf{u} \in L^\infty[0, T; L^2(\Omega(t))] \cap L^2[0, T; H^1_{0,\sigma}(\Omega(t))]$ satisfying (6).*

Outline of Proof. To establish the existence of this problem we use the technique of elliptic regularization. The basic idea behind this technique is to solve a sequence of approximate problems which are uniformly elliptic in space and time. After the existence to the regularized problems is established, we show that the solutions of these approximating problems converge to the desired solution in an appropriate limit.

The elliptic structure of the approximating problems is mathematically similar to the Laplacian operator. This structure allows us to prove existenceusing traditional theorems from Hilbert space theory. This technique is especially helpful for moving boundary problems because each of the approximating problems is analyzed using only classical Sobolev spaces defined on $\Omega(t) \times (0, T) \subset \mathbb{R}^{3+1}$. This approach bypasses the difficulties encountered by more traditional Galerkin methods for time dependent problems on cylindrical domains. Since the time variable is treated on an equal footing with the spatial variables for each of the approximating problems, the time-variability of the spatial function spaces is handled naturally, If one tried to use a traditional Galerkin truncation argument, one would immediately run into technical difficulty since a single orthonormal basis cannot capture the moving boundaries of the domain.

We first establish the existence of solutions to the following sequence of approximating problems. These approximate problems depend on a regularization parameter m, and we analyze these problems with the goal of showing that their solutions converge to the solution of the non-cylindrical flow problem when $m \to \infty$.

We wish to find a regularized velocity field $\mathbf{v}^m$ satisfying:

$$\int_0^T \left[\frac{1}{m}(\mathbf{v}_t^m, \boldsymbol{\phi}_t)_{\Omega(t)} - (\mathbf{v}^m, \partial_t\boldsymbol{\phi})_{\Omega(t)} + \nu((\mathbf{v}^m, \boldsymbol{\phi}))_{\Omega(t)} \right.$$

$$+ ((\mathbf{v}^m - U) \cdot \nabla\mathbf{v}^m, \boldsymbol{\phi})_{\Omega(t)} \tag{7}$$

$$\left. + (\mathbf{v}^m \cdot \nabla U, \boldsymbol{\phi})_{\Omega(t)} - (\mathbf{f}, \boldsymbol{\phi})_{\Omega(t)} \right] dt = (\mathbf{v}_0^m, \boldsymbol{\phi}(0)),$$

for any test function $\boldsymbol{\phi} \in \mathcal{D}_\sigma(\Omega_T)$.

To prove the existence of approximating solutions we study the following form:

$$a_m(\boldsymbol{\psi}, \boldsymbol{\phi}) = \int_0^T \left[\frac{1}{m}(\boldsymbol{\psi}_t, \boldsymbol{\phi}_t)_{\Omega(t)} - (\boldsymbol{\psi}, \boldsymbol{\phi}_t)_{\Omega(t)} + \nu((\boldsymbol{\psi}, \boldsymbol{\phi}))_{\Omega(t)} \right.$$

$$\left. + ((\boldsymbol{\psi} - U) \cdot \nabla\boldsymbol{\psi}, \boldsymbol{\phi})_{\Omega(t)} + (\boldsymbol{\psi} \cdot \nabla U, \boldsymbol{\phi})_{\Omega(t)} \right] dt$$

$$+ (\boldsymbol{\psi}(T), \boldsymbol{\phi}(T))_{\Omega(t)},$$

using assumption (A2) with α chosen sufficiently small, we can bound the nonlinear term and show this form is strongly positive over the space $L^2[0, T, H^1(\Omega(t))]$.

$$|a_m(\boldsymbol{\psi}, \boldsymbol{\psi})| \geq \int_0^T \left[\frac{1}{m} |\boldsymbol{\psi}_t, \boldsymbol{\phi}_t|_{\Omega(t)}^2 + C \, \|\boldsymbol{\psi}\|_{\Omega(t)}^2 \right] dt$$
$$+ \frac{1}{2} |\boldsymbol{\psi}(T)|_{\Omega(t)}^2 + \frac{1}{2} |\boldsymbol{\psi}(0)|_{\Omega(t)}^2,$$

$$|a_m(\boldsymbol{\psi}, \boldsymbol{\psi})| \geq C_m \, \|\boldsymbol{\psi}\|_{L^2[0,T,H^1(\Omega(t))]}^2 \, .$$

The existence of solutions to (7) follows directly from a variant of the nonlinear Lax-Milgram Theorem contained in Theorem 1.2

THEOREM 1.2. *Let $a(\boldsymbol{\phi}, \boldsymbol{\psi})$ be a form over the real, separable Hilbert space, $\mathcal{H}$, and let $\mathcal{L}(\boldsymbol{\phi})$ be a bounded linear functional on the same function space. Whenever:*

(i) the form $a(\boldsymbol{\psi}, \boldsymbol{\psi})$ is strongly positive, such that:

$$|a(\boldsymbol{\psi}, \boldsymbol{\psi})| \geq C \, \|\boldsymbol{\psi}\|_{\mathcal{H}}^2 \, ,$$

(ii) and the form $\boldsymbol{\psi} \to a(\boldsymbol{\psi}, \boldsymbol{\phi})$ is weakly continuous over $\mathcal{H}$, then there exists at least one $\boldsymbol{\psi} \in \mathcal{H}$ solving the problem:

$$a(\boldsymbol{\psi}, \boldsymbol{\phi}) = \mathcal{L}(\boldsymbol{\phi}).$$

(see [10, 11] and [12] for details).

With the existence of solutions to the approximating problems established, we obtain appropriate a priori estimates which are independent of m that allow passage to the limit.

The essential estimates are given by:

$$\int_0^T \frac{1}{m} |\mathbf{v}_t^m|_{\Omega(t)}^2 \, dt \leq C,$$
$$\int_0^T \|\mathbf{v}^m\|_{\Omega(t)}^2 \, dt \leq C, \tag{8}$$
$$|\mathbf{v}^m(T)|_{\Omega(T)}^2 \leq C.$$

These bounds allow passage to the limit in the linear terms, but to pass to the limit in the non-linear terms we need to establish an additional estimate in a strong topology. By defining the time-difference quotient:

$$\mathbf{v}_h^m = \frac{1}{h} \int_{t-h}^t \mathbf{v}^m(\mathbf{x}, s) ds,$$

and using the bounds above with some relatively straightforward calculations we can show:

$$\int_0^T |\mathbf{v}^m(t+h) - \mathbf{v}^m(t)|_{\Omega(t)}^2 \, dt \leq C\sqrt{h}. \tag{9}$$

Given the bounds obtained from (8) to (9), the sequence $\{\mathbf{v}^m\}_{m=1}^\infty$ is relatively compact in $L^2[0, T, L^2(\Omega(t))]$ [12].

Since $L^2[0, T; L^2(\Omega(t))]$ is a reflexive Banach space we can extract a subsequence $\mathbf{v}^{m_k}$ such that:

$$\lim_{m_k \to \infty} \int_0^T ((\mathbf{v}^{m_k} - U) \cdot \nabla \mathbf{v}^{m_k}, \boldsymbol{\phi})_{\Omega(t)} = \int_0^T ((\mathbf{v} - U) \cdot \nabla \mathbf{v}, \boldsymbol{\phi})_{\Omega(t)}.$$

Finally by passing to the limit $m \to \infty$, we obtain the existence of at least one weak solution to (6). The proof of Theorem 1.1 is now complete.

2. The Self-propelling Problem. A variant of the ambient flow problem which should be of particular interest to engineers hoping to create and control biomimetic vehicles is the self-propelling problem. Here, following San Martin et al. [7] we prescribe the body shape (which is assumed to deform), but we do not prescribe the trajectory of the body through the fluid. Of particular importance in these problems are the self-propelling conditions:

$$\int_{\mathcal{B}(t)} \rho^*(\mathbf{x}^*, t) \mathbf{w}^* d\mathbf{x}^* = 0, \qquad \int_{\mathcal{B}(t)} \rho^*(\mathbf{x}^*, t) (\mathbf{x}^*) \wedge \mathbf{w}^* d\mathbf{x}^* = 0, \qquad \forall t \geq 0.$$

Here: $\mathcal{B}(t) \subset \mathbb{R}^3$ is the volume occupied by the body at time t, ρ^* is the density of the body, and $\mathbf{w}^*$ is the Lagrangian velocity of the body in the body fixed frame.

These conditions ensure that no net linear or angular momentum is generated by the internal mechanics. The self-propelling conditions guarantee the internal forces generated by the body only change the shape of the body, and do not contribute to the overall momentum balance within the fluid. Essentially these conditions force the isolated body to obey conservation of linear and angular momentum locally. These conditions are assumed to hold in addition to the global conservation of momentum obeyed by the coupled fluid and body system. By imposing these conditions a priori we isolate the internal workings of the body and permit an analysis of the fluid-structure interaction problem with the internal mechanics treated as a black box. Furthermore, these conditions ensure that the motion of the body is indeed the natural motion of the body through the fluid, not the motion generated by a combination of shape-change and external (non-physical) motive forces.

Note carefully the domain of integration in these conditions. We expect from physical principles that in any locomotion problem where the body and fluid begin at rest the total linear and angular momentum of the fluid and body will be zero, but these conditions are more restrictive. Without these conditions, it is possible to obtain locomotion by introducing momentum sources within the body which are then transferred to the surrounding fluid. These conditions preclude such unnatural momentum sources, and also permit the boundary velocity to be decomposed into a rigid part and a part due to body deformation [7].

For engineers and physicists, these conditions yield the most natural approach to design and control problems for shape changing vehicles. Given a robotic system capable of deforming its shape, these conditions allow a direct analysis using the body geometry during locomotion rather than requiring an analysis of the internal mechanics. By working on the problem in this form it should be possible to design control systems and different locomotive 'gaits' without needing detailed specifications on the internal mechanical structure.

The self-propelling problem for a shape changing body is given by coupling the incompressible Navier-Stokes equations to Newton's laws for the position for the bary-center and the orientation of the body within the flow. Following the notation of [7] and using all the simplifications provided by the self-propelling conditions, the problem is to find the trajectory for the position of the bary-center and angular velocity of the body given by $(\boldsymbol{\xi}, \boldsymbol{\omega})$ along with the velocity and pressure fields of the ambient flow field.

Explicitly the self-propelling problem requires finding functions $(\boldsymbol{\xi}, \boldsymbol{\omega}, \mathbf{u}, p)$ satisfying the following system of coupled ordinary and partial differential equations:

$$\mathbf{u}_t - \nu \Delta \mathbf{u} + (\mathbf{u} \cdot \nabla)\mathbf{u} + \frac{1}{\rho}\nabla p = \mathbf{g}, \text{ in } \Omega(t), \tag{10}$$

$$\nabla \cdot \mathbf{u} = 0 \text{ in } \Omega(t), \tag{11}$$

$$\mathbf{u} = \frac{d\boldsymbol{\xi}(t)}{dt} + (\mathbf{x} - \boldsymbol{\xi}(t)) \wedge \boldsymbol{\omega}(t) + \mathbf{w} \text{ on } \delta\mathcal{B}(t), \tag{12}$$

$$\mathbf{u} = 0 \text{ on } \delta D, \tag{13}$$

$$\mathbf{u}(0) = \mathbf{u}_0, \text{ on } D - \mathcal{B}(0), \tag{14}$$

$$m\frac{d^2\boldsymbol{\xi}}{dt^2} = -\int_{\delta\mathcal{B}(t)} \boldsymbol{\sigma}(\mathbf{u}, p)\mathbf{n} dS, \tag{15}$$

$$\boldsymbol{\xi}(0) = (0,0,0), \quad \frac{d\boldsymbol{\xi}}{dt}(0) = (0,0,0), \tag{16}$$

$$\frac{d}{dt}(I(t)\boldsymbol{\omega}) = -\int_{\delta\mathcal{B}(t)} (\mathbf{x} - \boldsymbol{\xi}) \wedge \boldsymbol{\sigma}(\mathbf{u}, p)\mathbf{n} dS, \tag{17}$$

$$\boldsymbol{\theta}(0) = (0,0,0), \quad \boldsymbol{\omega}(0) = (0,0,0).$$

The solvability of this problem has been established in two spatial dimensions with no external body forces on the fluid, however the existence of solutions to this problem in three spatial dimensions with external forces is currently an open question. Since elliptic regularization allows such a simple treatment of the Navier-Stokes equations on non-cylindrical domains, we outline a proof for establishing the existence of strong and classical solutions to this problem. The approach uses the self-propelling assumptions to prove the existence of solutions to the fluid problem on a 'bundle' of different trajectories, and then uses classical ODE theory to deduce the local existence of a self-propelled trajectory.

We first assume that the hypotheses of Theorem 1.1 are satisfied while the trajectories remain in a ball compactly emdedded in the interior of D. By further assuming that enough additional regularity on the velocity and pressure fields can be obtained to ensure the hydrodynamic drag is well defined and continuous in time, we can pose the self-propelling problem as a system of first order, non-linear ordinary differential equations for the trajectory where the hydrodynamic force plays the role of a non-linear forcing term. Since the flow depends upon the whole trajectory of the body, this is a Volterra-like forcing term with a memory of the whole body trajectory. Since we are assuming the fluid problem is well posed for all trajectories within this ball, we can use the simpler topology of the space of trajectories to help obtain bounds on the fluid forcing. By establishing the uniform continuity of the fluid solutions with respect to perturbations in the trajectory of the body, we can then establish the existence of a fixed point (at least locally in time) for the body's trajectory. The existence of strong or even classical solutions to the self-propelling problem is then an easy consequence.

3. Discussion and Conclusion. We have shown how the technique of elliptic regularization may be applied to a pair of very general problems in animal locomotion. Furthermore this technique provides a new avenue for analyzing the solvability of the self-propelling problem. While the outline here is far from precise, the possibility of establishing rigorous results to the solvability of fluid structure interaction problems using the combination of elliptic regularization and classical results from ordinary differential equations opens the door to rigorous analysis justifying the use of simple control systems to handle flow control around shape changing bodies.

From the perspective of mathematical analysis, the idea of using the existence of fluid flows for a collection of prescribed boundary motions to help prove the existence of solution to fluid structure interaction problems, is to our knowledge completely new. This methodology breaks away from the more traditional approach of lifting to a fixed, reference fluid domain and conducting analysis on the coupled 'perturbed fluid'-structure system. We feel this approach has many potential benefits over the more traditional approach, not the least of which is the modularity of the analysis. Since this approach handles the Navier Stokes-equations directly, the analysis and results are more easily reconciled with current numerical schemes for solving fluid-structure interaction problems such as ALE methods. Further work on the regularity of the solutions to these problems is needed.

An investigation of how the regularity of the boundary impacts the regularity of the ambient flow field is important for both the construction and analysis of numerical methods specifically in animal locomotion, and more generally in fluid-structure interaction problems.

Acknowledgements. I would like to thank the workshop organizers for their hard work in preparation, and the workshop participants for their presentations and ideas. I would like to thank the IMA for hosting the Natural Locomotion in Fluids workshop. I would also like to thank my colleagues at the University of New Hampshire for their support and insight.

REFERENCES

[1] Agarwal RP, Meehan M, O'Regan D (2001) Fixed point theory and applications. Cambridge University Press, Cambridge/New York

[2] Childress S (1981) Mechanics of swimming and flying. Cambridge University Press, Cambridge

[3] Evans C (1998) Partial differential equations. Graduate studies in mathematics. AMS, Providence

[4] Galdi GP (1999) On the steady self-propelled motion of a body in a viscous incompressible fluid. Arch Ration Mech Anal 148:53-88

[5] Leveque RJ (2002) Finite volume methods for hyperbolic problems. Cambridge University Press, Cambridge/New York

[6] Lighthill J (1975) Mathematical biofluiddynamics. CBMS 17. SIAM, Philadelphia

[7] Martin JS, Scheid J-F, Takahashi T, Tucsnak M (2008) An initial boundary value problem modeling of fish-like swimming. Arch Ration Mech Anal 188:429-455

[8] Martin JS, Takahashi T, Tucsnak M A (2007) control theoretic approach to the swimming of microscopic organisms Quart Appl Math 65: 405-424

[9] Medeiros LA, Ferrel JL (1997) Elliptic regularization and Navier-stokes system. Mem Differ Equ Phys 12:165-177

[10] Salvi R (1985) On the existence of weak solutions of a nonlinear mixed problem for the Navier-Stokes equations in a time dependent domain. J Fac Sci Univ Tokyo Sect IA Math 32:213-221

[11] Salvi R (1988) On the Navier-stokes equations in non-cylindrical domains: one the existence and regularity. Math Z 199:153-170

[12] Salvi R (1994) On the existence of periodic weak solutions of Navier-stokes equations in regions with periodically moving boundaries. Acta Appl Math 37:169-179

[13] Simon J (1986) Compact Sets in the space $L^p(0, T, B)$. Annali Di Matematica Pura ed Applicata 146(1):65-96

[14] Silvestre AL (2002) On the slow motion of a self-propelled rigid body in a viscous incompressible fluid. J Math Anal Appl 274:203-227

[15] Takahashi T (2003) Analysis of strong solutions for the equations modeling the motion of a rigid-fluid system in a bounded domain. Adv Differ Equ 8: 1499-1532

[16] Temam R (1994) Navier Stokes equations: theory and numerical analysis. AMS Chelsea

[17] Wang ZJ (2000) Vortex shedding and frequency selection in flapping flight. J Fluid Mech 410:323-341

[18] Wang ZJ (2005) Dissecting insect flight. Annu Rev Fluid Mech 37:183-210

COMPARATIVE STUDIES REVEAL PRINCIPLES OF MOVEMENT ON AND WITHIN GRANULAR MEDIA

YANG DING*, NICK GRAVISH*, CHEN LI*, RYAN D. MALADEN*,
NICOLE MAZOUCHOVA†, SARAH S. SHARPE‡,
PAUL B. UMBANHOWAR§, AND DANIEL I. GOLDMAN(✉)¶

Abstract. Terrestrial locomotion frequently occurs on complex substrates such as leaf litter, debris, and soil that flow or solidify in response to stress. While principles of movement in air and water are revealed through study of the hydrodynamic equations of fluid motion, discovery of principles of movement in complex terrestrial environments is less advanced in part because describing the physics of limb and body interaction with such environments remains challenging. We report progress we have made in discovering principles of movement of organisms and models of organisms (robots) on and within granular materials (GM) like sand. We review current understanding of localized intrusion in GM relevant to foot and body interactions. We discuss the limb-ground interactions of a desert lizard, a hatchling sea turtle, and various robots and reveal that control of granular solidification can generate effective movement. We describe the sensitivity of movement on GM to gait parameters and discuss how changes in material state can strongly affect locomotor performance. We examine subsurface movement, common in desert animals like the sandfish lizard. High speed x-ray imaging resolves subsurface kinematics, while electromyography (EMG) allows muscle activation patterns to be studied. Our resistive force theory, numerical, and robotic models of sand-swimming reveal that subsurface swimming occurs in a "frictional fluid" whose properties differ from Newtonian fluids.

Key words. Locomotion, walking, running, crawling, swimming, lizard, robot, turtle, granular

AMS(MOS) subject classifications. Primary 1234, 5678, 9101112

1. Introduction. Discovery of biomechanical [2, 7] and neuromechanical [9, 39] principles of locomotion in a given environment requires comparative study of organisms in the environment, understanding the mechanics of interaction with the environment, and modeling of both organism and environment. For example, an understanding of organism flight has emerged through identification of effective wing kinematics [41] and experimental, computational, and analytic study of air flow patterns over real and model wing shapes. This extensive study of such aerial (and aquatic)

*School of Physics, Georgia Institute of Technology, Atlanta, GA 30332, USA. dingyang@gatech.edu

†School of Biology, Georgia Institute of Technology, Atlanta, GA 30332, USA.

‡Bioengineering Graduate Program, Georgia Institute of Technology, Atlanta, GA 30332, USA.

§Department of Mechanical Engineering, Northwestern University, Evanston, IL 60208, USA.

¶School of Physics, School of Biology, Bioengineering Graduate Program, Georgia Institute of Technology, Atlanta, GA 30332, USA. daniel.goldman@physics.gatech.edu
This work was supported by the Burroughs Wellcome Fund, NSF PoLS, NSF CMMI, ARL MAST CTA and DARPA.

S. Childress et al. (eds.), *Natural Locomotion in Fluids and on Surfaces*, IMA 155, DOI 10.1007/978-1-4614-3997-4_24,
© Springer Science+Business Media New York 2012

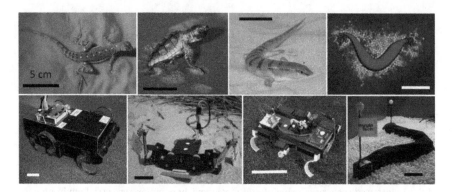

FIG. 1. *Organisms and models studied by our group to reveal principles of movement on and within GM. Top row: Zebra-tailed lizard, Loggerhead sea turtle (hatchling), Sandfish lizard, Numerical simulation of a sandfish. Bottom row: SandBot, FlipperBot, RoACH, Sandfish robot. All scale bars are approximately 5 cm long.*

locomotion has resulted in progress in the creation of robotic devices that can maneuver in fluid environments [5, 23].

Principles that govern the locomotion of animals that live in complex *terrestrial* environments are much less understood. For example, mountain goats bound across steep rubble-strewn slopes with agility human-made devices cannot currently match [36]. One reason for this gap in understanding is, unlike aerial and aquatic environments, common terrestrial environments like dirt, leaf litter, rubble, and sand are not yet adequately described by models at a level comparable to those that describe fluid-flow (e.g., the Navier-Stokes equations). Prediction of ground reaction forces is therefore challenging, and, consequently, quantitative discovery of principles of locomotion and design of devices (like robots) that operate effectively in such environments remains elusive.

Dry granular media (GM), e.g., the sands of deserts and beaches, are common examples of complex flowing substrates. Composed of collections of particles that interact through dissipative contact forces, these materials exhibit both solid- and fluid-like in response to stress. GM are good substrates with which to study terrestrial locomotion since they are readily controlled (by use of a fluidized bed [26], for example) and thus a range of repeatable initial conditions that mimic conditions of sand found in nature can be generated in the laboratory. A diversity of animals are important members of the ecosystems in sandy environments (examples in Fig. 1), and one can expect granular rheology to affect their locomotor strategies and performance. There have been descriptive studies of organism behavior (for examples, see [3, 33]) but fewer detailed studies [18, 22, 25] of above-ground biomechanics on GM. While many organisms move over the surface of sand, a large number also bury in it (e.g., many lizards, snakes, scorpions, spiders, crabs) and some even swim *within* the sand [33]. However, because of limited visualization tools, there have been even fewer detailed subsurface studies [35].

In this overview, we discuss progress our group has made towards discovering principles of locomotion in dry GM by comparative studies of animals and models (physical and numerical) on and within GM substrates. We illustrate how our modeling approaches have advanced descriptions of limb and body interaction with GM. We first review the relevant physics of localized intrusion in GM and then discuss locomotion above and below the surface.

2. Localized Intrusion of Granular Material. GM exhibit complex rheology [21] affected by both the properties of the particles (e.g., coefficient of friction, polydispersity, particle shape, etc.) and the compaction state of the medium. Compaction affects granular forces in nontrivial ways, but on average, closely packed GM resists larger stresses before flowing, compared to loosely packed GM [13, 40]. Compaction can be characterized by the volume fraction, ϕ, the ratio of solid volume to occupied volume within a region. In terms of the total mass (m), occupied volume (V), and the intrinsic particle density (ρ), $\phi = \frac{m}{\rho V}$. GM in nature such as on sand dunes [4] and beach environments [32] exist in a wide range of compaction states; using a fluidized bed and dry, round particles, GM can be prepared [13, 26] with compaction states ranging from loose ($\phi \approx 0.58$) to close ($\phi \approx 0.63$) with corresponding mechanical properties similar to natural sand.

The frictional nature of GM produces a yield force (F_{yield}), a threshold below which grains do not flow in response to forcing [34]. Above F_{yield} GM flow and, for low intrusion speeds, the force on the intruder is speed independent [42] unlike the case for fluids. As intrusion speed increases, inertial forces dominate frictional forces and typically vary as $F(v) \propto v^2$ [42]. Like the hydrostatic force in fluids, the average stress within GM increases approximately linearly with depth. Because the pressure drop across intruders is typically much smaller than the yield force, granular buoyancy in static granular beds is usually unimportant. Vertical intrusion of objects into GM often results in a penetration force linear in depth z and projected intruder surface area A, namely $F(z) = \alpha A z$. The resistance of GM to penetration α is a function of the material properties and packing state.

The physics of GM most relevant to our studies is that of localized forcing: The penetration and movement of feet, limbs, heads, or bodies [26, 30, 32]. Studies of localized forcing with horizontally and vertically translating intruders in initially homogeneous GM have been conducted [1, 10, 12, 40]. Much like in fluids, intruders moving horizontally through GM experience drag and lift forces. In GM however, these forces arise from normal and frictional forces on the intruder's surface, and are transmitted by "force chains" between particles in the bulk [10]. A previous study showed that the drag force on an intruder horizontally translated through GM depends less on intruder shape as compared to the drag force in a fluid [1]. Simulation revealed that in GM a band-shaped region of grains flow upward and forward in front of the intruder, with a volume roughly proportional to intruder cross-sectional area regardless of intruder shape; the drag force whose magnitude is set by the weight of the grains within the flowing region is thus insensitive to intruder shape [8]. For arbitrary intruder shape, we have discovered that both drag and lift forces can be approximated by decomposing the leading surface into small plates and summing the normal and tangential (frictional) forces on the plates. Since a portion of the flowing grains can also be pushed upward or downward by the leading surface, the intruder may experience a net positive or negative lift force depending on its shape [8].

3. Limbed Locomotion on Sand: Walking and Running. To run, walk, or crawl on a substrate, terrestrial animals generate forward thrust to advance and vertical lift to counter gravity. Unlike on rigid, non-slip ground, on yielding substrates like loosely packed sand animals like lizards [25] and turtles [31] intrude their feet into the substrate to generate sufficient forces to generate forward and vertical propulsion. This creates a dilemma: Deeper intrusion generates larger forces but increased foot penetration reduces stride length and increases drag on the limbs and the body. Deeper penetration also results in increased energy loss due to irreversible work done on the substrate.

Our study of a six legged robot, SandBot (Fig. 1), demonstrated the precariousness of moving on GM [26] and the importance of GM's finite yield stress. In our study SandBot ($\sim$30 cm, $\sim$2.3 kg) used an alternating tripod gait and relatively small, C-shaped legs (total area $\approx$ 15 cm^2) to walk on a fluidized bed trackway in which ϕ could be varied from 0.58 to 0.63. Since foot penetration forces are small at shallow depths, only when a large portion (>70%) of each leg intrudes into the substrate is sufficient vertical force produced to raise the body off the ground and move it forward. This is different from walking on hard ground (i.e., the inverted pendulum model [2]), because the foothold is deep relative to leg length and between steps the body is typically supported by the ground surface. SandBot can walk at speeds up to 30 cm/s ($\sim$1 bl/s) on sand or "swim" slowly ($\sim$1 cm/s). Only close to a particular set of limb kinematics (defined by stance duration, stance location, and duty factor) does it walk with little slip. Optimal limb kinematics are determined by forces generated during *rotational intrusion* into GM [27], and these forces differ from those generated during either pure vertical penetration [16] or horizontal drag [13]. During rotational intrusion, the maximum force occurs $\approx$40° before the angle at which the limb is deepest in the ground (90°).

With the optimal limb kinematics, SandBot's speed increased sublinearly with stride frequency for given volume fraction ϕ and increased with ϕ for given stride frequency, but for sufficiently low ϕ and/or high stride frequency, the speed was small ($\sim$1 cm/s). A model [26] revealed that the mechanism of effective movement on GM relies on solidification and explains the sublinear increase: Since penetration force increases as $\alpha A z$, the limb penetrates into the material to a depth governed by the body weight and the inertial forces needed to accelerate the body to limb velocity. At the depth where this force balance is achieved, the material under the limb solidifies and the robot "rotary" walks, with the limbs rotating relative to the solidified grains and moving the body forward by a distance determined by penetration depth and leg geometry. As ϕ decreases and/or stride frequency increases, penetration depth increases to the point where step length (inversely related to penetration depth) is smaller than leg length. Consequently, over consecutive steps the legs encounter previously disturbed ground which has reduced penetration resistance. The robot can no longer raise its body over a solid foothold, and instead "swims" forward slowly ($\sim 1\,\mathrm{cm/s}$) via drag on its legs moving through grains that are always locally fluidized.

Our study of SandBot provides broader insights into the principles governing effective locomotion on sand. Since surface penetration is required for locomotion on GM yet reduces stride length and causes increased drag, it may be advantageous for an animal (or a legged robot) to have large feet and long legs and use appropriate kinematics to reduce relative leg penetration while generating the required thrust. The zebra-tailed lizard (*Callisaurus draconoides*, $\sim 10\,\mathrm{cm}$, $\sim 10\,\mathrm{g}$, Fig. 1), a desert generalist, provides an excellent realization of these limb use principles [29]. It can run at speeds up to $\sim 4\,\mathrm{m/s}$ ($\sim 50\,\mathrm{bl/s}$) on both hard ground [19] and sand [18]. Compared to closely-related lizards of similar size, it has the longest hind limbs ($\approx 90\%$ body length) and the largest hind feet (total area $\approx 2\,\mathrm{cm^2}$) with extremely elongated toes [19]. While standing on loose sand, the foot only penetrates the surface by a few mm or 10% of vertical leg length ($\sim 3\,\mathrm{cm}$). High speed x-ray imaging revealed that during running on sand, the foot impacted the surface with a plantigrade foot posture to create the largest projected foot area and penetrates a maximal depth of $\approx 1.3\,\mathrm{cm}$ in stance; a majority ($>50\%$) of the leg remained above surface, making it possible for the animal to take long strides with reduced drag. Recently developed small robots like DASH and RoACH (Fig. 1, $\sim 10\,\mathrm{cm}$, $\sim 20\,\mathrm{g}$) are approaching small organisms in locomotor performance – like the zebra-tailed lizard, they have relatively large feet (total area $\approx 3\,\mathrm{cm^2}$) for their body weight, and can run at $\sim 10\,\mathrm{bl/s}$ on sand with appropriate foot design [28].

4. Limbed Locomotion on Sand: Crawling. Certain aquatically adapted organisms (e.g., mudskippers, sea lions and sea turtles) use paddle-like appendages (flippers) to crawl on deformable terrestrial materials [32, 37]. We have studied the terrestrial locomotion of sea turtles, animals whose evolutionary life history constrains them to come ashore and lay nests near sand dunes [6]. While crawling on sandy beaches using aquatically adapted limbs, a sea turtle rests most of its body weight on its flat plastron (belly), inserts its paddle-like flippers into sand, and pushes horizontally (and vertically) to generate enough thrust to overcome belly friction and accelerate its body. Both adults ($\sim$100 kg) and hatchlings ($\sim$20 g, Fig. 1) employ this strategy.

Since field studies can elicit behaviors not demonstrated in the laboratory [20], we conducted a study of locomotion performance and limb-ground interaction of hatchling Loggerhead Sea turtles (*Caretta caretta*) in the field on Jekyll Island, GA, USA [32]. A field-portable fluidized bed trackway allowed us to mimic the beach substrate in a controlled fashion by controlling the compaction state of the sand and the incline angle [26, 32]. High speed imaging under infrared lights revealed that during rapid runs (3 bl/s over the deformable sand–comparable to the speed on hard ground) the limbs exhibited minimal to no slip [32]; the animals did not paddle through sand. During each step forward speed increased to a maximum followed by a decrease to zero [32]. On sand, the flipper penetrated the material vertically and then the wrist bent in the fore-aft direction, maintaining solidified material behind the flipper during the thrust phase. This solidification process during forward movement results in a no-slip condition such that thrust forces stay below the material yield force ($F_{thrust} < F_{yield}$) and allows sea turtles to move effectively on sand.

To begin to systematically explore crawling on sand using paddle-like limbs, we developed a physical model of the sea turtle (FlipperBot, Fig. 1). Its performance is sensitive to belly friction, suggesting the importance of lift and forward thrust generation during a step. FlipperBot has a wrist that can be made free rotating or fixed. Our initial results show that a free rotating wrist allows solidification of material during forward motion, and improves locomotor performance. We attribute the performance increase to a decrease in the amount of material disturbed during a step by the flexible limb: This reduces the probability of encountering previously disturbed ground during the next step which, if it occurs, reduces forward progress.

5. Undulatory Swimming in a Frictional Fluid: Kinematics.
We now discuss the principles that allow organisms to "swim" within GM [3]. We used high speed x-ray imaging to study a small ($\sim$8 cm, 16 g) desert-dwelling lizard, the sandfish (Fig. 1), which inhabits the Saharan desert of Africa and moves within GM of different ϕ. We monitored the burial and swimming of the sandfish in the laboratory and found that once subsurface the animal no longer used limbs for propulsion. Instead it placed its limbs against its sides and executed an undulatory motion of the body with large amplitude axial oscillation, using the body to propel itself at speeds up to $\sim$2 bl/s. During bouts of steady swimming, the sandfish penetrated the GM at an angle of approximately 26° relative to the horizontal in low ϕ, and 19° in high ϕ states. During such bouts, the animal reached depths of 2–4 cm, as measured from the surface to the top of the back. However, the animal was capable of reaching depths greater than 10 cm.

Subsurface swimming kinematics were well characterized by a traveling sinusoidal wave propagating along the body from head to tail: $y = A \sin \frac{2\pi}{\lambda}(x + v_w t)$, with y the displacement from the mid-line of a straight animal, A the amplitude, λ the wavelength, f the wave frequency, $v_w = f\lambda$ the wave speed, t the time, and x the distance along a line joining the end points of the animal, pointing from the tail to the head, and parallel to the direction of motion. In both low and high ϕ GM (in both 0.3 mm and 3 mm glass beads) the ratio of A to λ was approximately 0.2. For each condition tested the animal increased its forward speed by increasing its undulation frequency [30, 31].

The forward speed of the sandfish, v_x, was less than the speed of the wave traveling along its body, and equal to the product of f and λ times a constant factor η, so that $v_x = \eta f\lambda$ with $\eta = 0.53 \pm 0.04$ (in 0.3 mm glass particles). Tracer particles placed in the GM revealed a backward flow of grains as the animal moved forward. Slipping while progressing is common to undulatory swimmers in deformable media over a wide range of length scales (e.g., eels and spermatozoa in fluids [2]) and is characterized by the wave efficiency, $\eta = v_x/v_w$ [15], defined as the ratio of the forward speed of the animal, v_x, to the velocity of the wave traveling down its body, v_w. Remarkably, η in 0.3 mm and 3 mm particles was approximately 0.5, and was independent of volume fraction ϕ despite differences in drag resistance of nearly a factor of two between the highest and lowest ϕ states.

6. Undulatory Swimming: Muscle Activity. To develop neuromechanical models [17, 39] of locomotion we must translate kinematics into forces (dynamics) and ultimately connect to the neuromuscular control system (including actuation and sensing). To learn how the sandfish generates force during sand-swimming, we studied the activation pattern in its epaxial musculature using electromyogram (EMG) recordings synchronized with high speed x-ray and visible light imaging [38]. We hypothesized that in such a frictional, highly dissipative environment the activation strategy would be dictated by the speed independent but depth dependent forces in GM.

To swim subsurface in GM, the sandfish generated an anterior-to-posterior wave of muscle activation. EMG onset occured at or just prior to the maximal convexity of the body (the point at which the muscle is maximally stretched), with more posterior muscles activated earlier in the muscle strain cycle. This timing pattern was in accord with the activation timing found for undulatory swimmers in water [11]. As a sandfish dove farther into the GM, EMG intensity increased with undulation number and, therefore, depth. For a given undulation number, EMG intensity was independent of speed; to move faster the sandfish propagated the wave faster. These results support the hypothesis that sandfish require a higher amount of muscle force with increasing depth but not speed. Despite 50% increase in resistance force measured in drag experiments, between low and high ϕ EMG intensity only changed a small amount (<20%) [38].

7. Modeling Sand-Swimming: Simulation, RFT, and Robots.
Challenged by the lack of constitutive equations for GM, we have modeled sand-swimming in three ways: Numerically, physically (robot), and analytically using a Resistive Force Theory (RFT). All modeling methods demonstrate that movement can be viewed as occurring within a "frictional" fluid where force is dominated by frictional contacts within the material locally flowing around the body. The models demonstrate the differences and similarities of swimming in GM compared to swimming in fluids.

We used a 50-segment numerical simulation to model the sandfish with relative angles between segments specified to generate a traveling wave as observed in the animal [31]; the center of mass position and inter-segment torques were unconstrained. The GM was modeled using an experimentally validated Discrete Element Model (DEM) [24] of $\sim 10^5$, 3 mm diameter particles (Fig. 1). Particle-particle and particle-intruder interactions included repulsive and viscous forces in the normal direction, and a frictional force in the tangential direction. The numerical model quantitatively reproduced kinematic features of the locomotion (e.g., speed vs. frequency). Examination of the flow of particles around the animal body supported the frictional fluid picture.

The DEM model also reproduced the observed muscle activation patterns: When constrained to dive at an angle, activation torque in the model increased with depth but was independent of speed. Force on the body was independent of undulation frequency which resulted in a frequency independent mechanical cost of transportation. The DEM model revealed only a small difference in average rectified motor torque between low and high ϕ media preparations for a given depth. The similarities between simulation and experiment therefore imply that the observed biological muscle activation pattern is dominated by the resistive forces of the GM.

Since the organism swims within a localized fluid, we used a RFT model [14, 30] originally developed for low Reynolds number (Re) swimmers to gain insight into swimming in the granular medium. In an RFT, the body of the organism is partitioned into infinitesimal elements along its length. When moving relative to the medium, each element experiences resistive thrust and drag forces. Resolving these forces into perpendicular and parallel components and balancing them by integrating forces over the length of the body (and head) predicts forward swimming speed at a given frequency. Since at biologically relevant swimming speeds (0–0.40 m/s) force is approximately independent of speed [30, 42], the force on an element could be characterized as a function of only the *direction* of the velocity relative to its orientation. As the entry angle of the animal was small (<30°), we approximated the motion of the animal as occurring in a horizontal plane.

Since comprehensive resistive force laws in GM were not available, we measured the forces on rods with cross-sections comparable to the animal body as the rods were dragged through GM at a fixed depth.

With these force laws, the RFT agreed well with the DEM model. The angular dependence of the force laws in GM resembles the forces generated in a Newtonian fluid at low Reynolds number: The perpendicular force increases and the parallel force decreases with the angle between the velocity of the rod and its longitudinal axis. However, while the functional forms of the forces in low Re can be approximated as sines and cosines, in GM, the functional forms are more complicated. Further the ratio of the average magnitude of perpendicular to parallel forces is larger in GM ($>3{:}1$) than in fluid ($\approx 2{:}1$). Consequently, thrust is relatively larger in GM compared to in fluids at low Reynolds number. The difference in force laws explains the higher η observed for sandfish (≈ 0.5) compared to non-inertial low Re swimmers in fluids (≈ 0.2). The RFT also suggests that the packing state does not affect η because thrust and drag scale similarly with changes in packing. An alternative explanation is that the material disturbed by the sandfish rapidly evolves to the same ϕ_c, the critical volume fraction [13, 40], and thus the body of the organism always moves within GM of the same ϕ. We have not yet determined which explanation is correct, and this is the subject of current study.

Finally, we developed and tested a physical model, the sandfish robot (Fig. 1). The robot consisted of six coupled motors and a passive block, each with a single joint that permitted angular excursions in the body plane, connected to each other by identical links to form the body [31]. We commanded the observed sandfish kinematics using an open-loop controller. Like the animal, the forward velocity of the robot monotonically increased with increasing frequency. However, for the robot $\eta = 0.34 \pm 0.03$, which was significantly below the values measured for the animal in experiment and predicted by the RFT and DEM simulation (with 50 segments). We found by simulating the robot that this smaller η was due to the robot's relatively few segments. Increasing the number of segments in simulation while keeping body length fixed caused η to increase, plateauing at a value equal to that from the animal ($\eta \approx 0.5$) when 15 or more segments were used.

All three models predict that η increases with increasing A, but we observed that the animal does not operate at high A. The RFT model shows that operating at large A comes at a cost: Since the animal's length is fixed, its wavelength λ decreases with increasing A. Because $v_x = \eta \lambda f$, as A increases, the competition between increasing η and decreasing λ results in a maximum in forward displacement per cycle at $A/\lambda \approx 0.2$. This finding is captured by all the models. The biological data reside close to the peak of the curve, indicating that the animal could be maximizing its sand-swimming speed in accord with the hypothesis that the sandfish's rapid burial and swimming behavior is an escape response [3].

8. Conclusions and Outlook. We have briefly described our efforts to discover principles of movement on and within GM using comparative studies of organisms and physical models. Discovery of principles of movement requires appropriate models of substrate interaction, which in GM include phase transitions, speed and depth dependent forces, and dependence on initial conditions (e.g., ϕ). Above ground, movement of some legged locomotors can be enhanced by maintaining solidification which suggests biological hypotheses for control of limb movement. Our walking and crawling models provide insights into the body and limb design and strategies for effective movement on the surface of sand. Subsurface swimming of the sandfish lizard and related models reveal that movement occurs within a "frictional" fluid whose properties dominate the locomotor pattern and activation strategy. Our RFT, numerical, and physical models predict kinematics and motor patterns for rapid sand-swimming. Our results are a step in developing interaction rules and principles of movement of organisms in even more complex terrestrial environments (flowing/solidifying materials composed of complex elements like rubble and leaf litter) and will enable creation of devices that can maneuver robustly in such terrains.

REFERENCES

[1] Albert I, Sample JG, Morss AJ, Rajagopalan S, Barabási AL, Schiffer P (2001) Granular drag on a discrete object: shape effects on jamming. Phys Rev E 64(6):61303

[2] Alexander RM (2003) Principles of animal locomotion. Princeton University Press, Princeton, USA

[3] Arnold EN (1995) Identifying the effects of history on adaptation: origins of different sand-diving techniques in lizards. J Zool 235(3):351–388

[4] Bagnold RA (1954) The physics of blown sand and desert dunes. Methuen and Co. Ltd, London, UK

[5] Baisch A, Wood R (2011) Design and fabrication of the Harvard ambulatory micro-robot robotics research, In: Pradalier C, Siegwart R, Hirzinger G (eds) Springer tracts in advanced robotics, vol 70. Springer, Berlin/Heidelberg, pp 715–730

[6] Bolten AB (2003) Variation in sea turtle life history patterns: neritic vs. oceanic developmental stages. In: Lutz PL, Musick JA, Wyneken J (eds) The biology of sea turtles, vol 2. CRC Press, Boca Raton, USA, pp 243–257

[7] Dickinson MH, Farley CT, Full RJ, Koehl MAR, Kram R, Lehman S (2000) How animals move: an integrative view. Science 288(5463):100–106

[8] Ding Y, Gravish N, Goldman DI (2011) Drag induced lift in granular media. Phys Rev Lett 106(2):028001

[9] Full RJ, Koditschek DE (1999) Templates and anchors: neuromechanical hypotheses of legged locomotion on land. J Exp Biol 2(12):3–125

[10] Geng J, Howell D, Longhi E, Behringer RP, Reydellet G, Vanel L, Clé ment E, Luding S (2001) Footprints in sand: the response of a granular material to local perturbations. Phys Rev Lett 87(3):35506

[11] Gillis GB (1998) Neuromuscular control of anguilliform locomotion: patterns of red and white muscle activity during swimming in the american eel Anguilla Rostrata. J Exp Biol 201:3245–3256

[12] Goldman DI, Umbanhowar P (2008) Scaling and dynamics of sphere and disk impact into granular media. Phys Rev E 77(2):021308–021321

[13] Gravish N, Umbanhowar PB, Goldman DI (2010) Force and flow transition in plowed granular media. Phys Rev Lett 105(12):128301

[14] Gray J, Hancock GJ (1955) The propulsion of sea-urchin spermatozoa. J Exp Biol 32(4):802–814

[15] Gray J, Lissman HW (1964) The locomotion of nematodes. J Exp Biol 41(1):135–154

[16] Hill G, Yeung S, Koehler SA (2005) Scaling vertical drag forces in granular media. Europhys Lett 72(1):137–143

[17] Holmes P, Koditschek D, Guckenheimer J (2006) The dynamics of legged locomotion: models, analyses, and challenges. Dynamics 48(2):207–304

[18] Irschick DJ, Jayne BC (1999) A field study of the effects of incline on the escape locomotion of a bipedal lizard, callisaurus draconoides. Physiol Biochem Zool 72(1):44–56

[19] Irschick DJ, Jayne BC (1999) Comparative three-dimensional kinematics of the hindlimb for high-speed bipedal and quadrupedal locomotion of lizards. J Exp Biol 202(9):1047–1065

[20] Irschick DJ, Herrel A, Vanhooydonck B, Huyghe K, Van Damme R (2005) Locomotor compensation creates a mismatch between laboratory and field estimates of escape speed in lizards: a cautionary tale for performance-to-fitness studies. Evolution 59(7):1579–1587

[21] Jaeger HM, Nagel SR, Behringer RP (1996) Granular solids, liquids, and gases. Rev Mod Phys 68(4):1259–1273

[22] Korff WL, McHenry MJ (2011) Environmental differences in substrate mechanics do not affect sprinting performance in sand lizards (Uma scoparia and Callisaurus draconoides). J Exp Biol 214:122–130

[23] Lauder G, Madden P (2006) Learning from fish: kinematics and experimental hydrodynamics for roboticists. Int J Autom Comput 3(4):325–335

[24] Lee J, Herrmann HJ (1993) Angle of repose and angle of marginal stability: molecular dynamics of granular particles. J Phys A Math Gen 26:373–383

[25] Lejeune TM, Willems PA, Heglund NC (1998) Mechanics and energetics of human locomotion on sand. J Exp Biol 201(13):2071–2080

[26] Li C, Umbanhowar PB, Komsuoglu H, Koditschek DE, Goldman DI (2009) Sensitive dependence of the motion of a legged robot on granular media. Proc Natl Acad Sci 106(9):3029–3034

[27] Li C, Umbanhowar PB, Komsuoglu H, Goldman DI (2010) The effect of limb kinematics on the speed of a legged robot on granular media. Exp Mech 50:1383–1393

[28] Li C, Hoover AM, Birkmeyer P, Umbanhowar PB, Fearing RS, Goldman DI (2010) Systematic study of the performance of small robots on controlled laboratory substrates. In: Proceedings of SPIE, Orlando, USA, vol 7679. pp 76790Z(1–13)

[29] Li C, Hsieh ST, Goldman DI, Multi-functional foot use during running in the zebra-tailed lizard (*Callisaurus draconoides*), Journal of Experimental Biology, in press

[30] Maladen RD, Ding Y, Li C, Goldman DI (2009) Undulatory swimming in sand: subsurface locomotion of the sandfish lizard. Science 325(5938):314–318

[31] Maladen RD, Ding Y, Kamor A, Umbanhowar PB, Komsuoglu H, Goldman DI (2011) Mechanical models of sandish locomotion reveal principles of high performance subsurface sand-swimming. J R Soc Interface 8(62):1332–1345

[32] Mazouchova N, Gravish N, Savu A, Goldman DI (2010) Utilization of granular solidification during terrestrial locomotion of hatchling sea turtles. Biol Lett 6(3):398–401

[33] Mosauer W (1932) Adaptive convergence in the sand reptiles of the Sahara and of California: a study in structure and behavior. Copeia 2:72–78

[34] Nedderman RM (1992) Statics and kinematics of granular materials. Cambridge University Press, Cambridge/New York, UK/USA

[35] Norris KS, Kavanau JL (1966) Burrowing of western shovel-nosed snake chionactis occipitalis hallowell and undersand environment. Copeia 4:650–664

[36] Playter R, Buehler M, Raibert M (2006) BigDog. In: Proceedings of SPIE, San Jose, USA, vol 6230. pp 62302O

[37] Renous S, Bels V (1993) Comparison between aquatic and terrestrial locomotions of the leatherback sea turle (Dermochelys coriacea). J Zool 230(3):357–378

[38] Sharpe SS, Ding Y, Goldman DI, Environmental interaction influences muscle activation strategy during sand-swimming in the sandfish lizard *Scincus scincus*. J Exp Biol (in review)

[39] Tytell ED, Hsu CY, Williams TL, Cohen AH, Fauci LJ (2010) Interactions between internal forces, body stiffness, and fluid environment in a neuromechanical model of lamprey swimming. Proc Natl Acad Sci 107(46):19832–19837

[40] Umbanhowar PB, Goldman DI (2010) Granular impact and the critical packing state. Phys Rev E 82(1):010301

[41] Vogel S (1994) Life in moving fluids. Princeton University Press, Princeton, USA

[42] Wieghardt K (1975) Experiments in granular flow. Annu Rev Fluid Mech 7:89–114

IMA LOCOMOTION WORKSHOP PARTICIPANTS

Mixed-Integer Nonlinear optimization: Algorithmic Advances and Applications

- Dhanapati Adhikari, Department of Mathematics, Oklahoma State University
- Silas Alben, School of Mathematics, Georgia Institute of Technology
- Alexander Alexeev, George W. Woodruff School of Mechanical Engineering, Georgia Institute of Technology
- Arezoo Ardekani, Department of Mathematics, Massachusetts Institute of Technology
- Nusret Balci, Institute for Mathematics and its Applications, University of Minnesota
- Tsevi Beatus, Department of Physics, Cornell University
- Gordon Berman, Lewis-Sigler Institute for Integrative Genomics, Princeton University
- Adam Boucher, Department of Mathematics, University of New Hampshire
- Luca Brandt, KTH Mechanics, Royal Institute of Technology (KTH)
- Kenny Breuer, Division of Engineering, Brown University
- Maria-Carme Calderer, School of Mathematics, University of Minnesota
- Chi Hin Chan, Institute for Mathematics and its Applications, University of Minnesota
- Xianjin Chen, Institute for Mathematics and its Applications, University of Minnesota
- Stephen Childress, Courant Institute of Mathematical Sciences, New York University
- Howie Choset, Department of Robotics, Carnegie Mellon University
- Luis Cisneros, Department of Physics, University of Arizona
- Fredric Cohen, Department of Molecular Biophysics and Physiology, Rush University Medical Center
- Itai Cohen, Department of Physics, Cornell University
- Aline Cotel, Department of Civil and Environmental Engineering, University of Michigan
- Darren Crowdy, Department of Mathematics, Imperial College London
- Zhenlu Cui, Department of Mathematics and Computer Science, Fayetteville State University
- John Dabiri, Graduate Aeronautical Laboratories and Bioengineering, California Institute of Technology
- Domenico D'Alessandro, Department of Mathematics, Iowa State University

S. Childress et al. (eds.), *Natural Locomotion in Fluids and on Surfaces*, IMA 155, DOI 10.1007/978-1-4614-3997-4,
© Springer Science+Business Media New York 2012

- Mark Denny, Hopkins Marine Station, Stanford University
- Antonio DeSimone, Department of Applied Mathematics, International School for Advanced Studies (SISSA/ISAS)
- Charles Doering, Department of Mathematics, University of Michigan
- Deborah Edmund, Department of Mechanical Engineering & Naval Architecture and Marine Engineering, University of Michigan
- Robert Eisenberg, Department of Molecular Biophysics and Physiology, Rush University Medical Center
- Jeff Eldredge, Mechanical & Aerospace Engineering, University of California, Los Angeles
- Acmae El Yacoubi, Department of Theoretical and Applied Mechanics (MAE), Cornell University
- Randy Ewoldt, Institute for Mathematics and its Applications, Massachusetts Institute of Technology
- Lisa Fauci, Department of Mathematics, Tulane University
- Frank Fish, Department of Biology, West Chester University
- Hermes Gadêlha, Centre for Mathematical Biology, Mathematical Institute, University of Oxford
- Eamonn Gaffney, Mathematical Institute, University of Oxford
- Daniel Goldman, School of Physics, Georgia Institute of Technology
- Michael Graham, Chemical and Biological Engineering Department, University of Wisconsin-Madison
- Andong He, Department of Mathematics, Pennsylvania State University
- Christel Hohenegger, Courant Institute of Mathematical Sciences, New York University
- David Hu, School of Mechanical Engineering, Georgia Institute of Technology
- Jifeng Hu, University of Minnesota
- Xianpeng Hu, Department of Mathematics, University of Pittsburgh
- YunKyong Hyon, Institute for Mathematics and its Applications, University of Minnesota
- Mark Iwen, Institute for Mathematics and its Applications, University of Minnesota
- Pieter Janssen, Department of Chemical and Biological Engineering, University of Wisconsin-Madison
- Srividhya Jeyaraman, Institute for Mathematics and its Applications, University of Minnesota
- Lijian Jiang, Institute for Mathematics and its Applications, University of Minnesota
- Mihailo Jovanovic, Department of Electrical and Computer Engineering, University of Minnesota
- Ning Ju, Department of Mathematics, Oklahoma State University
- Sunghwan (Sunny) Jung, Department of Engineering Science and Mechanics, Virginia Polytechnic Institute and State University

- Eva Kanso, Aerospace and Mechanical Engineering, University of Southern California
- Markus Keel, Institute for Mathematics and its Applications, University of Minnesota
- Scott Kelly, Mechanical Engineering and Engineering Science, University of North Carolina - Charlotte
- Hyejin Kim, Institute for Mathematics and its Applications, University of Minnesota
- Mimi Koehl, Department of Integrative Biology, University of California, Berkeley
- Pawel Konieczny, Institute for Mathematics and its Applications, University of Minnesota
- Arshad Kudrolli, Department of Physics, Clark University
- Amy Lang, Department of Aerospace Engineering and Mechanics, University of Alabama
- Ronald Larson, Department of Chemical Engineering, University of Michigan
- Eric Lauga, Department of Mechanical and Aerospace Engineering, University of California, San Diego
- Anita Layton, Department of Mathematics, Duke University
- Chiun-Chang Lee, Department of Mathematics, National Taiwan University
- David Lentink, Experimental Zoology Group, Wageningen University and Research Center
- Rachel Levy, Department of Mathematics, Harvey Mudd College
- Marta Lewicka, School of Mathematics, University of Minnesota
- Congming Li, Department of Applied Mathematics, University of Colorado
- Yongfeng Li, Institute for Mathematics and its Applications, University of Minnesota
- Zhi (George) Lin, Institute for Mathematics and its Applications, University of Minnesota
- Bin Liu, Courant Institute of Mathematical Sciences, New York University
- Chun Liu, Institute for Mathematics and its Applications, University of Minnesota
- Jerome Loheac, Université de Nancy I (Henri Poincaré)
- Ellen Longmire, Department of Aerospace Engineering and Mechanics, University of Minnesota
- Evelyn Lunasin, Department of Mathematics, University of Arizona
- Enkeleida Lushi, Courant Institute of Mathematical Sciences, New York University
- Krishnan Mahesh, Department of Aerospace Engineering and Mechanics, University of Minnesota

- Kara Maki, Institute for Mathematics and its Applications, University of Delaware
- Vasileios Maroulas, Institute for Mathematics and its Applications, University of Minnesota
- Hassan Masoud, Department of Mechanical Engineering, Georgia Institute of Technology
- Anna Mazzucato, Department of Mathematics, Pennsylvania State University
- Laura Miller, Department of Mathematics, University of North Carolina
- Michal Mlejnek, Department of Modeling and Simulation, Corning Incorporated
- Kamran Mohseni, Department of Aerospace Engineering Sciences, University of Colorado
- Yoichiro Mori, School of Mathematics, University of Minnesota
- Hoa Nguyen, Center for Computational Science, Tulane University
- Monika Nitsche, Department of Mathematics & Statistics, University of New Mexico
- Dongjuan Niu, School of Mathematical Sciences, Capital Normal University
- Clara O'Farrell, Department of Control and Dynamical Systems, California Institute of Technology
- Sarah Olson, Department of Mathematics, Tulane University
- Yizhar Or, Department of Mechanical Engineering, Technion-Israel Institute of Technology
- Cecilia Ortiz-Duenas, Institute for Mathematics and its Applications, University of Minnesota
- Hans Othmer, Department of Mathematics, University of Minnesota
- Neelesh Patankar, Department of Mechanical Engineering, Northwestern University
- Jifeng Peng, Department of Mechanical Engineering, University of Alaska
- Juan Restrepo, Department of Mathematics, University of Arizona
- Leif Ristroph, Department of Physics, Cornell University
- John Roberts, CSAIL, Massachusetts Institute of Technology
- David Saintillan, Department of Mechanical Science and Engineering, University of Illinois at Urbana-Champaign
- Fadil Santosa, Institute for Mathematics and its Applications, University of Minnesota
- Arnd Scheel, School of Mathematics, University of Minnesota
- William Schultz, Department of Mechanical Engineering, University of Michigan
- George Sell, School of Mathematics, University of Minnesota
- Tsvetanka Sendova, Institute for Mathematics and its Applications, University of Minnesota

- Shuanglin Shao, Institute for Mathematics and its Applications, Institute for Advanced Study
- Michael Shelley, Courant Institute of Mathematical Sciences, New York University
- Jian Sheng, Department of Aerospace Engineering and Mechanics, University of Minnesota
- Henry Shum, Mathematical Institute, University of Oxford
- Jake Socha, Department of Engineering Science and Mechanics, Virginia Polytechnic Institute and State University
- Saverio Spagnolie, Department of Mechanical and Aerospace Engineering, University of California, San Diego
- Panagiotis Stinis, School of Mathematics, University of Minnesota
- Wanda Strychalski, Department of Mathematics, University of California, Davis
- Susan Suarez, Department of Biomedical Sciences, Cornell University
- Vladimir Sverak, School of Mathematics, University of Minnesota
- Daniel Tam, Department of Mathematics, Massachusetts Institute of Technology
- Russ Tedrake, Electrical Engineering and Computer Science Department, Massachusetts Institute of Technology
- Jean-Luc Thiffeault, Department of Mathematics, University of Wisconsin-Madison
- Giordano Tierra Chica, Dpto. Ecuaciones Diferenciales y Análisis Numérico, University of Sevilla
- Edriss Titi, Department of Mathematics, University of California
- Chad Topaz, Department of Mathematics and Computer Science, Macalester College
- Marius Tucsnak, Departement de Mathématiques (IECN), Université de Nancy I (Henri Poincaré)
- Patrick Underhill, Department of Chemical and Biological Engineering, Rensselaer Polytechnic Institute
- Steven Vogel, Department of Biology, Duke University
- Shawn Walker, Courant Institute of Mathematical Sciences, New York University
- Changyou Wang, Department of Mathematics, University of Kentucky
- Jane Wang, Department of Theoretical and Applied Mechanics, Cornell University
- Qixuan Wang, School of Mathematics, University of Minnesota
- Paul Webb, School of Natural Resources and the Environment, University of Michigan
- Nathaniel Whitaker, Department of Mathematics & Statistics, University of Massachusetts
- Sijue Wu, Department of Mathematics, University of Michigan
- Wei Xiong, Institute for Mathematics and its Applications, University of Minnesota

- Jin Xu, Traditional Medicine Laboratory, Shanghai University of Traditional Chinese Medicine
- Sheng Xu, Department of Mathematics, Southern Methodist University
- Xiang Xu, Department of Mathematics, Pennsylvania State University
- Jeannette Yen, Department of Biology, Georgia Institute of Technology
- Tsuyoshi Yoneda, Institute for Mathematics and its Applications, University of Minnesota
- Jun Zhang, Department of Physics, New York University
- Weigang Zhong, Institute for Mathematics and its Applications, University of Minnesota
- Qiang Zhu, Department of Structural Engineering, University of California, San Diego

IMA ANNUAL PROGRAMS

1982–1983 Continuum and Statistical Approaches to Phase Transition
1983–1984 Mathematical Models for the Economics of Decentralized Resource Allocation
1984–1985 Continuum Physics and Partial Differential Equations
1985–1986 Stochastic Differential Equations and their Applications
1986–1987 Scientific Computation
1987–1988 Applied Combinatorics
1988–1989 Nonlinear Waves
Research Accomplishments
1989–1990 Dynamical Systems and their Applications
Research Accomplishments
1990–1991 Phase Transitions and Free Boundaries Research Accomplishments
1991–1992 Applied Linear Algebra Research Accomplishments
1992–1993 Control Theory and its Applications
1993–1994 Emerging Applications of Probability
1994–1995 Waves and Scattering
1995–1996 Mathematical Methods in Material Science
1996–1997 Mathematics of High Performance Computing
1997–1998 Emerging Applications of Dynamical Systems
1998–1999 Mathematics in Biology
1999–2000 Reactive Flows and Transport Phenomena
2000–2001 Mathematics in Multimedia
2001–2002 Mathematics in the Geosciences
2002–2003 optimization
2003–2004 Probability and Statistics in Complex Systems: Genomics, Networks, and Financial Engineering

2004–2005 Mathematics of Materials and Macromolecules: Multiple Scales, Disorder, and Singularities
2005–2006 Imaging
2006–2007 Applications of Algebraic Geometry
2007–2008 Mathematics of Molecular and Cellular Biology
2008–2009 Mathematics and Chemistry
2009–2010 Complex Fluids and Complex Flows
2010–2011 Simulating Our Complex World: Modeling, Computation and Analysis
2011–2012 Mathematics of Information
2012–2013 Infinite Dimensional and Stochastic Dynamical Systems and their Applications
2013–2014 Scientific and Engineering Applications of Algebraic Topology

IMA SUMMER PROGRAMS

1987 Robotics
1988 Signal Processing
1989 Robust Statistics and Diagnostics
1990 Radar and Sonar (June 18–29)
     New Directions in Time Series Analysis (July 2–27)
1991 Semiconductors
1992 Environmental Studies: Mathematical, Computational, and Statistical Analysis
1993 Modeling, Mesh Generation, and Adaptive Numerical Methods for Partial Differential Equations
1994 Molecular Biology
1995 Large Scale optimizations with Applications to Inverse Problems, Optimal Control and Design, and Molecular and Structural Optimization
1996 Emerging Applications of Number Theory (July 15–26) Theory of Random Sets (August 22–24)
1997 Statistics in the Health Sciences
1998 Coding and Cryptography (July 6–18)
     Mathematical Modeling in Industry (July 22–31)
1999 Codes, Systems, and Graphical Models (August 2–13, 1999)
2000 Mathematical Modeling in Industry: A Workshop for Graduate Students (July 19–28)
2001 Geometric Methods in Inverse Problems and PDE Control (July 16–27)
2002 Special Functions in the Digital Age (July 22–August 2)
2003 Probability and Partial Differential Equations in Modern Applied Mathematics (July 21–August 1)
2004 n-Categories: Foundations and Applications (June 7–18)
2005 Wireless Communications (June 22–July 1)

2006 Symmetries and Overdetermined Systems of Partial Differential Equations (July 17–August 4)
2007 Classical and Quantum Approaches in Molecular Modeling (July 23–August 3)
2008 Geometrical Singularities and Singular Geometries (July 14–25)
2009 Nonlinear Conservation Laws and Applications (July 13–31)

IMA "HOT TOPICS/SPECIAL" WORKSHOPS

- Challenges and Opportunities in Genomics: Production, Storage, Mining and Use, April 24–27, 1999
- Decision Making Under Uncertainty: Energy and Environmental Models, July 20–24, 1999
- Analysis and Modeling of Optical Devices, September 9–10, 1999
- Decision Making under Uncertainty: Assessment of the Reliability of Mathematical Models, September 16–17, 1999
- Scaling Phenomena in Communication Networks, October 22–24, 1999
- Text Mining, April 17–18, 2000
- Mathematical Challenges in Global Positioning Systems (GPS), August 16–18, 2000
- Modeling and Analysis of Noise in Integrated Circuits and Systems, August 29–30, 2000
- Mathematics of the Internet: E-Auction and Markets, December 3–5, 2000
- Analysis and Modeling of Industrial Jetting Processes, January 10–13, 2001
- Special Workshop: Mathematical Opportunities in Large-Scale Network work Dynamics, August 6–7, 2001
- Wireless Networks, August 8–10 2001
- Numerical Relativity, June 24–29, 2002
- Operational Modeling and Biodefense: Problems, Techniques, and Opportunities, September 28, 2002
- Data-driven Control and optimization, December 4–6, 2002
- Agent Based Modeling and Simulation, November 3–6, 2003
- Enhancing the Search of Mathematics, April 26–27, 2004
- Compatible Spatial Discretizations for Partial Differential Equations, May 11–15, 2004
- Adaptive Sensing and Multimode Data Inversion, June 27–30, 2004
- Mixed Integer Programming, July 25–29, 2005
- New Directions in Probability Theory, August 5–6, 2005
- Negative Index Materials, October 2–4, 2006
- The Evolution of Mathematical Communication in the Age of Digital Libraries, December 8–9, 2006
- Math is Cool! and Who Wants to Be a Mathematician?, November 3, 2006

- Special Workshop: Blackwell-Tapia Conference, November 3–4, 2006
- Stochastic Models for Intracellular Reaction Networks, May 11–13, 2008
- Multi-Manifold Data Modeling and Applications, October 27–30, 2008
- Mixed-Integer Nonlinear optimization: Algorithmic Advances and Applications, November 17–21, 2008
- Higher Order Geometric Evolution Equations: Theory and Applications from Microfluidics to Image Understanding, March 23–26, 2009
- Career Options for Women in Mathematical Sciences, April 2–4. 2009
- MOLCAS, May 4–8, 2009
- IMA Interdisciplinary Research Experience for Undergraduates, June 29–July 31, 2009
- Research in Imaging Sciences, October 5–7, 2009
- Career Options for Underrepresented Groups in Mathematical Sciences, March 25–27, 2010
- Physical Knotting and Linking and its Applications, April 9, 2010
- IMA Interdisciplinary Research Experience for Undergraduates, June 14–July 16, 2010
- Kickoff Workshop for Project MOSAIC, June 30–July 2, 2010
- Finite Element Circus Featuring a Scientific Celebration of Falk, Pasciak, and Wahlbin, November 5–6, 2010
- Integral Equation Methods, Fast Algorithms and Applications, August 2–5, 2010
- Medical Device-Biological Interactions at the Material-Tissue Interface, September 13–15, 2010
- First Abel Conference A Mathematical Celebration of John Tate, January 3–5, 2011
- Strain Induced Shape Formation: Analysis, Geometry and Materials Science, May 16–20, 2011
- Uncertainty Quantification in Industrial and Energy Applications: Experiences and Challenges, June 2–4, 2011
- Girls and Mathematics Summer Day Program, June 20–24, 2011
- Special Workshop: Wavelets and Applications: A Multi-Disciplinary Undergraduate Course with an Emphasis on Scientific Computing, July 13–16, 2011
- Special Workshop: Wavelets and Applications: Project Building Workshop, July 13–16, 2011
- Macaulay2, July 25–29, 2011
- Instantaneous Frequencies and Trends for Nonstationary Nonlinear Data, September 7–9, 2011

SPRINGER LECTURE NOTES FROM THE IMA

The Mathematics and Physics of Disordered Media
    Editors: Barry Hughes and Barry Ninham
    (Lecture Notes in Math., Volume 1035, 1983)

Orienting Polymers
    Editor: J.L. Ericksen
    (Lecture Notes in Math., Volume 1063, 1984)

New Perspectives in Thermodynamics
    Editor: James Serrin
    (Springer-Verlag, 1986)

Models of Economic Dynamics
    Editor: Hugo Sonnenschein
    (Lecture Notes in Econ., Volume 264, 1986)

THE IMA VOLUMES
IN MATHEMATICS AND ITS APPLICATIONS

Volume 1: Homogenization and Effective Moduli of Materials and Media
    Editors: Jerry Ericksen, David Kinderlehrer, Robert Kohn, and
    J.-L. Lions

Volume 2: Oscillation Theory, Computation, and Methods of Compensated
    Compactness
    Editors: Constantine Dafermos, Jerry Ericksen, David Kinderlehrer,
    and Marshall Slemrod

Volume 3: Metastability and Incompletely Posed Problems
    Editors: Stuart Antman, Jerry Ericksen, David Kinderlehrer, and
    Ingo Muller

Volume 4: Dynamical Problems in Continuum Physics
    Editors: Jerry Bona, Constantine Dafermos, Jerry Ericksen, and
    David Kinderlehrer

Volume 5: Theory and Applications of Liquid Crystals
    Editors: Jerry Ericksen and David Kinderlehrer

Volume 6: Amorphous Polymers and Non-Newtonian Fluids
    Editors: Constantine Dafermos, Jerry Ericksen, and David Kinder-
    lehrer

Volume 7: Random Media Editor: George Papanicolaou

Volume 99: Mathematics of Multiscale Materials
Editors: Kenneth M. Golden, Geoffrey R. Grimmett, Richard D. James, Graeme W. Milton, and Pabitra N. Sen

Volume 100: Mathematics in Industrial Problems, Part 10 by Avner Friedman

Volume 101: Nonlinear Optical Materials Editor: Jerome V. Moloney

Volume 102: Numerical Methods for Polymeric Systems Editor: Stuart G. Whittington

Volume 103: Topology and Geometry in Polymer Science
Editors: Stuart G. Whittington, De Witt Sumners, and Timothy Lodge

Volume 104: Essays on Mathematical Robotics
Editors: John Baillieul, Shankar S. Sastry, and Hector J. Sussmann

Volume 105: Algorithms for Parallel Processing
Editors:Robert S. Schreiber, Michael T. Heath, and AbhiramRanade

Volume 106: Parallel Processing of Discrete Problems Editor: Panos Pardalos

Volume 107: The Mathematics of Information Coding, Extraction, and Distribution
Editors: George Cybenko, Dianne O'Leary, and Jorma Rissanen

Volume 108: Rational Drug Design
Editors: Donald G. Truhlar, W. Jeffrey Howe, Anthony J. Hopfinger, Jeff Blaney, and Richard A. Dammkoehler

Volume 109: Emerging Applications of Number Theory
Editors: Dennis A. Hejhal, Joel Friedman, Martin C. Gutzwiller, and Andrew M. Odlyzko

Volume 110: Computational Radiology and Imaging: Therapy and Diagnostics
Editors: Christoph Börgers and Frank Natterer

Volume 111: Evolutionary Algorithms
Editors: Lawrence David Davis, Kenneth De Jong, Michael D. Vose and L. Darrell Whitley

Volume 112: Statistics in Genetics
Editors: M. Elizabeth Halloran and Seymour Geisser

Volume 113: Grid Generation and Adaptive Algorithms
Editors: Marshall Bern, Joseph E. Flaherty, and Mitchell Luskin

Volume 114: Diagnosis and Prediction Editor: Seymour Geisser

Volume 115: Pattern Formation in Continuous and Coupled Systems: A Survey Volume
Editors: Martin Golubitsky, Dan Luss, and Steven H. Strogatz

Volume 116: Statistical Models in Epidemiology, the Environment and Clinical Trials
Editors: M. Elizabeth Halloran and Donald Berry

Volume 117: Structured Adaptive Mesh Refinement (SAMR) Grid Methods
Editors: Scott B. Baden, Nikos P. Chrisochoides, Dennis B. Gannon, and Michael L. Norman

Volume 118: Dynamics of Algorithms
Editors: Rafael de la Llave, Linda R. Petzold, and Jens Lorenz

Volume 119: Numerical Methods for Bifurcation Problems and Large- Scale Dynamical Systems
Editors: Eusebius Doedel and Laurette S. Tuckerman

Volume 120: Parallel Solution of Partial Differential Equations
Editors: Petter Bjørstad and Mitchell Luskin

Volume 121: Mathematical Models for Biological Pattern Formation
Editors: Philip K. Maini and Hans G. Othmer

Volume 122: Multiple-Time-Scale Dynamical Systems
Editors: Christopher K.R.T. Jones and Alexander Khibnik

Volume 123: Codes, Systems, and Graphical Models
Editors: Brian Marcus and Joachim Rosenthal

Volume 124: Computational Modeling in Biological Fluid Dynamics
Editors: Lisa J. Fauci and Shay Gueron

Volume 125: Mathematical Approaches for Emerging and Reemerging Infectious Diseases: An Introduction
Editors: Carlos Castillo-Chavez with Sally Blower, Pauline van den Driessche, Denise Kirschner, and Abdul-Aziz Yakubu

Volume 126: Mathematical Approaches for Emerging and Reemerging Infectious Diseases: Models, Methods and Theory
Editors: Carlos Castillo-Chavez with Sally Blower, Pauline van den Driessche, Denise Kirschner, and Abdul-Aziz Yakubu

Volume 154: Mixed Integer Nonlinear Programming
    Editors: Jon Lee and Sven Leyffer

Volume 155: Natural Locomotion in Fluids and on Surfaces: Swimming,
    Flying, and Sliding
    Editors: Stephen Childress, Anette Hosoi, William W. Schultz, and
    Z. Jane Wang